Manufacturing Operations and Supply Chain Management

The LEAN Approach

David Taylor and David Brunt

THOMSON LEARNING™

Australia • Canada • Mexico • Singapore • Spain • United Kingdom • United States

Manufacturing Operations and Supply Chain Management

For more information, contact Thomson Learning, Berkshire House, 168–173 High Holborn, London, WC1V 7AA or visit us on the World Wide Web at: http://www.thomsonlearning.co.uk

---

British Library Cataloguing-in-Publication Data
A catalogue record for this book is available from the British Library

---

ISBN 1-86152-604-0

First edition published 2001 by Thomson Learning
Reprinted 2001 by Thomson Learning

Typeset by LaserScript Limited, Mitcham, Surrey
Printed in Great Britain by TJ International, Padstow, Cornwall

Contents

Dedication

To Chris and Alison for their support and encouragement.

List of Contributors

John Bicheno, Lean Enterprise Research Centre, Cardiff Business School, Cardiff, UK.

David Brunt, Lean Enterprise Research Centre, Cardiff Business School, Cardiff, UK.

Chris Butterworth, Supply Chain Manager, Corus (formerly British Steel Strip Products), Llanwern, UK.

Brian Daniels, Manager, Marketing and Commercial Services, Corus Engineering Steels, Rotherham, UK.

Ann Esain, Lean Enterprise Research Centre, Cardiff Business School, Cardiff, UK.

Peter Gallone, Logistics Manager, Tallent Engineering, Newton Aycliffe, UK.

Ann-Kristin Hahn, Research Student, University of Buckingham, UK.

Peter Hines, Professor of Supply Chain Management and Co-director of the Lean Enterprise Research Centre, Cardiff Business School, Cardiff, UK.

Matthias Holweg, Lean Enterprise Research Centre, Cardiff Business School, Cardiff, UK.

Daniel T. Jones, Professor, Co-director of Lean Enterprise Research Centre, Cardiff Business School, Cardiff, UK, and Chairman of International Car Distribution Programme (ICDP).

John S. Kiff, Lean Enterprise Research Centre, Cardiff Business School, Cardiff, UK and a Director of the International Car Distribution Programme (ICPD).

Jens Nießmann, Research Student, University of Buckingham, UK.

Nick Rich, Lean Enterprise Research Centre, Cardiff Business School, Cardiff, UK.

David Simons, Lean Enterprise Research Centre, Cardiff Business School, Cardiff, UK.

James Sullivan, Project Manager, British Strip Steel Products, Coventry, UK.

David Taylor, Lean Enterprise Research Centre, Cardiff Business School, Cardiff, UK.

Malcolm Taylor, Operations Manager, LDV Press Operations, Washwood Heath, UK.

Preface

This book has been developed as a result of work carried out in the Lean Processing Programme (LEAP) which ran from 1997 to 2000 and involved nine companies in the upstream automotive component supply chain in the UK. The programme was run by staff at the Lean Enterprise Research Centre at Cardiff Business School, together with staff from the participating companies.

The Lean Processing Programme was designed to extend Lean Thinking into this group of firms and their associated customer base. Over a three-year period it sought to make radical and incremental change both within and between the firms as well as at a network level. Specific improvements were made in terms of gaining a better understanding of customer requirements, an improved learning culture in the firms, faster reaction time, improved delivery performance, reduced new product time-to-market, better quality product, improved productivity and increased business opportunities.

The programme had two major objectives: first, to make improvements in business performance in the companies by applying lean principles and techniques; second, to further develop understanding of the methodologies and approaches required for lean improvement.

The aim of this book is to present the knowledge gained from the Lean Processing Programme and other related projects. It is not a theoretical textbook, but a collection of papers and cases which describe the practical approaches that were developed and applied with firms that were attempting lean transformation. It is our hope that readers, whether managers or students, will benefit from the experiences and reflections of the many people who, over a three-year period, endeavoured to develop lean systems within the upstream automotive supply chain in the UK. The book is, however, intended not just for readers concerned with the automotive industry, but for managers in any sector of manufacturing industry, as the principles and approaches described have general applicability.

David Taylor and David Brunt

Acknowledgements

The Lean Processing Programme was sponsored by the UK government's Engineering and Physical Science Research Council (Innovative Manufacturing Initiative) and a network of UK automotive/steel supply chain firms, namely:

British Steel Strip Products (now part of Corus Group Ltd[1]),
Thyssen Krupp Automotive Body Products,
TKA Chassis Camford,
Tallent Engineering Ltd,
GKN Auto-structures Ltd,
Steel and Alloy Processing Ltd,
LDV Ltd,
Wagon Automotive UK/USA.

We would like to thank all these organizations for their generous support and participation in the project.

The Lean Processing Programme was run by staff at the Lean Enterprise Research Centre at Cardiff Business School, together with Chris Butterworth and James Sullivan of British Steel Strip Products. We would like to thank all the research team members and the many managers and staff in the participating companies who contributed directly or indirectly to the production of this publication. We would also like to recognize the administrative assistance and support given by Shirlie Lovell and Sara Bragg at the Lean Enterprise Research Centre and Satinda Corr at British Steel Strip Products.

David Taylor and David Brunt

1 Corus Group was formed in 1999 by the merger of British Steel and Koninklijke Hoogovens of Holland.

Section 1
Introduction

Lean approaches and the Lean Processing Programme

1

David Taylor and David Brunt

THE DEVELOPMENT OF THE LEAN MOVEMENT

The Machine that Changed the World published in 1990[1], described the results of a five-year study of the world's automobile manufacturing industry. It showed clearly that certain Japanese automotive manufacturers and in particular Toyota were significantly ahead of established American and European car manufacturers in almost all of the key performance areas such as service levels, quality, productivity and time-to-market. The term 'lean production' was coined to describe the Toyota approach to manufacturing, which was contrasted to the 'mass production' approach of western manufacturers.

In the period since the publication of *The Machine that Changed the World*, a succession of firms around the world have improved their performance by the application of Lean principles and approaches. We have seen examples where throughput times and defects have been cut by 90 per cent, inventories reduced by three-quarters and space and unit costs slashed in half. All of this has been done at very little capital cost to the organizations involved and firms have begun to develop the flexibility necessary to meet their customers' needs. With performance improvements of this magnitude it has been possible for such companies to double output and profits with the same headcount.

WHAT IS LEAN PRODUCTION?

To quote from *The Machine that Changed the World*:

> Perhaps the best way to describe this lean production system is to contrast it with craft production and mass production, the two other methods humans have devised to make things.
>
> The craft producer uses highly skilled workers and simple but flexible tools to make exactly what the consumer asks for – one item at a time. Custom furniture, works of decorative art, and a few exotic sports cars provide current-day examples. We all love the idea of craft production, but the problem with it is obvious: goods produced by the craft method – as automobiles once were exclusively – cost too much for most of us to afford. So mass production was developed at the beginning of the twentieth century as an alternative.
>
> The mass-producer uses narrowly skilled professionals to design products made by unskilled or semiskilled workers tending expensive, single-purpose

machines. These churn out standardized products in very high volume. Because the machinery costs so much and is so intolerant of disruption, the mass-producer adds many buffers – extra supplies, extra workers, and extra space – to assure smooth production. Because changing over to a new product costs even more, the mass-producer keeps standard designs in production for as long as possible. The result: the consumer gets lower costs but at the expense of variety and by means of work methods that most employees find boring and dispiriting.

The lean producer, by contrast, combines the advantages of craft and mass production, while avoiding the high cost of the former and the rigidity of the latter. Toward this end, lean producers employ teams of multiskilled workers at all levels of the organization and use highly flexible, increasingly automated machines to produce volumes of products in enormous variety.

Lean production (a term coined by IMVP researcher John Krafcik) is 'lean' because it uses less of everything compared with mass production – half the human effort in the factory, half the manufacturing space, half the investment in tools, half the engineering hours to develop a new product in half the time. Also, it requires keeping far less than half the needed inventory on site, results in many fewer defects, and produces a greater and ever growing variety of products.

Perhaps the most striking difference between mass production and lean production lies in their ultimate objectives. Mass producers set a limited goal for themselves – 'good enough,' which translates into an acceptable number of defects, a maximum acceptable level of inventories, a narrow range of standardized products. To do better, they argue, would cost too much or exceed inherent human capabilities.

Lean producers, on the other hand, set their sights explicitly on perfection: continually declining costs, zero defects, zero inventories, and endless product variety. Of course, no lean producer has ever reached this promised land – and perhaps none ever will, but the endless quest for perfection continues to generate surprising twists.

For one, lean production changes how people work but not always in the way we think. Most people – including so-called blue-collar workers – will find their jobs more challenging as lean production spreads. And they will certainly become more productive. At the same time, they may find their work more stressful, because a key objective of lean production is to push responsibility far down the organizational ladder. Responsibility means freedom to control one's work – a big plus – but it also raises anxiety about making costly mistakes.

Similarly, lean production changes the meaning of professional careers. In the West, we are accustomed to think of careers as a continual progression to ever higher levels of technical know-how and proficiency in an ever narrower area of specialization as well as responsibility for ever larger numbers of subordinates – director of accounting, chief production engineer, and so on.

Lean production calls for learning far more professional skills and applying these creatively in a team setting rather than in a rigid hierarchy. The paradox is that the better you are at teamwork, the less you may know about a specific, narrow specialty that you can take with you to another company or to start a new business. What's more, many employees may find the lack of a steep career ladder with ever more elaborate titles and job descriptions both disappointing and disconcerting.

If employees are to prosper in this environment, companies must offer them a continuing variety of challenges. That way, they will feel they are honing their skills and are valued for the many kinds of expertise they have attained. Without these continual challenges, workers may feel they have reached a dead end at an early point in their career. The result: they hold back their know-how and commitment, and the main advantage of lean production disappears.

This sketch of lean production and its effects is highly simplified, of course. Where did this new idea come from and precisely how does it work in practice? Why will it result in such profound political and economic changes throughout the world? In this book we provide the answers.

(Womack, Jones and Roos (1990) *The Machine That Changed the World*, pp. 12–14)

The Machine That Changed the World not only identified and described the principal differences between Lean production, Mass production and its predecessor Craft production (Table 1), but also provided factual benchmarking of the relative performance of all of the world's car manufacturers at the end of the 1980s. Many aspects of industrial performance were measured and compared. Figure 1 illustrates some of the key findings.

The data shows that the best Japanese producers were clearly out-performing manufacturers in other parts of the world in most if not all of the key performance criteria, although by the end of the 1980s a number of American manufacturers had started to copy and adapt many of the lean production approaches particularly in relation to car assembly. This was in part due to competitive necessity – Japanese

Table 1 Three eras of motor car production

	Craft	*Mass*	*Lean*
● Workforce	Highly skilled in design, machine operations and fitting. Apprenticeship for workers.	Interchangeable workers (division of labour). Improvement responsibility – Industrial engineer and foreman	Flexible teams work the process. Little management layers. Improvement responsibility throughout the organization
● Organization	Extremely decentralized but concentrated in one city. Most parts and design from small machine shops. Coordination by owner/entrepreneur.	Vertical integration. Central organization – design, engineering and production in one place.	Network of suppliers with design and engineering capability. Improvement along supply chain.
● Tools	General-purpose machine tools.	Dedicated machines.	General purpose.
● Product	Very low production volume – 1000 or fewer per year. No two exactly alike.	High volume. Long product life cycle.	Ever-decreasing model life cycles. Niche models possible.

Source: Summarized from *The Machine that Changed the World* (1990).

	Japanese in Japan	*Japanese in North America*	*American in North America*	*All Europe*
Performance:				
Productivity (hours/veh.)	16.8	21.2	25.1	36.2
Quality (assembly defects/100 vehicles)	60.0	65.0	82.3	97.0
Layout:				
Space (sq. ft./vehicle/year)	5.7	9.1	7.8	7.8
Size of Repair Area (as % of assembly space)	4.1	4.9	12.9	14.4
Inventories (days for 8 sample parts)	0.2	1.6	2.9	2.0
Work Force:				
% of Work Force in Teams	69.3	71.3	17.3	0.6
Job Rotation (0 = none, 4 = frequent)	3.0	2.7	0.9	1.9
Suggestions/Employee	61.6	1.4	0.4	0.4
Number of Job Classes	11.9	8.7	67.1	14.8
Training of New Production Workers (hours)	380.3	370.0	46.4	173.3
Absenteeism	5.0	4.8	11.7	12.1
Automation:				
Welding (% of direct steps)	86.2	85.0	76.2	76.6
Painting (% of direct steps)	54.6	40.7	33.6	38.2
Assembly (% of direct steps)	1.7	1.1	1.2	3.1

Figure 1a Summary of Assembly Plant Characteristics, Volume Producers, 1989 (Averages for Plants in Each Region)

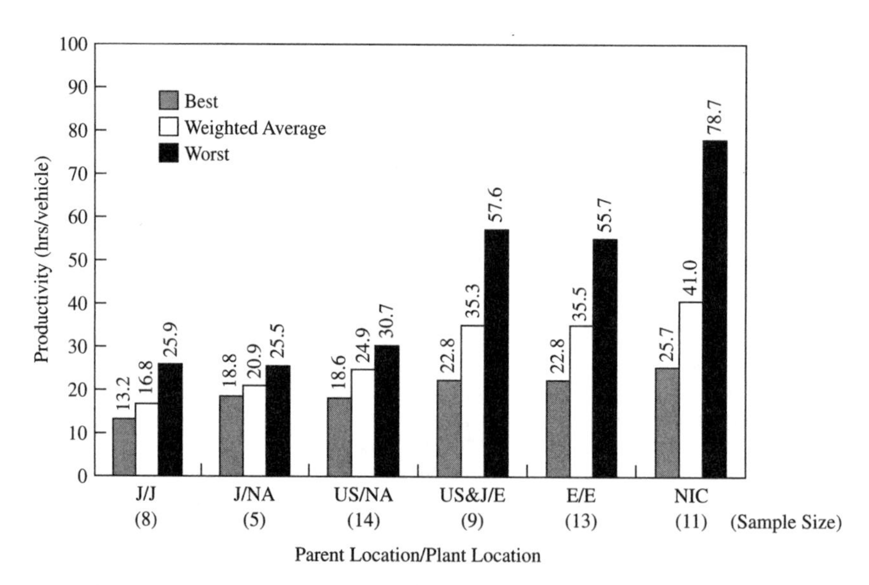

Note: Volume producers include the American "Big Three"; Fiat, PSA, Renault, and Volkswagen in Europe; and all of the companies from Japan.

J/J = Japanese-owned plants in Japan.
J/NA = Japanese-owned plants in North America, including joint venture plants with American firms.
US/NA = American-owned plants in North America.
US&J/E = American- and Japanese-owned plants in Europe.
E/E = European-owned plants in Europe.
NIC = Plants in newly industrializing countries: Mexico, Brazil, Taiwan, and Korea.

Figure 1b Assembly Plant Productivity, Volume Producers, 1989

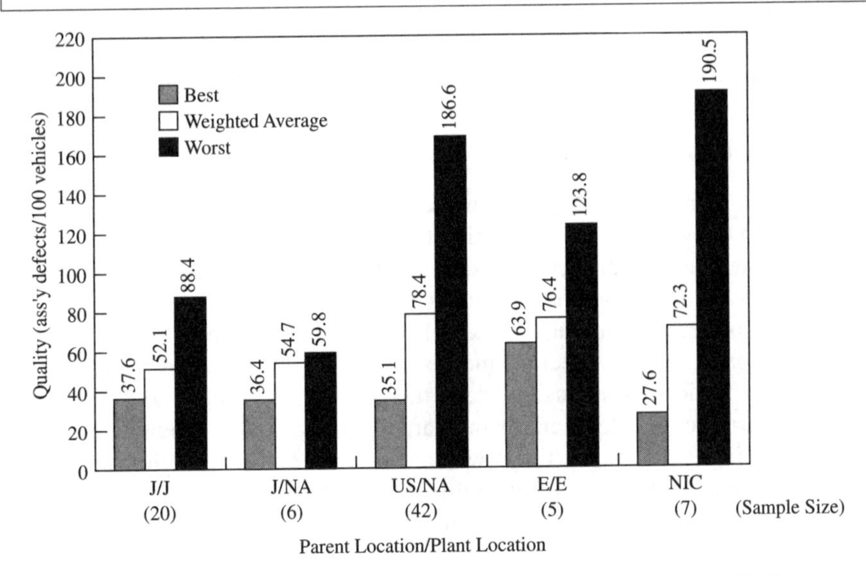

Note: Quality is expressed as the number of defects per 100 cars traceable to the assembly plant, as reported by owners in the first three months of use. The reports only include cars sold in the United States.

Figure 1c Assembly Plant Quality, Volume Producers, 1989

Figure 1 Benchmark data comparing world car producers.

Source: J.P. Womack, D.T. Jones and D. Roos *The Machine That Changed The World* (New York: Rawson Associates 1990, pp. 85–92).

producers had moved from virtually zero market share in 1950 to about 30 per cent share of the world's car market in 1989 and western manufacturers had to respond, and partly due to increasing knowledge of the Japanese approach to production – largely brought about by the development of Japanese car assembly plants in North America, either independently or as joint ventures with American manufacturers. However, the data in Figure 1 also clearly shows that at the start of the 1990s, European manufacturers lagged very significantly behind the Japanese in virtually every aspect that was measured.

Since 1990 further development of lean production systems has continued in America and European auto assemblers have also started to adopt many of the lean principles, with impetus coming from the development of Japanese implants such as Toyota, Nissan and Honda in the UK. The 1990s have also witnessed development of lean approaches to production and supply in a number of other industry sectors in western Europe and North America

LEAN THINKING

At the same time as these practical applications of lean approaches were being explored by companies, the theoretical base of the lean approach was being further developed and defined by researchers in the USA and the UK. Womack and Jones published the results of much of this conceptual and practical work in 1996, in their book *Lean Thinking.*[2]

The stated aim of Lean Thinking is to address the practical issues facing companies and managers who wanted to pursue the lean approach and attempt the transformation from mass to lean. To quote from the preface of the book:

> ***From Lean Production to Lean Enterprise***
>
> We realized that we needed to concisely summarize the principles of 'lean thinking' to provide a sort of North Star, a dependable guide for action for managers striving to transcend the day-to-day chaos of mass production. This summary was hard for most readers to construct because the Japanese originators of lean techniques worked from the bottom up. They talked and thought mostly about specific methods applied to specific activities in engineering offices, purchasing departments, sales groups, and factories: dedicated product development teams, target pricing, level scheduling, cellular manufacturing. Although they wrote whole books describing specific techniques and a few high-level philosophic reflections as well (such as the memoirs of Taiichi Ohno), the thought process needed to tie all the methods together into a complete system was left largely implicit. As a result, we met many managers who had drowned in techniques as they tried to implement isolated bits of a lean system without understanding the whole.
>
> After interactions with many audiences and considerable reflection, we concluded that lean thinking can be summarized in five principles: precisely specify value by specific product, identify the value stream for each product, make value flow without interruptions, let the customer pull value from the producer, and pursue perfection. By clearly understanding these principles and then tying them all together, managers can make full use of lean techniques and maintain a steady course. These principles and their application are the subject of Part I of this book.
>
> With regard to the conversion process, we knew of one heroic example, the original lean leap by Toyota immediately after World War II – but only in sketchy outline. What was more, our most striking benchmark examples in *Machine* were the 'greenfield' plants started from scratch by Japanese auto firms in the West in the 1980s. These were critical achievements because they blew away all the claims, so prevalent up to that time, that, to work, lean production somehow depended on Japanese cultural institutions. Greenfields, however – with new bricks and mortar, new employees, and new tools – bore little resemblance to the long-established 'brownfields' most managers were struggling to fix. Our readers wanted a detailed plan of march suited to their reality and one that would apply in any industry.
>
> We therefore resolved to identify firms in a range of industries in the leading industrial countries that had created or were creating lean organizations from mass-production brownfields. Observing what they had done seemed to be our best hope of discovering the common methods of becoming lean. In doing this, we did not want a survey to discover average practice but rather to concentrate on the outliers – those organizations recently moving far beyond convention to make a true leap into leanness.
>
> But where to find them? We knew the motor vehicle industry well, but we wanted examples from across the industrial landscape, including service organizations. In addition, we wanted examples of small firms to complement

household-name giants, low-volume producers to contrast with high volume auto-makers, and 'high-tech' firms to compare against those with mature technologies.

In the end, through a lot of hard digging and some good fortune, we tapped into networks of lean thinking executives in North America, Europe, and Japan, and gained hands-on experience from a personal investment in a small manufacturing company. Over a four-year period, we interacted with more than fifty firms in a wide range of industries and gained a deep understanding of the human exertions needed to convert mass-production organizations to leanness. We describe our findings and prescribe a practical plan of action in Part II of this book.

(Source: Womack and Jones (1996) *Lean Thinking*, pp. 10–11)

Lean Thinking provides the high-level vision of the lean company and indeed the lean supply chain and indicates the major steps that are involved for firms that wish to move towards that vision.

THE LEAN ENTERPRISE RESEARCH CENTRE, CARDIFF UNIVERSITY

In the early 1990s, two research centres were established. In the USA the Lean Enterprise Institute was established by Jim Womack, while in the UK the Lean Enterprise Research Centre was set up by Professors Dan Jones and Peter Hines at Cardiff University. The aim of both these centres is to assist companies in the transition to a lean approach. In so doing, much of the work has been concerned with the development and testing of tools and techniques to translate the lean vision into practice, as well as to further develop the theory of lean production and lean supply chain management.

More specifically the aims of the Lean Enterprise Research Centre at Cardiff (LERC) are as follows:

- developing pioneering, leading-edge lean thinking tools and techniques;
- helping organizations achieve world-class performance through the application of lean thinking principles and techniques;
- disseminating lean thinking knowledge through a broad range of education programmes and management courses, and through communicating to the broader management population.

The mission is to develop a detailed knowledge of lean management principles and techniques, and identify new methods of applying them for the benefit of businesses and other organizations.

Since its establishment in 1993, LERC has been involved in a wide variety of research projects aimed at developing lean supply chains. All projects undertaken by LERC are of an applied nature whereby concepts and methodologies are developed and tested in conjunction with industrial or commercial organizations. To date, over 30 projects have been carried out in a variety of sectors, including automotive, capital equipment manufacture, electrical distribution, food retailing, telecommunications and public sector administration.

Increasingly, the research at LERC has been looking beyond the production environment to examine the challenges and requirements of applying lean principles to the management of the whole supply chain from raw material sources to end consumer.

One of the most significant projects to date is the Lean Processing Programme, a three-year project involving nine companies in the upstream automotive component supply chain, the results of which form the basis of this book.

THE LEAN PROCESSING PROGRAMME (LEAP)

LEAP was a three-year research project which commenced in April 1997. It was funded jointly by the UK government (through the Engineering and Physical Science Research Council's 'Industrial Manufacturing Initiative') and the nine participating companies (Figure 2). The objective of the programme was to develop a lean supply chain from the major raw material supplier, British Steel Strip Products (BSSP), through two steel service centres, to six first-tier component suppliers which provide sub-assemblies and components to the major car manufacturers in the UK. The automobile original equipment maufacturers (OEMs) were not active participants in the programme other than in the role of customers.

The motivation to engage in incremental and fundamental change in the way in which business was conducted throughout the supply chain was heightened by the wave of consolidations that were taking place between auto assemblers in the mid-1990s, the globalization of automotive supply chains and competition from low-cost European economies. The convergence of these changes also included a tightening of pressure on first-tier suppliers to meet new and ever-challenging measures of customer service. A common denominator in this process to change within and beyond the organization was the rallying call to 'lean'.

The research team comprised six research staff from the Lean Enterprise Research Centre, two managers seconded full-time from BSSP and staff from each of the other eight companies seconded for specific projects as required. Additionally, a steering

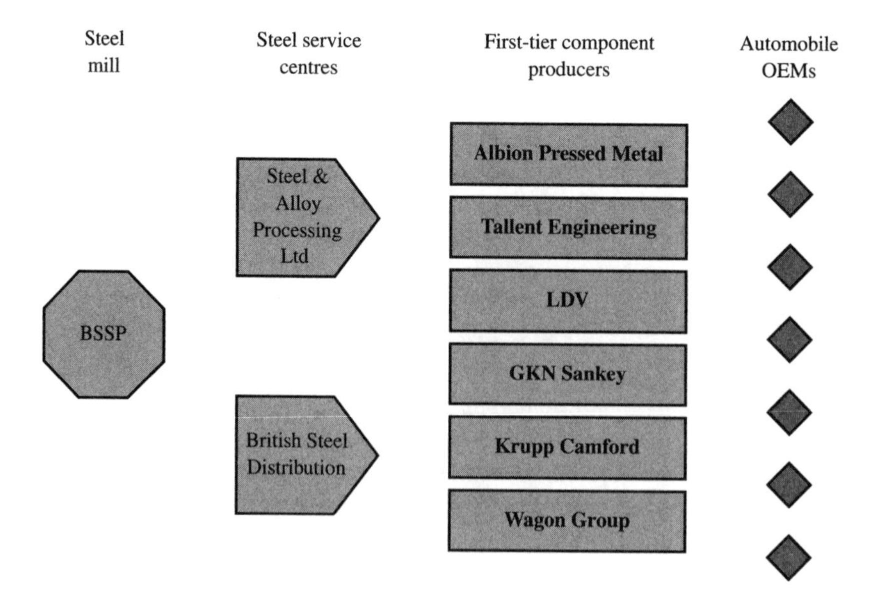

Figure 2 Companies in the Lean Processing Programme.

group was established comprising the managing director or operations director from each of the participating companies. The role of the steering group was not only to oversee and direct the project but also to become practically involved in the lean transformation process. The total project cost over three years was in the order of £2 million.

Project background and justification

Between 1994 and 1997 Professor Peter Hines of LERC carried out the Dantotsu Benchmarking Project, which compared the performance of the supplier systems used by Toyota in Japan with equivalent systems used by Toyota in the UK. The work showed that the performance of typical automotive component suppliers in the UK was well behind that of the best in Japan as exemplified by Toyota (Figure 3). The LEAP project set out to address these and other gaps in performance.

Importantly, however, the Dantotsu research[3] also showed that relatively little work had been done in Japan on improving the upstream elements of the automotive supply chain, particularly from component suppliers back to the producers of steel and other raw materials. Figure 3b shows that the competitive advantage of Toyota primarily originated from the company itself and its various component suppliers, rather than from the raw material supply. The LEAP project, therefore, provided an opportunity not just to catch up with the Japanese at the component supplier level, but also to forge ahead in terms of the performance of raw material production and supply. This was deemed to be a particularly significant opportunity as raw materials account for 44 per cent of total value adding in terms of auto production (Figure 3c).

The LEAP project aimed to address a number of issues that had hitherto been neglected.

- In terms of supply chain management, most previous research had concentrated on the downstream elements of supply chains. That is, from manufacturers, through the distribution channels to the consumer. Where upstream issues had been addressed, it was usually in terms of improving linkages between two adjacent elements, for example from manufacturer to first-tier supplier. As far as was known, LEAP was the first major research initiative anywhere in the world to comprehensively address improvement of an entire upstream supply chain (i.e. from raw material to original equipment manufacturer).
- Furthermore, in terms of the 'Lean movement', most research and practical applications had been set in the context of improving the manufacturing operations of the individual firm or plant, or of improving the links between manufacturer and first-tier supplier. For the first time, the LEAP project took a comprehensive look at the issues and challenges of applying lean principles to multi-echelons of a major supply chain.

Aims of LEAP

The overall aim of LEAP was 'to create competitive advantage in the UK upstream automotive industry by the application of Lean Thinking'.

The stated objectives of the programme were to make step-change improvements for each participating company in:

	Toyota Japan Suppliers	Toyota UK Suppliers
Quality Failure	5.1 ppm	1,250 ppm
Value Added per Employee	£106,000	£40,000
Late Deliveries	400 ppm	107,100 ppm

Figure 3a Gaps Between Automotive Suppliers Japan v UK c 1996

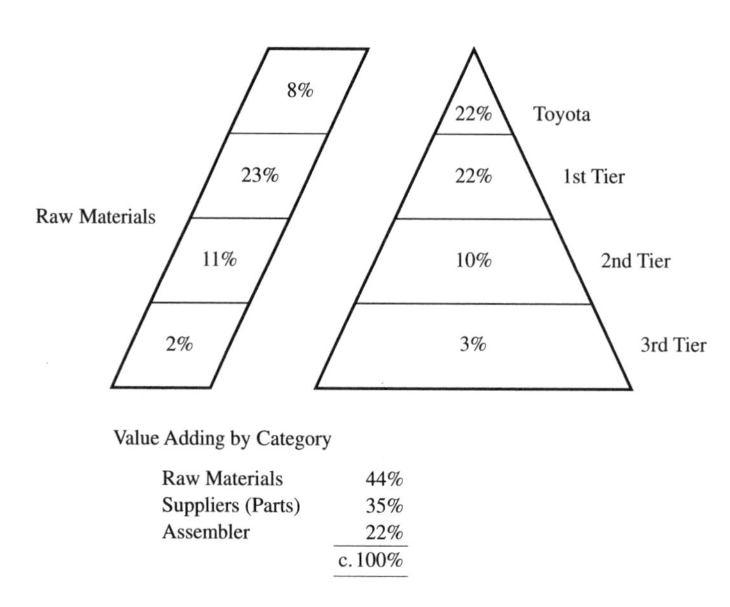

Value Adding by Category

Raw Materials	44%
Suppliers (Parts)	35%
Assembler	22%
	c. 100%

Figure 3b Value Adding in the Toyota Supplier System in Japan

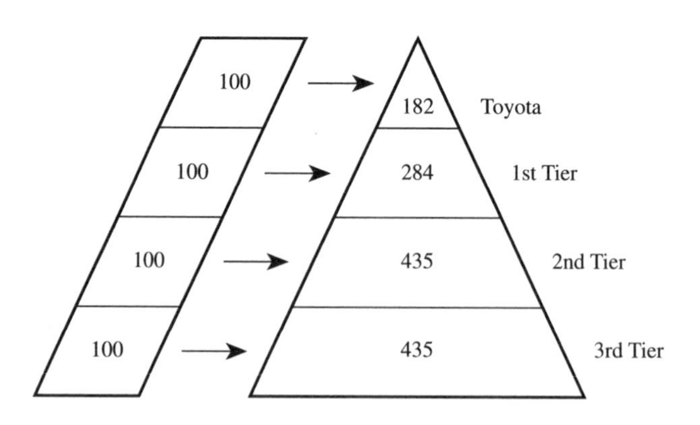

100 = Index for UK equivalent
Based on Value Added per Shop Floor Worker

Figure 3c Productivity Advantage in Toyota Supplier System in Japan

Figure 3 Comparative performance of component suppliers to Toyota in Japan and the UK in the mid-1990s.

- lead time;
- quality;
- customer performance;
- time-to-market;
- Company/supply chain culture.

These objectives were translated into the specific quantified targets shown in Table 2.

A three-phase approach was adopted for the project.

Phase 1 (April 1997 to December 1997) involved situational analysis in each of the nine individual companies using the techniques of waste analysis and value stream mapping techniques developed at LERC. Target improvement projects were then agreed between the company management and the researchers and a variety of initiatives commenced internal to each company.

Phase 2 (January 1998 to April 1998) involved waste analysis and mapping of selected value streams along the complete supply chain. This involved stringing together relevant internal mapping data from each company and the collection of additional data on intercompany activities such as transport or information transfer. A series of potential intercompany supply chain improvement projects were then identified and presented to the LEAP Steering Group. This group then selected the projects deemed to be most beneficial.

Phase 3 (May 1998 to April 2000) was devoted to implementing improvement projects both within individual companies and across the supply chain as a whole. A wide variety of projects were carried out, primarily by company staff with guidance in lean methodologies and approaches from the university research team.

At the same time as achieving quantifiable improvements in the steel to automotive supply chain, a second major objective of the LEAP project was to apply, test and further refine the lean methodology for supply chain improvement which had been developed at Cardiff University in the period from 1994 to 1997.

OBJECTIVES AND STRUCTURE OF THE BOOK

During the course of the Lean Processing Programme (LEAP), a wide variety of projects and initiatives were undertaken both within individual companies and across the wider supply chain.

Table 2 LEAP quantified targets

Improvement area	*British Steel Strip Products*	*Distributors*	*Metal processors*	*Total for the three tiers*
1. Lead-time reduction	40%	40%	40%	40%
2. Time-to-market reduction	30%	30%	30%	30%
3. Stock reduction	10%	20%	20%	15%
4. Quality improvement	50%	20%	50%	45%
5. Productivity improvement	5% per annum	5% per annum	5% per annum	15% over 3 years
6. Increased mutual business	30%	30%	30%	30%

An important requirement of the LEAP brief was to disseminate the research findings as widely as possible both within the UK automotive sector and beyond. Consequently during the three years of the project a significant number of papers were written for publication in trade journals, academic journals and the business press. Presentations were also made at conferences in Britain and abroad to audiences of practitioners and academics (see Appendix 1). This book reproduces a number of these papers, which have been selected in order to try to give a balanced view of both the theory and practice of lean development.

The book has a number of aims:

- to describe a representative sample of the activities and projects carried out in the course of the LEAP project and in so doing provide an overview of what was a major industry/academic collaborative research initiative;
- to provide a series of case studies to illustrate the lean approach to situational analysis, operational improvement and organizational transformation;
- to provide insights into the methodological developments in terms of lean operations and supply chain management that have taken place during the course of the LEAP project.

The book is divided into seven sections.

In Section 1, following this introduction, is a chapter by Dan Jones which gives a high-level overview of the historical development and current state of approaches to car manufacturing and supply chain systems. Readers are commended to this chapter as it sets the strategic context for the work in the LEAP project and focuses on the macro-level picture before subsequent chapters become involved in the detail. Chapter 2 should also be read in conjunction with Chapter 26, the last chapter in the book, which looks to the future and the challenge of building and delivering cars to order. This is perhaps the ultimate challenge for the lean producer and Chapter 26 reports on a new research project which explores the characteristics of the automotive supply chain that will be needed to meet this goal.

Section 2: The Lean Approach To Supply Chain Improvement contains two papers, which describe the theoretical underpinning of the lean approach that has been developed at the Lean Enterprise Research Centre. The first paper, 'The seven value stream mapping tools' by Hines and Rich (1997), effectively describes the toolkit with which the LEAP researchers embarked at the start of the project.

The second paper, 'Value stream management', published in 1998 by Hines and the other members of the LEAP team, describes the further development of these tools and the need to consider the wider corporate context. This paper reflects the knowledge gained in the initial phases of the LEAP.

The third paper in this section describes the results of a benchmarking study comparing UK and Japanese suppliers to Toyota manufacturing plants and highlights the gaps in performance which formed part of the justification for the LEAP project.

Section 3: Analysing Supply Chain Performance and Identifying Waste gives case studies of the situational analysis carried out in the various LEAP companies using the suite of mapping tools described in Section 1. Each chapter describes one or more of the mapping and analysis tools that were used to create a quantified picture of the value stream performance and relates the method and results of application in specific situations.

Section 4: From Analysis To Improvement describes a number of cases to show how the tools of lean situational analysis can and indeed must be linked to identifiable improvement initiatives within companies and across multi-levels of the supply chain.

Section 5: Achieving Organizational Change contains three papers which consider the critical issue of the approaches that were applied or developed in the LEAP project to move individuals, departments, companies and the supply chain partners towards the adoption of the Lean philosophy and techniques.

Section 6: Lean Applications in Other Situations describes lean initiatives in new product development and supplier association development and automotive after-sales which ran alongside the main thrust of LEAP which focused on the production and supply of existing components and parts. During the course of LEAP a number of Lean initiatives were started in other sections of the companies involved as a result of the successes within LEAP; one such initiative is described in Chapter 21.

Section 7: Towards the Future contains two papers which highlight the importance of linking lean operational improvements to corporate strategy and indeed to supply chain strategy and describes the methods that are available within the Lean environment with which to approach this task. Chapter 25, 'From current to future state', is in many ways the culmination of the LEAP project. It gives an overview of the complete steel to automotive supply chain and summarizes its current characteristics and performance using a number of key measures. It then goes on to describe the vision of the 'lean' steel to automotive supply chain that has been developed by the research team. It is thus a statement of how good we believe this supply chain could be if the lean principles were fully adopted. The final chapter in the book, 'Building cars to order – can current automotive supply systems cope?', reports some of the preliminary results of the '3DayCar' research programme which commenced in 1999 and focuses attention once again on the future and the major challenges facing automotive supply chain systems as we move into the twenty-first century.

Note regarding use of the book

Most of the chapters in the book are based on papers that have been presented or published elsewhere during the course of the LEAP project. As such, the papers were written as 'stand-alone articles', often with an introduction giving the context of the LEAP project and possibly some background as to Lean principles and techniques.

As editors, we have asked individual authors to modify some the original articles to avoid too much repetition, particularly in relation to the context of the LEAP project which we have described in this introduction. The book is designed as a collection of readings and case studies that practitioners or academics can dip into and read as stand-alone articles. There is therefore no need to read each chapter in the order presented, although we would suggest that Section 1 is read first to provide a framework for the subsequent chapters. Obviously, as the book has been designed in this way, a degree of overlap is inevitable between some of the articles but the editors hope that this does not detract too much from your reading of the cases.

NOTES

1 *The Machine That Changed The World.* Womack, J.L., Jones, D.T. and Roos, D. (1990) New York: Rawson Associates.
2 Womack, J.P. and Jones, D.T. (1996) *Lean Thinking Banish Waste and Create Wealth in your Corporation.* New York: Simon & Schuster.
3 A fuller account of the Dantotsu research is given in Chapter 5 of this book.

The Lean transformation 2

Daniel T. Jones

THE ORIGINS OF LEAN THINKING

The starting point for this book is Lean Thinking. Lean Thinking is an articulation of the core principles behind the Toyota Production System (TPS), acknowledged to be the most efficient in the world today.[1] The project reported in this book set out to see whether these principles could be applied to the upstream supply chain from automotive assembler through component supplier to steel production. Even Toyota in Japan has only had limited success in extending TPS to the production of raw materials.

Although the term Lean Thinking was coined only a decade ago it has a long history.[2] The key principle of linking every value-creating step in a continuous sequence was evident in Colt's armoury in Hertford, Connecticut in the USA back in 1855.[3] This was probably one of the earliest examples of what we would now call single-piece flow – every machine needed to make a rifle was lined up in process sequence. Each component was moved one at a time from machine to machine until it was complete.

This logic reached its peak in Henry Ford's first large assembly plant in Highland Park in 1915, making the famous Model T.[4] There every machine making every component that was eventually assembled into a finished car was also lined up in process sequence (Figure 1). The whole plant was completely synchronized and the time it took to make a complete car was little more than the stacked lead time of the processing steps for the longest lead-time component – a matter of hours. For this system Henry Ford coined the term 'Flow Production'.

However, from then on industrial history took a different direction. It became clear that customers wanted more choice and product variety than Ford offered with the Model T. This posed a challenge for Flow Production, in which all the tools were configured to make a single product with no tool changes. General Motors chose instead to organize their production processes by separate departments specializing in different activities. The machines in these departments would be kept busy by ensuring there were always batches of work waiting to be done. Instead of following the product flow, batches of different products would wander from department to department until ready for final assembly. This allowed engineers to concentrate on designing faster machines, optimized to make large batches. Henry Ford also followed this route in his next major plant in River Rouge in 1931, and the age of what he called 'Mass Production' began (Figure 2).

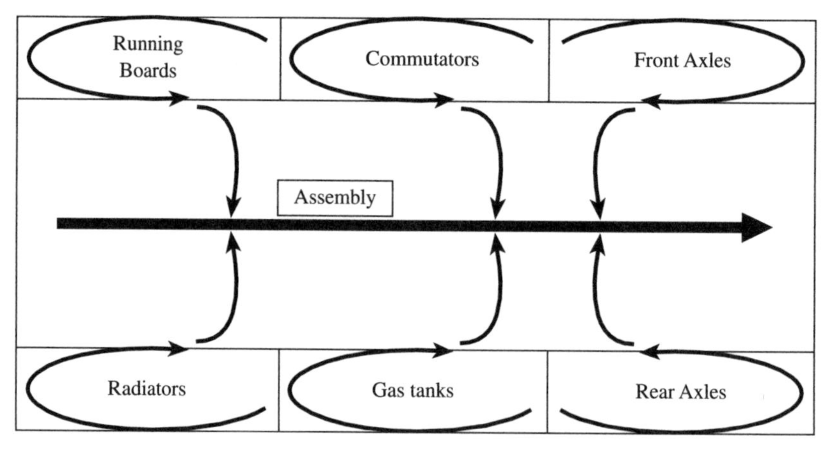

Figure 1 Ford Highland Park Plant.

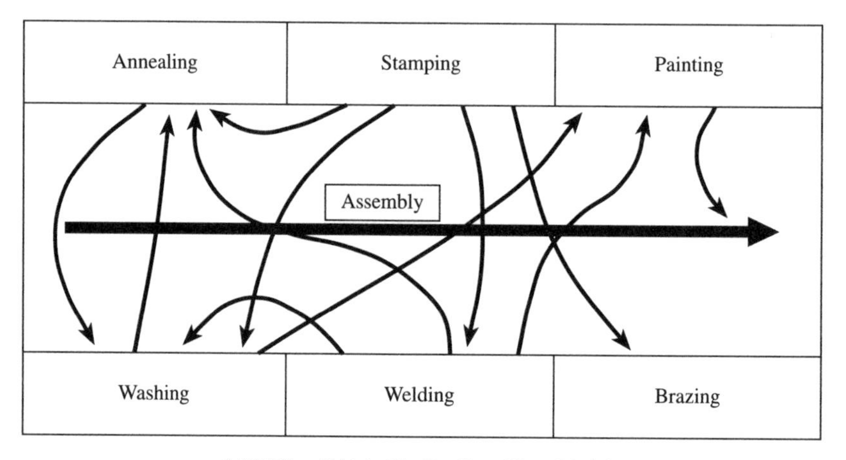

Figure 2 Ford River Rouge Plant.

Gradually over time, what started out as internal parts making shops became self-standing businesses in their own right, supplying several assembly plants. These firms were in turn supplied by a growing network of sub-suppliers and raw materials processors. This vertical de-integration continued until recently, when GM and Ford spun off their component operations into Delphi and Visteon respectively. One might describe the eventual result of Mass Production as a 'Spaghetti world' of bewildering complexity, in which every firm is focused on optimizing their own operations and protecting themselves from their customers and suppliers (Figure 3). As a result it takes many months rather than hours to build a car from raw material to the dealer.

Mass Production worked well when demand grew rapidly – and everyone could sell every car they made. Once markets became saturated the inherent limitations of Mass Production began to surface. Long lead times mean all production decisions are based on forecasts rather than customer orders. Forecasts a long time out are

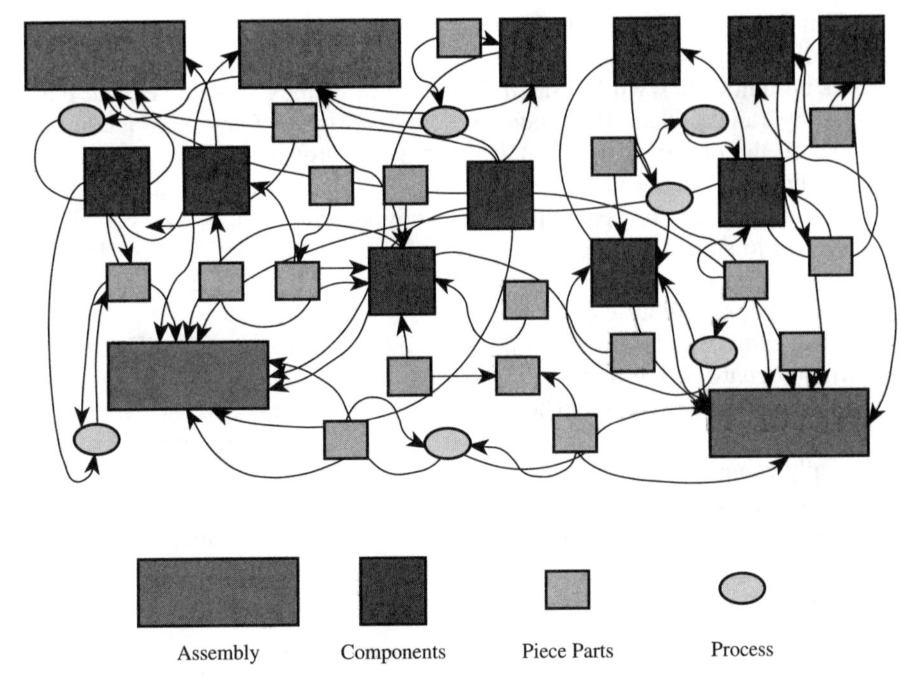

Figure 3 Spaghetti world.

notoriously unreliable, leading to too much stock of the wrong products and long replenishment cycles for the critical missing products. Sales have to be made from stock, which may need discounting to shift in the marketplace. And it is the eventual disposal of these products that in turn forms the basis for the next round of forecasts and production decisions.

Close examination of the flow of products through this 'Spaghetti world' reveals huge amounts of wasted time and effort, or '*Muda*' that adds absolutely no value for the end customer. The identification and step-by-step elimination of this *Muda* is one of the key objectives of Lean Thinking. This *Muda* is not confined to the physical product flow. Mass Production needs ever more expensive and complicated forecasting, planning, scheduling, and supplier coordination and marketing systems to keep it running. Indeed, as the research in this book shows, some of the biggest obstacles to eliminating *Muda* are in the way information is processed up the value stream. Complicated scheduling systems and batch processing of information turn out to compound the *Muda* in the system.

THE REDISCOVERY OF LEAN PRINCIPLES

Across the Pacific the founders of Toyota, Sakichi Toyoda and his son Kiichiro, were working on their own version of Flow Production in the 1930s. They formulated the two key principles of what later became the Toyota Production System: automatic machine and line stopping whenever a mistake is made so no bad parts will be passed forward to interrupt the downstream flow (which they called *Jidoka*), and a pull system so that only parts actually needed are made (called *Just-in-Time*). Later on, the third pillar involving levelling the workload in a mixed model production flow was added (called *Heijunka*).

It was not until after World War II that these principles were linked and put into operation by Taiichi Ohno, the production chief at Toyota. Ohno was determined to overcome all the obstacles to producing a range of products in low volumes, using simple equipment laid out in process sequence. His 20-year experiment started in the engine plant, and then moved to pressing, body welding and assembly. Only when Ohno needed to spread TPS to the supply base in the early 1970s was TPS written down for the first time, though it took another decade before these principles were published in books and articles.

Japanese carmakers began to realize that Toyota was pursuing a different path to them when it continued to make profits during the first oil shock of 1973. Until then the other Japanese carmakers had been following the Total Quality route to improving performance – and quality in particular. TPS on the other hand focused on dramatically compressing time to squeeze out *Muda*, improve quality and only make just what the customer ordered. Japanese exports grew to such an extent that by 1981 alarm bells were ringing and quotas were imposed on Japanese cars entering the USA and the European Community. By 1990 benchmarking studies had identified lean production as the underlying reason for the superior competitiveness of Japanese carmakers, and Toyota in particular.[5]

The successful transfer of lean production to Japanese-owned plants in North America and Europe triggered a massive catch-up movement by all their competitors. This began in engine and assembly plants and later spread to first-tier component suppliers. It soon became apparent that copying what Toyota was doing now was not the answer, and that individual elements of TPS could not work in isolation. A more fundamental transformation was needed that mirrored the earlier reconfiguration from batch to flow at Toyota in the 1950s and 1960s and in its suppliers in the 1970s. This in turn required a deeper understanding of the principles behind TPS. By observing firms that had followed Ohno's example and made the transition it was possible to articulate the principles of Lean Thinking and to describe the action path to implement it.[6]

The general manufacturing, aerospace and electronics industries began to implement Lean Thinking during the 1990s.[7] The clothing, grocery and fast-moving consumer goods (FMCG) industries pursued parallel paths through the Quick Response (QR) and Efficient Consumer Response (ECR) movements respectively.[8] By the end of the 1990s the construction industry began to show an interest, as did the healthcare industry.[9] In 1999 we saw the first evidence that raw materials producers were ready to go down the lean path.[10] Meanwhile Toyota continued to make profits and to build additional plants across the globe, even during the recession in Japan and South East Asia, which exposed critical weaknesses in Nissan and other Japanese carmakers.

RESEARCHING LEAN THINKING IN THE AUTO INDUSTRY

The MIT *International Motor Vehicle Programme* (IMVP) laid the groundwork for our detailed understanding of the impact of Lean Thinking on the automobile industry.[11] In 1993 the Lean Enterprise Research Centre was set up at Cardiff University Business School to continue this work in the UK. Its first project was the *Lean Enterprise Benchmarking Project* (1992–94), which measured the performance of automotive component plants making brakes, seats, exhausts and wiring harnesses across the world.[12] The results were broadly similar to IMVP findings on assembly plants – exposing big differences in productivity, quality and lead times.

For suppliers hoping to supply the new assembly plants being opened by Nissan, Toyota and Honda this posed a huge challenge. It became clear that continuous improvement of current operations was not going to be enough. A more radical reconfiguration from batch to single-piece-flow production, accompanied by the other TPS principles, was going to be necessary to reduce defects from 1500 to 5 per million parts and to deliver exact quantities on time to assembly plant customers without relying on large stocks of finished goods.

The results caused a considerable stir and led to several initiatives by the industry and the UK government to improve the competitiveness of the supply base. These included the establishment of the Industry Forum by the Society of Motor Manufacturers and Traders, to train engineers to lead shop-floor lean transformations, under the guidance of Japanese master engineers.

At the other end of the automotive value stream the *International Car Distribution Programme* (ICDP 1992–2003) carried out the first detailed simulation ever undertaken of the new car supply system and of the after-market parts and service systems. Calibrated by benchmarking both dealers and assemblers across Europe, these simulation models revealed the huge potential for reducing the costs of distribution while at the same time giving the customer exactly what they wanted. They also enabled us to analyse the key steps that would be necessary to make the transition from a 'stock push' to 'customer pull' distribution system.[13]

Car assemblers and importers in the UK responded to the initial findings by reducing the lead times for cars ordered directly from the factory and by pulling most of the stock of unsold cars back from dealers to central distribution centres. This gave dealers a much better chance of obtaining the car the customer wanted – exact matches rose from 25 per cent to 76 per cent, between 1992 and 1997 and lost sales from not being able to supply the right car or an acceptable alternative fell from 20 to 6 per cent. At the same time overall stock in the system fell from 58 to 40 days' supply. This is still some way from a 'customer pull' system. Some manufacturers, such as Rover, Volvo and Renault, have at various times tried true customer order systems in the UK, with mixed success to date.

The parts simulation work showed how the application of Lean Thinking principles could also make a huge difference to the operation of parts distribution, and hence also to the ability of dealers to service and repair cars on time. By shortening order lead times from suppliers and out to dealers, by centralizing inventory management and supplying dealers from local distribution centres it is possible to cut total system parts stock by 75 per cent. This also cuts errors and improves availability, so that dealers can get the job completed right first time on time.

The results of the *Lean Processing Project* (LEAP 1997–2000) which looked at the upstream value stream from supplier back to steel maker are reported in this book. Through detailed value stream mapping the project showed how critical the management of information is to reducing *Muda* and improving responsiveness.[14] As we also discovered, in the grocery industry orders are amplified as they pass from firm to firm, making it impossible to synchronize operations without large amounts of stock and long lead times.[15]

The final ongoing research being conducted at LERC is the *3DayCar Project* (1999–2001). This is trying to bring together all the findings so far to build a picture of the complete automotive value stream. One of the significant early findings, reported in the last chapter in this book, is that the highly complex order processing systems within the assembler represent the biggest obstacle to improving system performance. The detailed mapping of these systems reveals that nothing less than a complete reconfiguration of the ordering and production planning systems will be enough to move to a real customer order-driven system.

Figure 4 brings together representative findings from all these studies to give us a picture of the automotive value stream from hot rolling at the steel mill through to the end customer in 1998. It shows that the production lead time in this example was some 93 days and the lead time for a customer ordering a car directly from the factory was 42 days. It should be noted that only one-third of cars were actually ordered from the factory, the remaining two-thirds being ordered by dealers to sell from stock. Both of these lead times would have been significantly longer a decade ago. Going further back upstream from hot rolling to raw material extraction would add several more months to the production lead time.

Figure 5 shows how these lead times might fall if most of the changes identified in our research were implemented. It shows a production lead time for a customer-ordered car of 22 days from the hot mill and an order lead time of 14 days. A vast majority of customers are prepared to wait for 14 days for the car to be built to their specification. A detailed costing of the benefits from implementing Lean Thinking across the whole automotive value stream remains to be completed by the 3DayCar project. However, there is no doubt that the savings run into several million pounds for each value stream.

There is still a lot to be done to work out the details of what it would take to implement all these changes, but there is no doubt that this lean value stream is within our grasp in the next few years. The biggest obstacles to making this happen will probably not be physical or system changes but changes in the attitudes and behaviour of everyone involved, from the manager of the steel plant through the production planner at the assembler to the dealer salesperson. Now that we have a vision of the

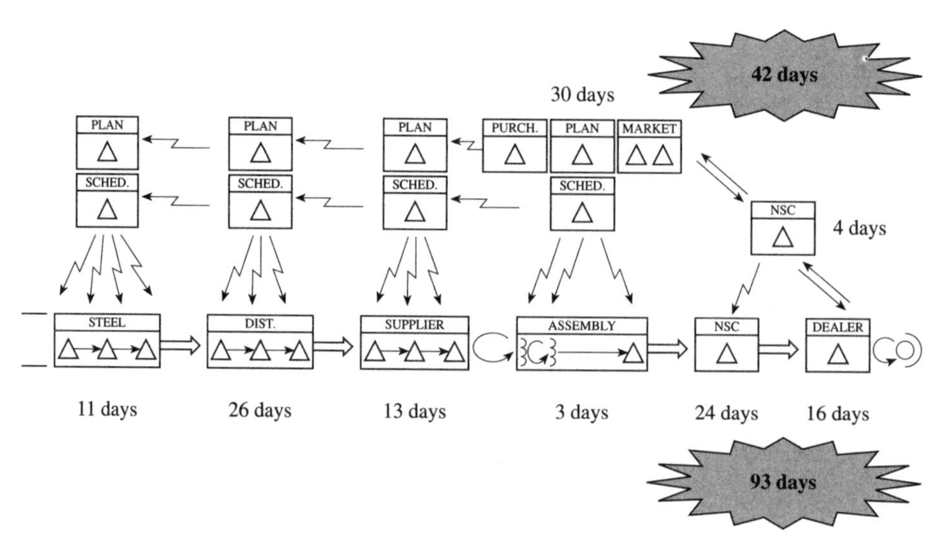

Figure 4 Auto value stream circa 1998: the current state.

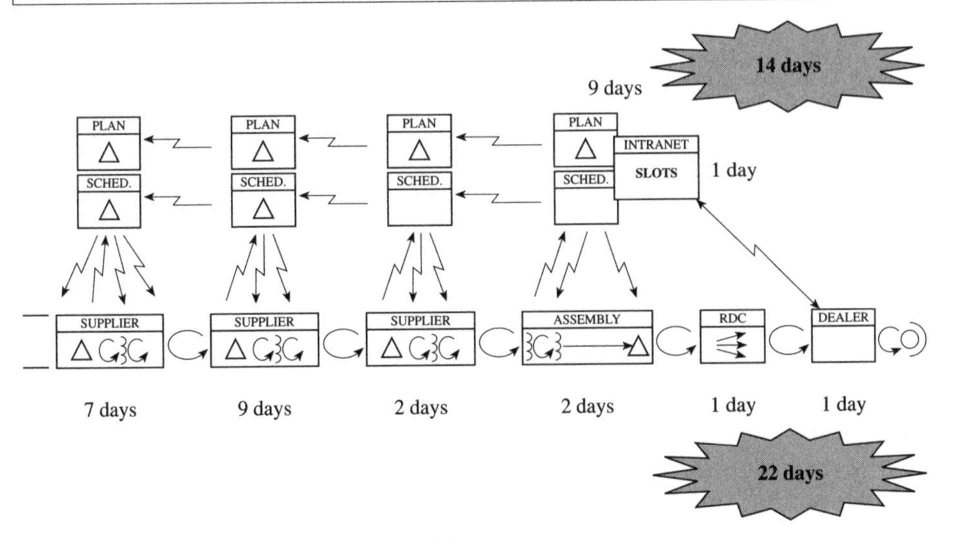

Figure 5 Auto value stream circa 2002? The future state.

possible future state of the whole system it is becomes possible to begin to align incentives and structures to optimize the value stream as a whole. This will be one of the most important challenges over the next decade.

NOTES

1 See J.P. Womack and D.T. Jones (1996) *Lean Thinking: Banish Waste and Create Wealth in your Corporation*, London and New York: Simon & Schuster. The first part of this chapter draws on speeches by Womack and Jones to Lean Summits in the USA and Europe in 1998.
2 The term lean production entered the public domain in J. P. Womack *et al.* (1990) *The Machine that Changed the World*. London & New York: Rawson Macmillan.
3 See D.A. Hounschell (1984) *From the American System to Mass Production, 1800–1932*. Baltimore and London: Johns Hopkins University Press.
4 See H. L. Arnold and F. L. Fauroute (1915) *Ford Methods and the Ford Shops*. New York: The Engineering Magazine Company.
5 Summarized in J. P. Womack *et al., The Machine that Changed the World*, op. cit.
6 See J. P. Womack and D. T. Jones, *Lean Thinking*, op. cit.
7 Between 1995 and 1999 attendance at the Lean Summits grew from 150 to 1200 and Lean Summits were held in the USA, the UK, Brazil, Turkey, France and Germany – see www.lean.org.
8 See www.ecrnet.org.
9 See for instance the 'Egan Report' titled *Rethinking Construction*, Department of the Environment, Transport and the Regions, London, July 1998, www.leanconstruction.org and for healthcare see www.ihi.org.
10 The *Wall Street Journal* of 10 May 1999 reported on the Alcoa's new Toyota-style production system.
11 Reported in J. P. Womack *et al., The Machine that Changed the World*, op.cit.
12 *World-wide Benchmarking Competitiveness Study*, Andersen Consulting, London, 1994.
13 See G. Williams, European New Car Supply and Stocking Systems, 1997, ICDP Report, Solihull, 1998 and J. S. Kiff, T. Chieux and D. Simons, *Parts Supply Systems in the Franchised After-Market,* ICDP Report, Solihull, 1998, available from www.icdp.net.
14 See J. Shook and M. Rother, *Learning to See*, Lean Enterprise Institute, Boston, 1998 and P. Hines *et al., Value Stream Management: Strategy and Excellence in the Supply Chain,* Financial Times Prentice Hall, London, 2000.
15 See D. T. Jones and D. Simons, Future Directions for the Supply Side of ECR, in D. Corsten and D. T. Jones, eds, *Academic Perspectives on the Future of the Consumer Goods Industry*, ECR Europe Academic Partnership, Brussels, 2000.

Section 2

The Lean Approach to Supply Chain Improvement

The seven value stream mapping tools 3

Peter Hines and Nick Rich

INTRODUCTION

Work carried out in the first Supply Chain Development Programme (SCDP I), together with early work in the second programme (SCDP II), has shown that in order fully to understand the different value streams [1] in which the sponsors operate, it is necessary to map these intercompany and intracompany value-adding processes. These value-adding processes make the final product or service more valuable to the end customer than otherwise it would have been. The difference between the traditional supply or value chain and the *value stream* is that the former includes the complete activities of all the companies involved, whereas the latter refers only to the specific parts of the firms that actually add value to the specific product or service under consideration. As such the value stream is a far more focused and contingent view of the value-adding process.

At present, however, there is an ill-defined and ill-categorized toolkit with which to understand the value stream, although several workers (e.g.[2–5]) have developed individual tools. In general these authors have viewed their creations as *the* answer, rather than as a part of the jigsaw. Moreover, these tools derive from functional ghettos and so, on their own, do not fit well with the more cross-functional toolbox required by today's best companies. It is the purpose of this paper to construct a typology or total jigsaw to allow for an effective application of sub-sets of the complete suite of tools. The tools themselves can then effectively be applied, singularly or in combination, contingently to the requirements of the individual value stream.

WASTE REMOVAL INSIDE COMPANIES

The rationale underlying the collection and use of this suite of tools is to help researchers or practitioners to identify waste in individual value streams and, hence, find an appropriate route to removal, or at least reduction, of this waste. The use of such waste removal to drive competitive advantage inside organizations was pioneered by Toyota's chief engineer, Taiichi Ohno, and *sensei* Shigeo Shingo [6,7] and is oriented fundamentally to productivity rather than to quality. The reason for this is that improved productivity leads to leaner operations which help to expose

This paper was first published in IJOPM Vol. 17 No. 1 1997 pp. 46–64 and is reproduced with kind permission of the publisher.

further waste and quality problems in the system. Thus the systematic attack on waste is also a systematic assault on the factors underlying poor quality and fundamental management problems [8].

In an internal manufacturing context, there are three types of operation that are undertaken according to Monden [9]. These can be categorized into:

1. non-value adding (NVA);
2. necessary but non-value adding (NNVA); and
3. value-adding (VA).

The first of these is pure waste and involves unnecessary actions which should be eliminated completely. Examples would include waiting time, stacking intermediate products and double handling.

Necessary but non-value adding operations may be wasteful but are necessary under the current operating procedures. Examples would include: walking long distances to pick up parts; unpacking deliveries; and transferring a tool from one hand to another. In order to eliminate these types of operation it would be necessary to make major changes to the operating system such as creating a new layout or arranging for suppliers to deliver unpacked goods. Such change may not be possible immediately.

Value-adding operations involve the conversion or processing of raw materials or semi-finished products through the use of manual labour. This would involve activities such as: sub-assembly of parts, forging raw material and painting body work.

The seven wastes

There are seven commonly accepted wastes in the Toyota production system (TPS):

1. overproduction;
2. waiting;
3. transport;
4. inappropriate processing;
5. unnecessary inventory;
6. unnecessary motion;
7. defects.

Overproduction is regarded as the most serious waste as it discourages a smooth flow of goods or services and is likely to inhibit quality and productivity. Such overproduction also tends to lead to excessive lead and storage times. As a result defects may not be detected early, products may deteriorate and artificial pressures on work rate may be generated. In addition, overproduction leads to excessive work-in-progress stocks which result in the physical dislocation of operations with consequent poorer communication. This state of affairs is often encouraged by bonus systems that encourage the push of unwanted goods. The pull or *kanban* system was employed by Toyota as a way of overcoming this problem.

When time is being used ineffectively, then the waste of *waiting* occurs. In a factory setting, this waste occurs whenever goods are not moving or being worked on. This waste affects both goods and workers, each spending time waiting. The ideal state should be no waiting time with a consequent faster flow of goods. Waiting time for workers may be used for training, maintenance or *kaizen* activities and should not result in overproduction.

The third waste, *transport*, involves goods being moved about. Taken to an extreme, any movement in the factory could be viewed as waste and so transport minimization rather than total removal is usually sought. In addition, double handling and excessive movements are likely to cause damage and deterioration with the distance of communication between processes proportional to the time it takes to feed back reports of poor quality and to take corrective action.

Inappropriate processing occurs in situations where overly complex solutions are found to simple procedures such as using a large inflexible machine instead of several small flexible ones. The over-complexity generally discourages ownership and encourages the employees to overproduce to recover the large investment in the complex machines. Such an approach encourages poor layout, leading to excessive transport and poor communication. The ideal, therefore, is to have the smallest possible machine, capable of producing the required quality, located next to preceding and subsequent operations. Inappropriate processing occurs also when machines are used without sufficient safeguards, such as *poke-yoke* or *jidoka* devices, so that poor quality goods are able to be made.

Unnecessary inventory tends to increase lead time, preventing rapid identification of problems and increasing space, thereby discouraging communication. Thus, problems are hidden by inventory. To correct these problems, they first have to be found. This can be achieved only by reducing inventory. In addition, unnecessary inventories create significant storage costs and, hence, lower the competitiveness of the organization or value stream wherein they exist.

Unnecessary movements involve the ergonomics of production where operators have to stretch, bend and pick up when these actions could be avoided. Such waste is tiring for the employees and is likely to lead to poor productivity and, often, to quality problems.

The bottom-line waste is that of *defects* as these are direct costs. The Toyota philosophy is that defects should be regarded as opportunities to improve rather than something to be traded off against what is ultimately poor management. Thus defects are seized on for immediate *kaizen* activity.

In systems such as the Toyota production system, it is the continuous and iterative analysis of system improvements using the seven wastes that results in a *kaizen*-style system. As such, the majority of improvements are of a small but incremental kind, as opposed to a radical or breakthrough type.

WASTE REMOVAL INSIDE VALUE STREAMS

As the focus of the value stream includes the complete value adding (and non-value adding) process, from conception of requirement back through to raw material source and back again to the consumer's receipt of product, there is a clear need to extend this internal waste removal to the complete supply chain. However, there are difficulties in doing this. These include lack of visibility along the value stream and lack of the tools appropriate to creating this visibility. This paper aims to help researchers and practitioners remedy such deficiencies. The waste terminology described above has been drawn from a manufacturing environment, specifically from the automotive industry, and from a Japanese perspective. As a result some translation of the general terminology will be required to adapt it to a particular part of the value stream and to particular industries in non-Japanese settings. Consequently, a contingency approach is required to some extent.

Such an approach has been the subject of considerable previous work at the Lean Enterprise Research Centre. This would include the application by Hines [10] of the *kyoryoku kai* (supplier association) to a range of UK-based industry sectors and the introduction by Jones [11] of the Toyota production system philosophy to a warehouse environment. Jones has shown that the seven wastes required rewording to fit an after-market distribution setting. He therefore retitled the seven wastes as:

1. faster-than-necessary pace;
2. waiting;
3. conveyance;
4. processing;
5. excess stock;
6. unnecessary motion; and
7. correction of mistakes.

THE SEVEN VALUE STREAM MAPPING TOOLS

The typology of the seven new tools is presented in terms of the seven wastes already described. In addition the delineating of the overall combined value stream structure will be useful and will also be combined as shown in the left-hand column in Table I. Thus, in order to make improvements in the supply chain it is suggested here that at least an outline understanding of the particular wastes to be reduced must be gained before any mapping activity takes place.

At this point it should be stressed that several of the seven mapping tools were already well-known before the writing of this paper. At least two can be regarded as new, and others will be unfamiliar to a wide range of researchers and practitioners. Until now, however, there has been no decision support mechanism to help choose the most appropriate tool or tools to use.

The tools themselves are drawn from a variety of origins as shown in Table 2. These origins include engineering (tools 1 and 5), action research/logistics (tools 1 and 6) operations management (tool 3), and two that are new (tools 4 and 7). As can be seen, they are generally from specific functional ghettos and so the full range of tools will not be familiar to many researchers, although specific tools may be well-known to individual readers. Each of these is reviewed in turn before a discussion is undertaken of how they can be selected for use.

THE TOOLS

Process activity mapping

As noted above, process activity mapping has its origins in industrial engineering. Industrial engineering comprises a group of techniques that can be used to eliminate from the workplace waste, inconsistencies and irrationalities, and provide high-quality goods and services easily, quickly and inexpensively [12]. The technique is known by a number of names in this context, although process analysis is the most common [13].

Table 1 The seven stream mapping tools

	Mapping tool						
Wastes/structure	*Process activity mapping*	*Supply chain response matrix*	*Production variety funnel*	*Quality filter mapping*	*Demand amplification mapping*	*Decision point analysis*	*Physical structure (a) volume (b) value*
Overproduction	L	M		L	M	M	
Waiting	H	H	L		M	M	
Transport	H						L
Inappropriate processing	H		M	L		L	
Unnecessary inventory	M	H	M		H	M	L
Unnecessary motion	H	L					
Defects	L			H			
Overall structure	L	L	M	L	H	M	H

Notes: H = High correlation and usefulness
M = Medium correlation and usefulness
L = Low correlation and usefulness

Table 2 Origins of the seven value stream mapping tools

Mapping tool	*Origin of mapping tool*
(1) Process activity mapping	Industrial engineering
(2) Supply chain response matrix	Time compression/logistics
(3) Production variety funnel	Operations management
(4) Quality filter mapping	New tool
(5) Demand amplification mapping	Systems dynamics
(6) Decision point analysis	Efficient consumer response/logistics
(7) Physical structure mapping	New tool

There are five stages to this general approach:

1. the study of the flow of processes;
2. the identification of waste;
3. a consideration of whether the process can be rearranged in a more efficient sequence;
4. a consideration of a better flow pattern, involving different flow layout or transport routeing; and
5. a consideration of whether everything that is being done at each stage is really necessary and what would happen if superfluous tasks were removed.

Process activity mapping involves the following simple steps: first, a preliminary analysis of the process is undertaken, followed by the detailed recording of all the items required in each process. The result of this is a map of the process under consideration (see Figure 1). As can be seen from this process industry example, each step (1—23) has been categorized in terms of a variety of activity types (operation, transport, inspection and storage). The machine or area used for each of these activities is recorded, together with the distance moved, time taken and number of people involved. A simple flow chart of the types of activity being undertaken at any one time can then be made. These are depicted by the darker shade boxes in Figure 1.

Next the total distance moved, time taken and people involved can be calculated and recorded. The completed diagram (Figure 1) can then be used as the basis for further analysis and subsequent improvement. Often this is achieved through the use of techniques such as the 5W1H (asking: *Why* does an activity occur? *Who* does it? On *which* machine? *Where? When?* And *How?*). The basis of this approach is therefore to try to eliminate activities that are unnecessary, simplify others, combine yet others and seek sequence changes that will reduce waste. Various contingent improvement approaches can be mapped similarly before the best approach is selected for implementation.

#	Step	Flow	Machine	Dist (M)	Time (Min)	People	Operation	Transport	Inspect	Store	Delay	Comments
1	Raw material	S	Reservoir				O	T	I	S	D	Reservoir/Additives
2	Kitting	O	Warehouse	10	5	1	O	T	I	S	D	
3	Delivery to lift	T		120		1	O	T	I	S	D	
4	Offload from lift	T			0.5	1/2	O	T	I	S	D	
5	Wait for mix	D	Mix area		20		O	T	I	S	D	
6	Put in cradle	T		20	2	1/2	O	T	I	S	D	
7	Pierce/Pour	O	Mix area 12		0.5	1	O	T	I	S	D	
8	Mix (blowers)	O			20	1/2	O	T	I	S	D	Base material, blow & additives
9	Test #1	I			30	1+1	O	T	I	S	D	Sample/Test
10	Pump to storage tank	T	Store tank	100		1	O	T	I	S	D	Dedicated reservoir
11	Mix in storage tank	O	Store tank		10	1	O	T	I	S	D	
12	I.R. test	I			10	1+1	O	T	I	S	D	Stamp & approve
13	Await filling	D			15		O	T	I	S	D	Longer if screen late
14	To filler head	T		20	0.1	1	O	T	I	S	D	
15	Fill/Top/Tighten	O	Filler head		1	1+1	O	T	I	S	D	1 unit
16	Stack	T	Pallet	3	0.1	1	O	T	I	S	D	1 unit
17	Delay to fill 1 pallet	D			30		O	T	I	S	D	
18	Strap pallet	O			2	1	O	T	I	S	D	
19	Transfer to store	T		80	2	1	O	T	I	S	D	
20	Await truck	D	Store		540		O	T	I	S	D	Batch 360/Queue 180
21	Pick/Move by fork lift	T		90	3	1	O	T	I	S	D	Fork lift
22	Wait to fill full load	D	Lorry		30	1+1	O	T	I	S	D	1 Operator, 1 Haulier
23	Await shipment	D	Lorry		60	1	O	T	I	S	D	1 Haulier
	Total		23 Steps	443	781.2	25	6	8	2	1	6	
	Operators				38.5	8						
	% Value adding				4.93%	32%						

Figure 1 Users and non-users of DFM.

Supply chain response matrix

The origin of the second tool is the time compression and logistics movement and goes under a variety of names. It was used by New [2] and by Forza *et al.* [3] in a textile supply chain setting. In a more wide-ranging work, Beesley [4] applied what the termed 'time-based process mapping' to a range of industrial sectors including automotive, aerospace and construction. A similar, if slightly refined, approach was adopted by Jessop and Jones [5] in the electronics, food, clothing and automotive industries.

This mapping approach, as shown in Figure 2, seeks to portray in a simple diagram the critical lead-time constraints for a particular process. In this case it is the cumulative lead time in a distribution company, its suppliers and its downstream retailer. In Figure 2 the horizontal measurements show the lead time for the product both internally and externally. The vertical plot shows the average amount of standing inventory (in days) at specific points in the supply chain.

In this example the horizontal axis shows the cumulative lead time to be 42 working days. The vertical axis shows that a further 99 working days of material are held in the system. Thus a total response time in this system of 141 working days can be seen to be typical. Once this is understood, each of the individual lead times and inventory amounts can be targeted for improvement activity, as was shown with the process activity mapping approach.

Production variety funnel

The production variety funnel is shown in Figure 3. This approach originates in the operations management area [14] and has been applied by New [2] in the textiles

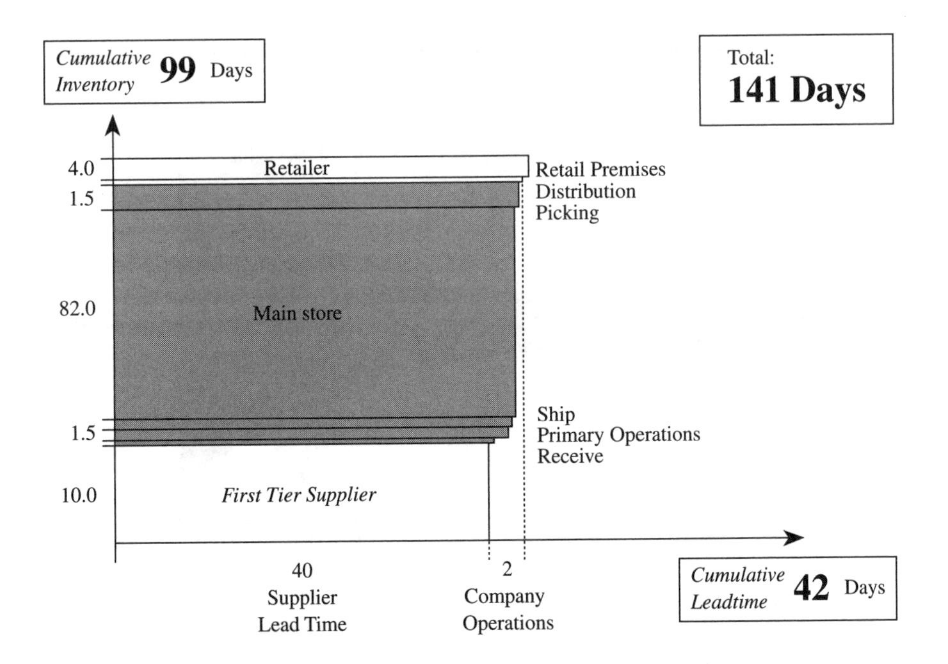

Figure 2 Supply chain response matrix – a distribution example.

industry. A similar method is IVAT analysis which views internal operations in companies as consisting of activities that conform to I, V, A or T shapes [15]:

- 'I' plants consist of unidirectional, unvarying production of multiple identical items such as a chemical plant.
- 'V' plants consist of a limited number of raw materials processed into a wide variety of finished products in a generally diverging pattern.
- 'V' plants are typical in textiles and metal fabrication industries.
- 'A' plants, in contrast, have many raw materials and a limited range of finished products with different streams of raw materials using different facilities; such plants are typical in the aerospace industry or in other major assembly industries.
- 'T' plants have a wide combination of products from a restricted number of components made into semi-processed parts held ready for a wide range of customer-demanded final versions; this type of site is typical in the electronics and household appliance industries.

Such a delineation using the production variety funnel (Figure 3) allows the mapper to understand how the firm or the supply chain operates and the accompanying complexity that has to be managed. In addition, such a mapping process helps potential research clients to understand the similarities and differences between their industry and another that may have been more widely researched. The approach can be useful in helping to decide where to target inventory reduction and making changes to the processing of products. It is also useful in gaining an overview of the company or supply chain being studied.

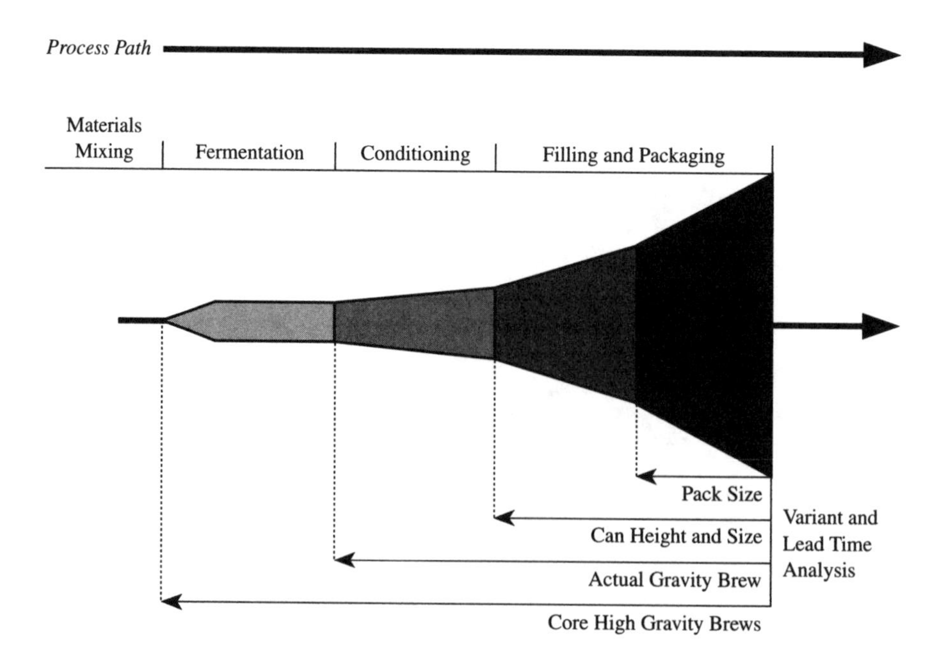

Figure 3 Production variety funnel – a brewing industry case.

Quality filter mapping

The quality filter mapping approach is a new tool designed to identify where quality problems exist in the supply chain. The resulting map itself shows where three different types of quality defect occur in the supply chain (see Figure 4):

1. The first of these is the *product* defect. Product defects are defined here as defects in goods produced that are not caught by in-line or end-of-line inspections and are therefore passed on to customers.
2. The second type of quality defect is what may be termed the *service* defect. Service defects are problems given to a customer that are not directly related to the goods themselves, but rather are results of the accompanying level of service. The most important of these service defects are inappropriate delivery (late or early), together with incorrect paper work or documentation. In other words, such defects include any problems that customers experience which are not concerned with production faults.
3. The third type of defect is *internal scrap*. Internal scrap refers to defects produced in a company that have been caught by in-line or end-of-line inspection. The in-line inspection methods will vary and can consist of traditional product inspection, statistical process control or through poke-yoke devices.

Each of these three types of defect are then mapped latitudinally along the supply chain. In the automotive example given (Figure 4), this supply chain is seen to consist of distributor, assembler, first-tier supplier, second-tier supplier, third-tier supplier and raw material source. This approach has clear advantages in identifying where defects are occurring and hence in identifying problems, inefficiencies and wasted effort. This information can then be used for subsequent improvement activity.

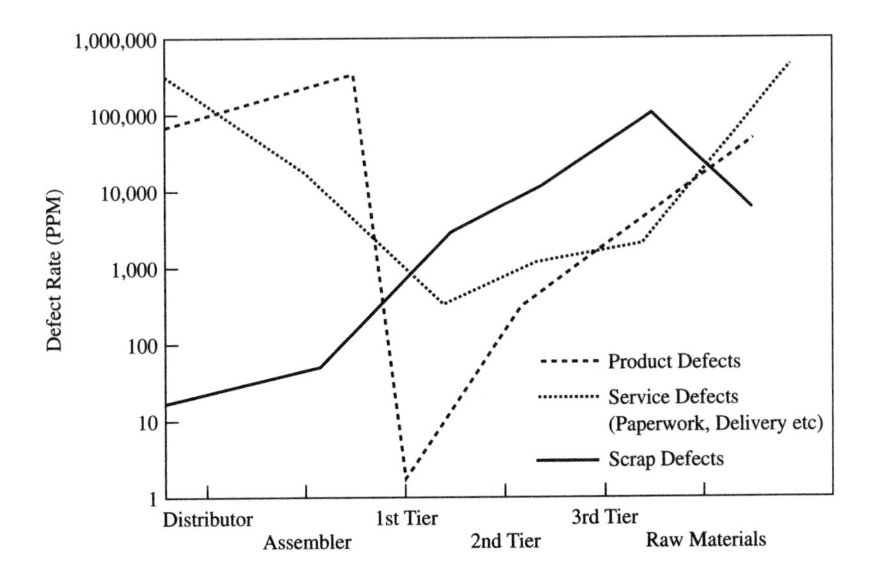

Figure 4 The quality filter mapping approach – an automotive example.

Demand amplification mapping

Demand amplification mapping has its roots in the systems dynamics work of Forrester [16] and Burbidge [17]. What is now known as the 'Forrester effect' was first described in a *Harvard Business Review* article in 1958 by Forrester [16]. This effect is linked primarily to delays and poor decision making concerning information and material flow. The Burbidge effect is linked to the 'law of industrial dynamics' which states:

> if demand is transmitted along a series of inventories using stock control ordering, then the amplification of demand variation will increase with each transfer [17].

As a result, in unmodified supply chains generally excess inventory, production, labour and capacity are found. It is then quite likely that on many day-to-day occasions manufacturers will be unable to satisfy retail demand even though on average they are able to produce more goods than are being sold. In a supply chain setting, manufacturers therefore have sought to hold in some cases sizeable stocks to avoid such problems. Forrester [16] likens this to driving an automobile blindfolded with instructions being given by a passenger.

The use of various mapping techniques loosely based on Forrester and Burbidge's pioneering work is now quite common (e.g. [18]) and in at least one case the basic concept has even been developed into a game called The Beer Game [19] which looks at the systems-dynamics within a retail brewing situation [19]. The basis of the mapping tool in the supply chain setting is given in Figure 5. In this instance an FMCG food product is being mapped along its distribution through a leading UK supermarket retailer. In this simple example two curves are plotted. The first, in the lighter shading, represents the actual consumer sales as recorded by electronic point-of-sale data. The second, and darker, curve represents the orders placed to the supplier to fulfil this demand. As can be seen, the variability of consumer sales is far lower than it is for supplier orders. It is also possible subsequently to map this product further upstream. An example may be the manufacturing plant of the cleaning products company or even the demand they place on their raw material suppliers.

This simple analytic tool can be used to show how demand changes along the supply chain in varying time buckets. This information then can be used as the basis for decision making and further analysis to try to redesign the value stream configuration, manage the fluctuations, reduce the fluctuation or to set up dual-mode solutions where regular demand can be managed in one way and exceptional or promotional demand can be managed in a separate way [20].

Decision point analysis

Decision point analysis is of particular use for 'T' plants or for supply chains that exhibit similar features, although it may be used in other industries. The decision point is the point in the supply chain where actual demand pull gives way to forecast-driven push. In other words, it is the point at which products stop being made according to actual demand and instead are made against forecasts alone [21]. Thus, with reference to Figure 6 an example from the FMCG industry the decision point can be at any point from regional distribution centres to national distribution centres through to any point inside the manufacturer or indeed, at any tier in the supply chain [22].

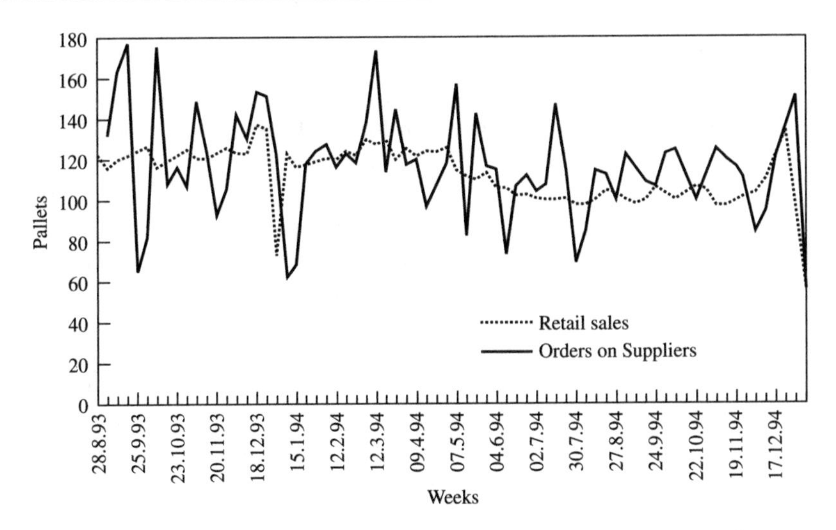

Figure 5 Demand amplification mapping – an FMCG food product example.

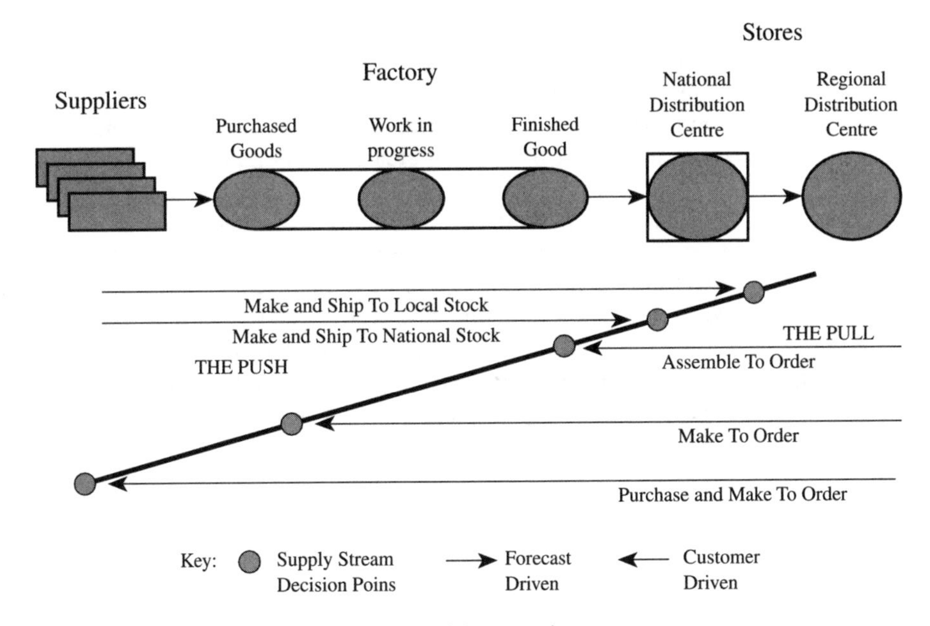

Figure 6 Decision point analysis – an FMCG example.

Gaining an understanding of where this point lies is useful for two reasons:

1. In terms of the present, with this knowledge it is possible to assess the processes that operate both downstream and upstream from this point. The purpose of this is to ensure that they are aligned with the relevant pull or push philosophy.
2. From the long-term perspective, it is possible to design various 'what if' scenarios to view the operation of the value stream if the decision point is moved. This may allow for a better design of the value stream itself.

Physical structure

Physical structure mapping is a new tool which has been found to be useful in understanding what a particular supply chain looks like at an overview or industry level. This knowledge is helpful in appreciating what the industry looks like, understanding how it operates and, in particular, in directing attention to areas that may not be receiving sufficient developmental attention.

The tool is illustrated in Figure 7 and can be seen to be split into two parts, namely: volume structure and cost structure. The first diagram (Figure 7(a)) shows the structure of the industry according to the various tiers that exist in both the supplier area and the distribution area, with the assembler located at the middle point. In this simple example, there are three supplier tiers as well as three mirrored distribution tiers. In addition, the supplier area is seen to include raw material sources and other support suppliers (such as tooling, capital equipment and consumable firms). These two sets of firms are not given a tier level as they can be seen to interact with the assembler as well as with the other supplier tiers.

The distribution area in Figure 7 includes tiers as well as a section representing the after-market (in this case for spare parts), as well as various other support organizations providing consumables and service items. This complete industry map therefore captures all the firms involved, with the area of each part of the diagram proportional to the number of firms in each set.

The second diagram maps the industry in a similar way with the same sets of organizations. However, instead of linking the area of the diagram to the number of firms involved, it is directly linked to the value-adding process (or, more strictly to the cost-adding process). As can be noted in this automotive case, the major cost adding occurs in the raw material firms, the first-tier suppliers and the assembler, respectively. The distribution area is not seen to be a major cost-adding area.

The basis for use of this second figure, however, is that it is then possible to analyse the value adding required in the final product as it is sold to the consumer. Thus value analysis or function analysis tools employed by industrial engineers can be focused at the complete industry or supply chain structure [23,24]. Such an approach may result in a redesign of how the industry itself functions. Thus, in a way

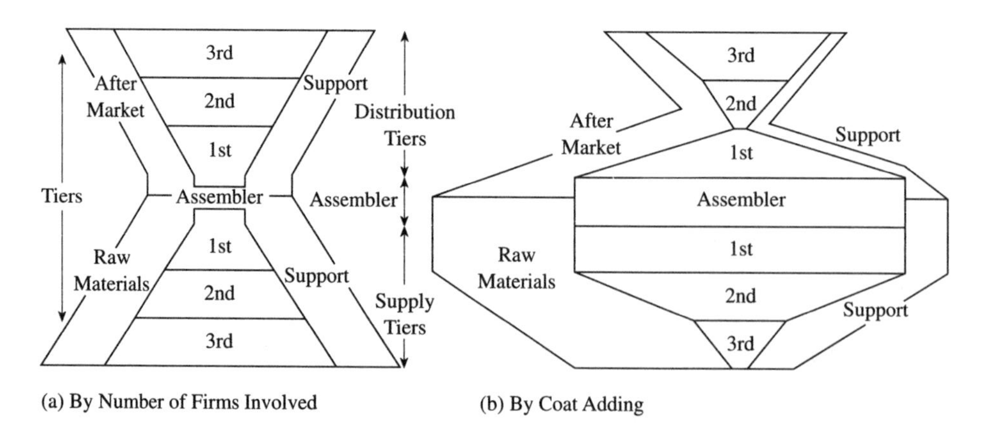

Figure 7 Physical structure mapping – an automotive industry example.

similar to the application of the process activity mapping tool discussed above, attempts can be made to try to eliminate activities that are unnecessary, to simplify others, combine yet others and to seek sequence changes that will reduce waste.

USING THE TOOLKIT

The use of this toolkit at this stage should not be confined to any particular theoretical approach to ultimate implementation. Thus the options can be left open at this stage over whether to adopt a kaizen or business process re-engineering approach once the tools have been used [25,26]. It is the authors' belief that the framework does not constrain this choice process.

Value stream mapping tools

The specific focusing of which tools to use in what circumstances is done using a simplified version of the value stream analysis tool (VALSAT) [27]. The first part of this process is to identify the specific value stream to be reviewed. Second, through a series of preliminary interviews with managers in the value stream, it is necessary to identify the various wastes that exist in the value stream that managers believe can and should be removed (reference should be made to the earlier discussion of the seven wastes). In addition, it is important to gain the views of these managers on the importance of understanding the complete industry structure, irrespective of which wastes are to be removed.

This selection of tools is achieved by giving the interviewees a written overview of each of the wastes as well as an explanation of what is meant by the industry structure. At this stage, if necessary, descriptions of the seven wastes may be reworded in terms more appropriate to the industry under consideration. For instance, in the health-care industry the concept of overproduction may not have great value. However, to call this potential waste 'doing things too early' instead may be more useful in getting the interviewees to relate the concept to their own situations.

Once this has been done, the reworded seven wastes and the account of the overall structure are recorded as row eight in the VALSAT in Table 1 diagram, or as eight rows in area A of Figure 8. Comparison of Table 1 and Figure 8 will show that the former is a simplification of the latter but with sections A, B and C already completed. Thus using this VALSAT method, the different approaches to identifying how these eight variables can be mapped has already been completed by the addition of the seven value stream mapping tools (B). In addition, area C of Figure 8 has already been completed as the correlation between tools and wastes was completed within the main body of Table 1.

At this point it is informative to ask the firm or firms involved to identify for each of the eight wastes/structure (D) the benchmark company in the sector. In other words, by opening these discussions at this stage it forces the firm to think about which of their competitors is best at reducing particular wastes and managing their complete supply/distribution chain. This knowledge may then lead on to more formal benchmarking with these companies, if this is felt to be appropriate, or at least a good focus for subsequent mapping activities.

The next stage (E), therefore, is to ascertain the individual importance weighting of the seven wastes and the overall industry structure. This is achieved most

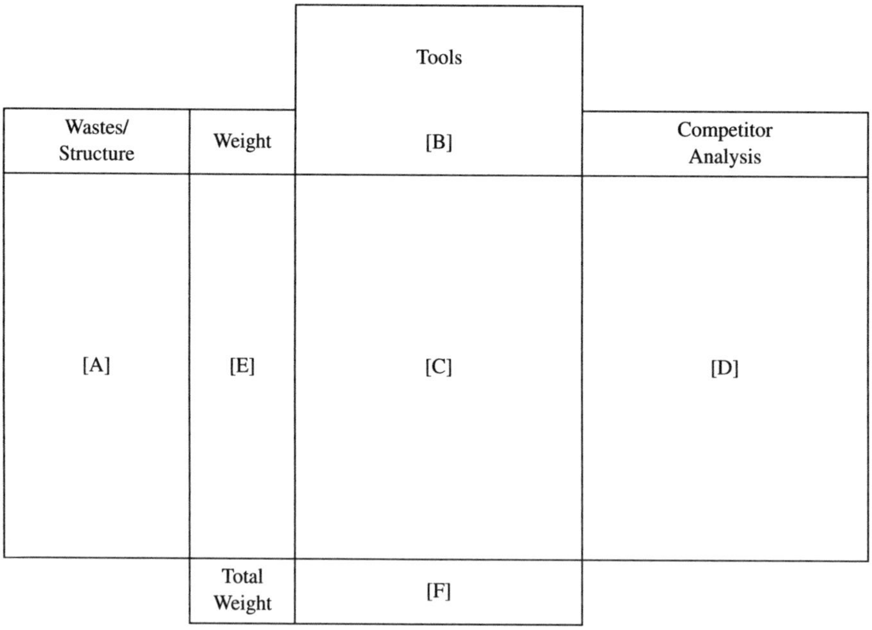

		Tools	
Wastes/ Structure	Weight	[B]	Competitor Analysis
[A]	[E]	[C]	[D]
	Total Weight	[F]	

Figure 8 Using the VALSAT approach to select effective value stream mapping tools.

effectively by allocating a total of 40 points for the eight factors and asking the interviewee to apportion these on the basis of an importance rating between the factors, with the proviso that no one factor can attract more than ten points. If there is more than one interviewee, then the different scores may be aggregated and rebased to total 40 points.

The last arithmetical stage of this approach is to create total weights for each tool. In effect, what is being done here is to give a rating to each tool in terms of how useful it is in identifying the various wastes designated as of most importance by the organization or organizations. This is achieved by giving each of the different correlations given in Table 1 a score. Thus, high correlations are equivalent to nine points; medium three points; and low, to a single point. Then, for each correlation, a total importance score is calculated. This is achieved by multiplying the weighting of each waste/structure by the correlations. Thus, referring to the correlations in Table 1, if the weighting for overproduction is six points the usefulness of the tools in addressing overproduction will be:

- six for process activity mapping (6×1);
- 18 for supply chain response matrix (6×3);
- zero for production variety funnel (6×0);
- six for quality filter mapping (6×1);
- 18 for demand amplification mapping (6×3);
- 18 for decision point analysis (6×3); and
- zero for physical structure mapping (6×0).

This type of calculation is then applied to each of the other rows so that scores are recorded for each individual correlation. Once this is complete the total scores for each column are then summed and recorded in the total weight section, or ‘F’ in Figure 8. The columns which have the highest scores are those that contain the most appropriate tools. As a rule of thumb it is useful to choose more than one tool. Indeed, as a final check, the more important two or three wastes/structure should have been addressed by tools with which they are highly correlated or failing this by at least two tools with which they have a medium correlation. This will ensure that each waste/structure is covered adequately in the mapping process.

This section therefore will have assisted the reader to identify which tool or tools to use. However, once the tools have been run it may be that some unexpectedly high wastes have become apparent. For instance, the demand amplification mapping tool may have been employed to identify unnecessary inventory and waiting. However, as the tool also has a medium correlation with overproduction, it will be useful in identifying such waste if it exists but was not recognized at first by the managers involved. This backflushing may therefore identify some unanticipated but potential improvement areas and, hence, lead to some breakthroughs.

After this mapping process is complete, the researcher will be able to use each individual tool with its associated benefits to undertake more detailed analysis of the value stream with a view to its improvement. As stated above, it is not the purpose of this paper to convert other researchers to a *kaizen* or business process re-engineering approach in this subsequent work. However, the various mapping tools described will help with whichever approach is chosen. In general, the removal of non-value adding waste is best done using a *kaizen* approach, whereas the removal of necessary but non-value adding waste requires a more revolutionary strategy wherein the application of business process re-engineering may be more appropriate.

CASE EXAMPLE

For the reader to gain a better understanding of the approach, a brief case example will be reviewed. The company involved is a highly profitable leading industrial distributor with over 60,000 products and an enviable record for customer service.

After undertaking preliminary discussions it was decided to focus on the upstream value stream to the point at which goods are available for distribution by the firm. Nine products were chosen based on a Pareto analysis from one particular value stream, namely: the lighting product range. Within this range, interviews with key cross-functional staff showed that unnecessary inventory, defects, inappropriate processing and transport were the most serious wastes in the system. In order to understand this in more detail, and using the mapping correlation matrix (Table 1), it was decided to adopt five of the tools:

1. process activity mapping;
2. supply chain response matrix;
3. quality filter mapping;
4. demand amplification mapping; and
5. decision point analysis.

The on-site mapping work was carried out over a three-day period and proved that each of the tools was of value in analysing the selected value streams. An example of the effective interplay of the tools was that the supply-chain response matrix suggested, as the key priority for the firm, supplier lead-time reduction. However, when the data from the quality filter mapping were added, it was found that the real issue was on-time delivery rather than lead-time reduction. Thus, if the supply-chain response matrix had been used on its own, it might have resulted in shorter lead time, but would have exacerbated the true problem of on-time delivery.

The work assisted the firm to conclude that, although it did not need to change, there was plenty of room for improvement, particularly regarding the relatively unresponsive suppliers. As a result, attention had been paid to the setting-up of a cross-function-driven supplier association [10], with six key suppliers in one product group area for the purpose of supplier co-ordination and development. In this supplier association there is an awareness-raising programme, involving ongoing benchmarking, of why change is required. In addition, education and implementation are being carried out using methods such as vendor-managed inventory, due date performance, milk rounds, self-certification, stabilized scheduling and EDI.

The company has found the ongoing mapping work to be very useful, and one senior executive noted that 'the combination of mapping tools has provided an effective means of mapping the [company's] supply chain, concentrating discussion/action on key issues'. Another described the work as 'not rocket science but down to earth common sense which has resulted in us setting up a follow-up project which will be the most important thing we do between now and the end of the century'. Indeed, a conservative estimate of the savings that could be reaped is in excess of £10m per year as a result of this follow-on work.

CONCLUSION

This chapter has outlined a new typology and decision-making process for the mapping of the value stream or supply chain. This general process is grounded in a contingency approach as it allows the researcher to choose the most appropriate methods for the particular industry, people and types of problem that exist. The typology is based around the identification of the particular wastes the researcher/company/value stream members wish to reduce or eliminate. As such, it allows for an extension of the effective internal waste-reduction philosophy pioneered by leading companies such as Toyota. In this case, however, such an approach can be widened and so extended to a value stream setting. This extension capability lies at the heart of creating lean enterprises, with each of the value stream members working to reduce wasteful activity both inside and between their organizations.

ACKNOWLEDGEMENT

Hines, P. and Rich, N. (1997) 'The seven value stream mapping tools', International Journal of Operations & Productions Management, 17(1), 46–64. Reproduced by kind permission of the publisher.

REFERENCES

(1) Womack, J. and Jones, D., (1994) From lean production to the lean enterprise, *Harvard Business Review*, March–April, pp. 93–103.

(2) New, C., (1993) The use of throughput efficiency as a key performance measure for the new manufacturing era, *The International Journal of Logistics Management*, Vol. 4 No. 2, pp. 95–104.

(3) Forza, C., Vinelli, A. and Filippini, R., (1993) Telecommunication services for quick response in the textile apparel industry, *Proceedings of the 1st International Symposium on Logistics*, The University of Nottingham, pp. 119–26.
(4) Beesley, A., (1994) A need for time-based process mapping and its application in procurement, *Proceedings of the 3rd Annual IPSERA Conference*, University of Glamorgan, pp. 41–56.
(5) Jessop, D. and Jones, O., (1995) Value stream process modelling: a methodology for creating competitive advantage, *Proceedings of the 4th Annual IPSERA Conference*, University of Birmingham.
(6) Japan Management Association, (1985) Kanban: *Just-in-Time at Toyota*. Cambridge, MA: Productivity Press.
(7) Shingo, S., (1989) *A Study of the Toyota Production System from an Industrial Engineering Viewpoint*. Cambridge, MA: Productivity Press.
(8) Bicheno, J., *34 for Quality*, Picsie Books, Buckingham, 1991.
(9) Monden, Y., *Toyota Production System: An Integrated Approach to Just-in-Time*, 2nd ed, Industrial Engineering and Management Press, Norcross, GA, 1993.
(10) Hines, P., *Creating World Class Suppliers: Unlocking Mutual Competitive Advantage*, Pitman Publishing, London, 1994.
(11) Jones, D., Applying Toyota principles to distribution, *Supply Chain Development Programme I, Workshop #8 Workbook*, Britvic Soft Drinks Ltd, Lutterworth, 6–7 July 1995.
(12) Ishiwata, J., *Productivity through Process Analysis*, Productivity Press, Cambridge, MA, 1991.
(13) Practical Management Research Group, *Seven Tools for Industrial Engineering*, PHP Institute, Tokyo, 1993.
(14) New, C., The production funnel: a new tool for operations analysis, *Management Decision*, Vol. 12 No. 3, 1974, pp. 167–78.
(15) Macbeth, D. and Ferguson, N., *Partnership Sourcing: An Integrated Supply Chain Approach*, Pitman Publishing, London, 1994.
(16) Forrester, J., Industrial dynamics: a major breakthrough for decision makers, *Harvard Business Review*, July–August 1958, pp. 37–66.
(17) Burbidge, J., Automated production control with a simulation capability, *Proceedings IFIP Conference Working Group 5–7*, Copenhagen, 1984.
(18) Wikner, J., Towill, D. and Naim, M., Smoothing supply chain dynamics, *International Journal of Production Economics*, Vol. 22, 1991, pp. 23–48.
(19) Senge, P., *The Fifth Discipline: The Art and Practice of the Learning Organization*, Doubleday Currency, New York, NY, 1990.
(20) James, R., Promotions update, *Supply Chain Development Programme I, Workshop 8*, Britvic Soft Drinks Limited, Lutterworth, 6–7 July 1995.
(21) Hoekstra, S. and Romme, S. (eds), *Towards Integral Logistics Structure Developing Customer-Oriented Goods Flows*, McGraw-Hill, New York, NY, 1992.
(22) Rich, N., Supply stream 'responsiveness' project, *Supply Chain Development Programme I Workshop #6*, Tesco Stores Limited, Hertford, 25–26 January 1995.
(23) Miles, L., *Techniques of Value Analysis and Engineering*, McGraw-Hill, New York, NY, 1961.
(24) Akiyama, K., *Function Analysis: Systematic Improvement of Quality and Performance*, Productivity Press, Cambridge, MA, 1989.
(25) Imai, M., *Kaizen: The Key to Japan's Competitive Success*, McGraw-Hill, New York, NY, 1986.
(26) Hammer, M., Re-engineering work: don't automate, obliterate, *Harvard Business Review*, July–August 1990, pp. 104–12.
(27) Hines, P., Rich, N. and Hittmeyer, M., Competing against ignorance: advantage through knowledge, *International Journal of Physical Distribution & Logistics Management*, 1996.

4 Value stream management[1]

Peter Hines, Nick Rich, John Bicheno, David Brunt, David Taylor, Chris Butterworth and James Sullivan

INTRODUCTION

Value Stream Management is a strategic and operational approach to the data capture, analysis, planning and implementation of effective change within the core cross-functional or cross-company processes required to achieve a truly lean enterprise. This paper describes the method in detail. This involves a summary of the previous Value Stream Mapping approach together with a discussion of the weaknesses of this previous approach. Subsequently the new approach is described involving a strategic review of a business or supply chain's activities, the delimitation of key processes and the mapping of these processes. A discussion of how to analyse and synthesize these data is followed by a section on an approach to planning strategic and operational change together with a framework in which to do this. Some observations on the benefits and limitations of the new approach are then summarized.

The removal of wastes within a supply chain setting has been highlighted explicitly or implicitly as perhaps the most important task facing the modern day logistician [1,2,3]. In order to do this, the first stage adopted by a number of researchers has been to develop and apply diagnostic tools either within internal (one company) or extended supply chains [4,5,6,7]. In general, such tools have proved useful in their original settings but have often failed to provide a wider applicability particularly if used on their own. As a result, the Value Stream Mapping approach was developed [8]. This approach, described in more detail below, sought to take a cross functional approach employing a contingently selected toolkit from a variety of functional and academic disciplines in order to diagnose supply chain wastes.

Two years of empirical testing of the method have shown that it can be highly effective in diagnosing waste and helping to pinpoint how radical or incremental improvement programmes may yield considerable advantage [9,10,11,12]. However, this testing also showed a number of shortcomings of the approach which will be reviewed below. Chief among these was what could be described as 'shop floor myopia' where improvement activities could often only be envisaged within a shop floor setting, and often then only at an operational level. As a result, work within the Lean Processing Programme at Cardiff Business School has focused on how the method may be improved to provide a more holistic approach addressing both the strategic and operational areas of complete supply chains. This paper will develop the earlier Value Stream *Mapping* approach into a more complete Value Stream *Management* methodology.

VALUE STREAM MAPPING

Value Stream Mapping was initially developed in 1995 with an underlying rationale for the collection and use of the suite of tools as being ‘to help researchers or practitioners to identify waste in individual value streams and, hence, find an appropriate route to [its] removal’ [8]. The approach requires the researcher to identify the severity of a series of wastes that exist generically within a supply chain and to choose, apply and then analyse the output from a series of appropriate contingent tools as illustrated in Table 1.

The choice of wastes is based around Ohno’s seven wastes (with the addition of the importance of knowing the overall supply chain structure) shown in the left hand column of Table 1 [13,14,15]. The seven initial tools were drawn from a variety of functional or academic backgrounds with each adding a unique dimension to the mapping process. The tools, as illustrated in the top row of Table 1, are drawn from industrial engineering (Process Activity Mapping) [16,17], time compression/ logistics (Supply Chain Response Matrix) [4,5,6,7], operations management (Production Variety Funnel) [4,5,18], systems dynamics (Demand Amplification Mapping) [19,20], Efficient Consumer Response (Decision Point Analysis) [21] with Quality Filter Mapping and Physical Structure Mapping being new tools [8].

In general a combination of between three and six tools has been applied when it has been judged that the chosen tools are useful in understanding the detailed waste that exists. Further details of the selection process can be found in Hines and Rich, 1997 [8]. After the selection process, the tools are applied to the appropriate internal or external supply chain. Subsequent to this, the researcher is able to use each individual tool with its individual associated benefits to undertake more detailed analysis of the value stream with a view to improvement. The development of an action plan requires the use of the so-called ‘Chocolate Box’, a correlation of the

Table 1 Selection matrix for the seven Value Stream Mapping tools

Wastes/Structure	*Process activity mapping*	*Supply chain response matrix*	*Production variety funnel*	*Mapping tool Quality filter mapping*	*Demand amplification mapping*	*Decision point analysis*	*Physical structure (a) volume (b) value*
Overproduction	L	M		L	M	M	
Waiting	H	H	L		M	M	
Transport	H						L
Inappropriate processing	H		M	L		L	
Unnecessary inventory	M	H	M		H	M	L
Unnecessary motion	H	L					
Defects	L			H			
Overall structure	L	L	M	L	H	M	H

Notes: H = High correlation and usefulness
M = Medium correlation and usefulness
L = Low correlation and usefulness

existing wastes and particular tools that might be employed to make radical or incremental improvements to the existing situation. The reason for the name is that where a highly rated waste correlates with a very useful tool for addressing the particular waste, the relevant adjacent square is shaded dark with lighter shades where lower ranked wastes correlate with respectively useful tools. Blank squares occur where there is no correlation. As a result when viewing the resultant 'Chocolate Box' the eye is drawn to the dark (chocolates) in the matrix (box) as their application is likely to yield most benefit to the value stream under consideration. Subsequent to this, various of the relevant improvement tools can be implemented. The stages of this Value Stream Mapping approach are illustrated in Table 2.

PROBLEMS, CONSTRAINTS AND ISSUES WITHIN VALUE STREAM MAPPING

As of January 1998, the Value Stream Mapping approach had been applied in over 30 different value streams. These value streams are drawn from a wide range of environments spanning the automotive component industry, capital equipment manufacture, electrical distribution, food retailing, telecommunications and public sector administration. However, the most intensive and detailed use of the method has been within the Lean Processing (LEAP) Programme, a three year government and industry funded research programme in the UK upstream automotive industry. In the latter instance it has been used successfully to map the complete value stream network stretching from six automotive component companies back through their steel service centres to British Steel Strip Products (Figure 1). The reason for the use of the method is to provide the basis for decision making about the development and implementation of appropriate inter- and intra-company improvement approaches to be applied to this important part of the automotive industry.

However, during the LEAP Programme and other projects [9,10,11,12,22,23] in spite of the widespread benefits a number of problems, constraints and issues have

Table 2 Stages in Value Stream Mapping

1. Understand Wastes Within the Supply Chain	4. Develop a Framework for Implementation
2. Select Appropriate Mapping Tools	5. Implement Change Programme
3. Undertake Mapping Activity	

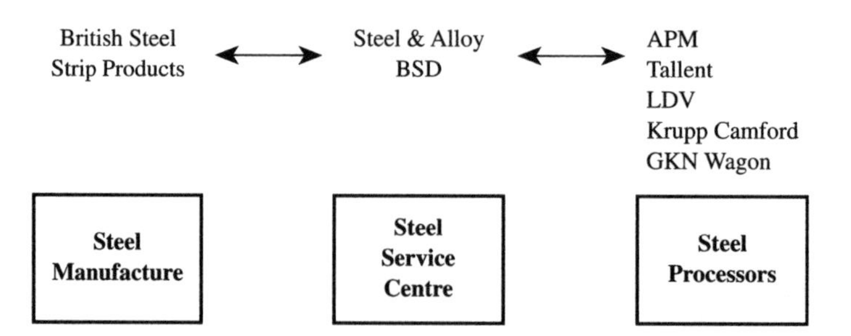

Figure 1 The Leap network.

arisen when using the Value Stream Mapping approach. These can be divided into those specifically connected with the method, those regarding the general environment of use and those involving wider limitations of its use. These are summarized in Table 3.

Four problems or difficulties have been identified or encountered specifically when using Value Stream Mapping. The first of these is that there are a range of other wastes that were found within a supply chain setting such as wasted energy (such as lighting, heating) and the waste of human potential when human resources are under-used or their value and contribution is not recognized. The second problem was that it soon became apparent that the seven initial tools did not cover every eventuality and were weaker in some circumstances, such as in mapping information flows [11,12]. In addition, it was felt that there were often lower level tools that could be employed on occasion as required. The method did not explicitly allow for the use of this including such methods as shop-floor downtime analysis.

The third area concerned the method's concentration on value streams or those linked activities that went towards producing a single or closely linked group of products or services [1]. The weakness of this approach was that where value streams met, for instance, at the purchasing department of one company, absolute capacity constraints were not identified as these points were rarely a constraint on any one particular value stream but were possibly key bottlenecks if all other value streams were considered. The last criticism in this area was that it was often difficult for the company or companies involved to understand how the mapping method was translated into a particular 'Chocolate Box' and why one method was being preferred to another. Often the problem here revolved around the use of tacit rather than

Table 3 Problems encountered when applying Value Stream Mapping

A. Those specifically associated with the Value Stream Mapping Method
1. There may be other wastes that are not incorporated into the existing scheme such as wasted energy or wasted human potential.
2. There may be other tools which could usefully be added either at a top level or as a follow up at a more detailed level.
3. Due to the concentration on value streams, areas of overlap between value streams were not well covered such as finite capacity planning.
4. A lack of understanding of how the method moves from mapping outcomes to the 'Chocolate Box' and the final implementation approach.

B. Those regarding the general environment of use
1. A lack of understanding of what becoming lean means and how you might go about doing so.
2. A lack of a formal education step in the process at either senior or operational levels.
3. Presently much of the subjective data is lost within the formal analysis stage.
4. The method is at present generally very time intensive.

C. Wider limitations of the method
1. A lack of understanding of a particular firm's position in a supply chain and the implications of their actions.
2. A lack of linkage to corporate strategy and the wider market environment.
3. A lack of review of other key processes in different business and supply chain environment such as new product development and maintenance in order to avoid 'shop floor myopia'.
4. A lack of understanding about human resource issues such as the appropriate internal or external culture, language and relationships required by organizations.

codified knowledge [24,25], hence somewhat leaving the researched company a little in the dark.

The second type of problems encountered were concerned with the general environment in which the Value Stream Mapping approach was being used. The first major issue here is that the method has been employed by research teams that have a detailed knowledge of what lean production and the lean enterprise are, how they should work and why they should operate in such a set way. However, in several cases the companies being studied have not initially shared this knowledge and hence have found it difficult to grasp some of the concepts being put forward as a result of the mapping. This is compounded by the lack of formal education steps within Value Stream Mapping at all levels of the company meaning that the tacit knowledge gaps were often never completely closed hence lowering the likely buy-in of company directors or employees in subsequent programmes.

Another general weakness of the method was that at least 50 per cent of the useful information that was collected during the mapping process was subjective, informal or even by way of participant observation [26]. As such, the rather positivist weighted formal Value Stream Mapping approach did not readily find a way of incorporating more naturalist related research findings [27]. As a result in some cases very useful subjective data was not always used to its full and found little influence in the subsequent 'Chocolate Box'. The last general constraint of the approach is that, although very thorough, it is generally quite time consuming to undertake.

The last area of concern about Value Stream Mapping lay within its wider limitations. The first of these lay in many companies' or value streams' lack of understanding of their position within the supply chain and in particularly the results of their actions on other members of their supply chain. For instance in the value streams studied within the LEAP programme a small and sometimes also insignificant change (for them) by a vehicle manufacturer in their schedule could yield major problems and costs not only to their automotive component suppliers but also their steel service centres and British Steel Strip Products. However, if the complete supply chain was not educated or mindful of these problem, progress to correct them would be hard at best.

The second wider limitation of Value Stream Mapping was that its use was not explicitly linked to the corporate strategy and the wider business environment in which a company was operating within. As such a danger of the mapping was that it would be done in isolation of the needs of the company, potentially yielding impressive percentage improvements but be unrelated to what the business or supply chain needs to do to be successful. Linked to this point is that where applied Value Stream Mapping activity has far too often focused on the order fulfilment or supplier integration process ignoring other key processes such as product development or human resource management. Perhaps a consequence of the lack of attention to human resource issues is that the researchers have on some occasions under-estimated the need to understand the cultural environment and soft system issues that could reduce the impact or totally halt the mapping or change process [9].

As a result of these 12 problems, constraints and issues it was felt by the LEAP research team that a further development of the initial Value Stream Mapping approach was required that would seek to reduce the impact of these weaknesses while building on the many strengths of the initial method.

THE NEW METHODOLOGY: VALUE STREAM MANAGEMENT

The new Value Stream Management method is a strategic and operational approach designed to help a company or complete supply chain achieve a lean status. It has its antecedents grounded in the Value Stream Mapping approach [8] but seeks to overcome some of the problems and drawbacks of this earlier approach. Value Stream Management also incorporates various education and policy deployment [28] stages to make it a far better basis for ongoing company or supply chain development. The new approach can be divided into twenty individual and consecutive stages as illustrated in Table 4, each of which will be outlined below.

The first stage of the method is to gain an overall understanding of the company mission, the general customer environment and their needs as well as the strategic direction set by the individual company or companies in the supply chain. This is achieved through the use of a formal presentation from the company followed up by semi-structured interviews with senior executives either collectively or individually. During this stage the researcher is trying to gauge what the exact strategy is, whether it appears likely to be successful, the changing demands of the market place and indeed whether the company or companies involved have a formal strategic planning process. This information is also supplemented with an understanding of what important change programmes are already in place or planned in the near future.

As a result of these initial discussions the researcher attempts within a non structured interview format to understand what the key processes are within the company or supply chain and whether these are indeed recognized and actively managed at the time of interview. Within this context, the term process refers to the main cross-company or cross-supply chain activities that are responsible for achieving the key deliverables required by customers. As Dimancescu *et al.* note, 'for many companies, aligning organizational competencies around a vital few core

Table 4 The Top Level Strategic Analysis Stage

1. Understand Company Mission, Customer Environment & Needs and Strategic Direction
2. Delimit key processes
   a) Customer Facing such as Order Fulfilment
   b) Non Customer Facing such as Supplier Integration
3. Understand existing Roles and Responsibilities
4. Understand existing Organizational Structure
5. Lean Enterprise Education for Senior Managers
6. Develop Top Level 'Big Picture' Cartoons of Key Processes
7. Define Products and Processes to be Mapped in Detail
8. Appoint Senior Level Steering Board
9. Appoint Process Champion for Each Key Process to Lead Mapping Activity
10. Lean Enterprise Education for Process Champions
11. Understand Specific Operating Environment and Wastes
12. Select Second Level Mapping Tools by Process
13. Undertake Detailed Mapping
14. Identify Areas where Further Analysis is Required
15. Undertake Third Level Analysis
16. Analyze Objective and Subjective Data and Develop Timed Implementation Plan
17. Develop Key Control Metrics by Process Area
18. Educate and Train Implementation Action Teams
19. Undertake Implementation
20. Measure Progress Against Plan

processes has become a competitive strategy. By doing so, the most visionary business leaders are recognizing that it is processes, not functions or departments, that deliver customer value and satisfaction. These processes have two purposes: One is to 'qualify' your company [or supply chain] in the eyes of a buyer; the other is to demonstrate a capability that will "win an order".' [28].

Although the single processes that cut across organisational boundaries may be labelled with a variety of names, they fall into a two generic categories, namely that they are customer facing or non-customer facing. Customer facing processes are those that directly interact with customers and are what companies or supply chains are usually directly measured on. Examples of these include the traditional Japanese 'QCD' of Quality, Cost and Delivery [29] but may also include other areas such as New Product Development or Environmental Control. In an American or European setting the original QCD nomenclature is usually changed to fit a western culture into Quality, Cost Management and Order Fulfilment.

In addition to the customer facing processes there are usually a whole series of internal or upstream processes that need to be satisfactorily managed in order to allow the customer facing processes to meet customer requirements. Examples of non customer facing processes are Supplier Integration, Human Resource Management and Information Management, each of these being areas that in general do not attract detailed day-to-day customer attention but are nevertheless equally critical to the company or supply chain.

For the majority of companies between four and ten such key processes can be identified and can be seen to cut across company divides, functions and traditional management domains. A far from exhaustive list of such processes is illustrated in Table 5. The researcher during this second Value Stream Management stage therefore seeks to identify the key processes within the language presently in use within the relevant supply chain and attempts to give a broad definition for each key process.

The third and fourth stages of Value Stream Management also involve interviews with senior management and seek to gain a greater understanding of the company or supply chain by delimiting the existing roles and responsibilities and organization structure. This knowledge enables the researcher to understand who is trying to manage what within what operating structure. It also helps to give the basis of later analysis why some areas of the company or supply chain are not performing as well as they might. During each of these first stages, although a great deal of factual information may be collected, the most important information may be in the form of subjective or 'off-the-cuff' comments perhaps made over coffee or lunch breaks. Indeed, to obtain a richer vein of this information it is often useful to employ a passive researcher during the interview process who does not take part in the formal

Table 5 A range of frequently encountered key processes

1. Supplier Integration	8. Human Resource Management	14. Kaikaku
2. Order Fulfilment	9. Training & Education	15. Order Acquisition
3. New Product Introduction	10. Quality	16. Legal Control
4. New Product Development	11. Cost Management	17. Public Relations
5. Research & Development	12. Information Management	18. Maintenance
6. Environmental Control	13. Kaizen	19. Promotions Management
7. Strategic Management		20. New Facility Development

or informal questioning but merely observers the reactions, voice inflections and body language of the interviewees [26].

After these initial top management interviews are completed possibly during a one day workshop, the next stage is to educate the senior team in what lean production, the lean enterprise and lean thinking is [1,28,30,31,32]. This can usually be achieved during a half day seminar that ensures that the top managers understand the central tenets of lean thinking and can therefore understand why and to some degree how the lean message applies to them. At the end of this education phase it is possible to return to a discussion of the company or supply chain under consideration and to debate with the senior staff involved which of the key processes defined in stage two are the most critical or problematic at the present time. At this point either a key few or all of these processes can then be mapped at a 'big picture' cartoon level as illustrated in a very simple form in Figure 2. This big picture cartoon seeks to consolidate the top managers understanding of what the process is, who is involved, how long it takes, where problems presently are as well as giving some early brainstorming of what improvements might be beneficially applied.

A rough schematic is produced for each key process either on a large flip chart or by the use of post-it notes along a wall. To the initial process stage and department (X and Y axis) configuration a flow line or simplistic critical path may be added [33] together with the total process time, split into distinct sections if this level of detail is known. To this may be added, by way of post-it notes, examples of existing problems and possible improvement areas. Depending on the detail of the cartoon produced this activity should take between an hour and half a day for each map produced. This top level analysis will also help the senior managers to gauge which of the key processes

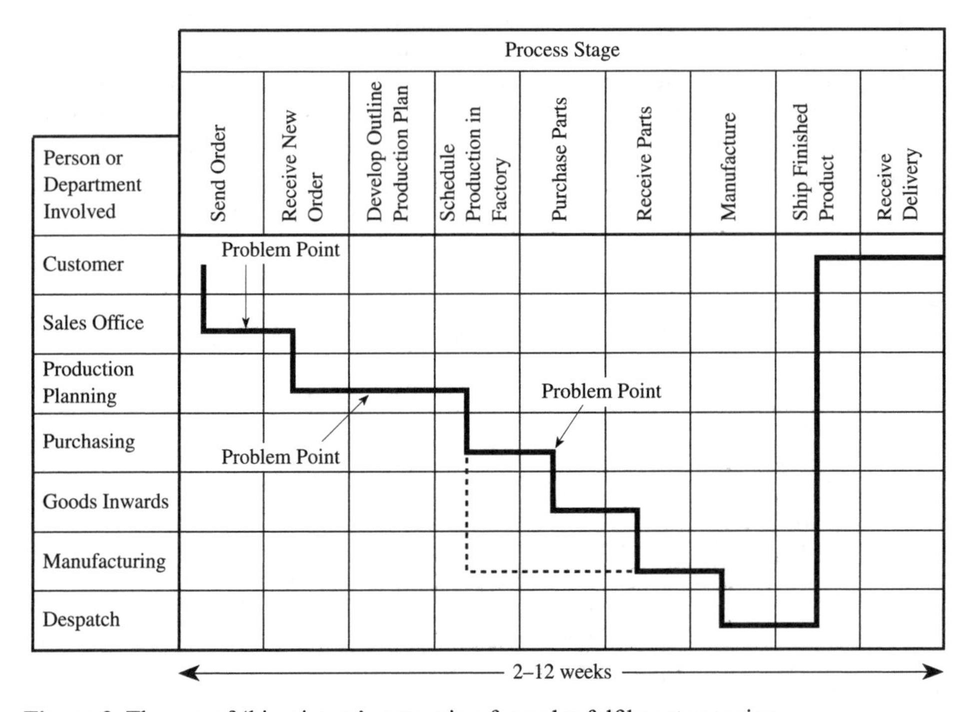

Figure 2 The use of 'big picture' cartooning for order fulfilment mapping.

needs to be understood in further detail in order to make either radical or incremental change. As a result of these discussions the products to be mapped as well as the specific processes to be mapped in detail will be defined. In the normal course of events most companies will find that the order fulfilment process is their most critical day-to-day activity and will almost certainly call for further more detailed analysis.

In general, it is useful to define a number of specific value streams within this order fulfilment and other process areas before choosing which product flows to map in detail. Value streams are specific groups of activities within a company or wider supply chain that involve the production of a similar range of products or services employing a similar routing of tasks and activities through the supply chain. An example of this from the LEAP programme would be one automotive component manufacturer who is making two specific types of products. In one part of the factory they are making complex sub-assemblies involving simpler metal pressings welded together and assembled with a range of bought in components. These products are shipped to vehicle manufacturers and have a specific and distinctive internal and external supply chain. In another part of the factory the company is bending and welding a metal tubing product that it sells to another metal processing company involving the purchase of steel from a different supply source with no use of other purchased components. In this case there are clearly two different value streams even though there may be perhaps over 100 variants of the first value stream and a dozen of the latter. In this case the researchers chose to map one complex sub-assembly and one of the manipulated tubing products as the respective value streams were clearly different. Further detailed mapping of several very similar products within a particular value stream at this stage of analysis are unlikely to prove to be a wise investment of research time.

In order to achieve an effective product mapping, analysis and ultimate change programme, it is essential to have an adequate management structure in place for this change process. The Value Stream Management method employs the Three Tier Management approach in order to achieve this [28]. Depicted schematically (Figure 3), the management system consists of three levels (excluding in this example the Cardiff outside facilitation group):

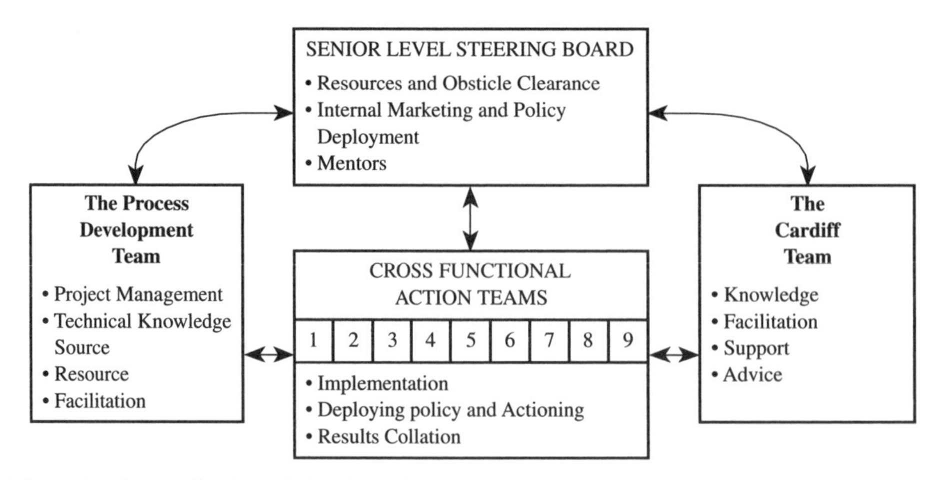

Figure 3 The application of the three-tier system of management.

1. The Senior Level Steering Board
2. The Process Development Team
3. The Cross Functional Action Teams.

This internal tiering method serves to translate top-level (static) objectives, sometimes referred to as the organization's scorecard, into competitively critical (motivational) performance gaps, which, if not narrowed, will diminish an enterprise's competitive vitality, and finally into (dynamic) targeted actions that are executed in short time frames in order to close the gap.

The purpose of creating a senior level steering board is to identify the resources required for a programme of work and to make these available. A major problem of making change in companies is that the people designated to make this change are often also still required to undertake their normal activities, clearly constraining their ability to contribute to the change process. However, if senior managers can take these constraints into account they can provide a more effective resource to the change process. In addition the steering board also seeks to clear obstacles either in the subsequent mapping or implementation stages. At this point various obstacles may be encountered such as the non cooperation of one person in providing data for the detailed maps. Thus as a last resort there needs to be recourse to this top level board to unblock such problems.

The senior level steering board will also market the change programme internally and develop outline policies for the individual process teams to deploy. This activity may even become the subject of a process team in itself and will set the type of overview scorecard described above as a guiding direction for all of the activities in the company and supply chain. A last role for the steering board is to mentor individual implementation projects when these are applied later on in the Value Stream Management approach. The board itself is formed from the existing directors of the business and/or the designated champions of the various key processes defined by the management group earlier (stage 2). Where the leaders of these process groups are different from the original senior group taking part in stages 1–7, the education stage 5 needs to be repeated for any staff who did not take part in the earlier seminar.

At this stage the focus of the Value Stream Management activity moves from the senior level steering board to the different process development teams and their respective champions (Figure 3). The eleventh stage of Value Stream Management is to identify the working environment in which the change is to take place as well as the generic type of wastes that may exist. As discussed above, the original Value Stream Mapping approach was applied primarily within a shop floor order fulfilment environment and relied on Ohno's traditional seven wastes [13,14,15] to understand where value was not being added. However, the LEAP research team soon found this to be overly limiting as there were a number of other wastes within a range of other operating environments.

In all, the LEAP team were able to identify twelve wastes (including Ohno's original seven) that were applicable to the manufacturing or production environment; these are illustrated in Table 6. The original seven wastes are well documented elsewhere and are well defined by Bicheno [14] within this working environment. However, the new five wastes although frequently encountered were not being fully recognized as waste within the early Value Stream Mapping work. As later tool selection depended on an understanding of the wastes, this meant that this selection was on occasion not being optimized. The new wastes that were identified included wasted power and energy (for instance when machines were left switched on even when not in operation) and the waste of human potential (where people were either being under-utilized in their daily routines or where their full potential was not being realized as they were for example not consulted about improvements within their own working environments). In addition environmental waste or waste caused by pollution due to, for instance, chemical discharges was identified as was the waste of unnecessary overhead (too large factory, too many supervisors, fork lift trucks or office staff) as well as inappropriate design of either the product or manufacturing process itself.

Linked to the identification of these different wastes was the fact that the 12 defined wastes were being described with shop floor manufacturing language that meant very little to, for instance, a design engineer or food retailer. As a result the LEAP team have developed a list of seven generic work environments in which the twelve wastes can be applied with adequate re-naming of each waste; these are illustrated in Table 7. In each of these working environments the wastes need to be appropriately re-named so that people working in these different environments can identify the waste more easily (although, of course some of the existing names can adequately be used in other environments). An example of this is the re-naming of the twelve wastes for the warehouse and distribution environment as illustrated in Table 8.

The twelfth stage of Value Stream Management is to choose the appropriate mapping tools for the different processes under consideration. At this stage a different toolkit will be used as the basis for selection for the different processes under consideration; thus different tools are required to map the Human Resource Management process and the order fulfilment process for instance. However, for the present discussion attention will be maintained on the supply chain oriented order fulfilment process. Ten tools have been identified for the initial mapping of the order

Table 6 The twelve value stream wastes within a component and assembly production manufacturing environment

1. Overproduction	5. Unnecessary Inventory	9. Human Potential
2. Waiting	6. Unnecessary Motions	10. Environmental Pollution
3. Transporting	7. Defects	11. Unnecessary Overhead (including training)
4. Inappropriate Processing	8. Power and Energy	12. Inappropriate Design

Table 7 The seven work environments

1. Component and Assembly Production Manufacturing	3. Office and Information Systems	5. Warehouse and Distribution
		6. New Product Development
2. Complete Supply Chain	4. Process Manufacturing	7. Service and Retail

Table 8 The twelve value stream wastes within a warehousing and distribution environment

1. Doing Things Too Early	6. Unnecessary Ergonomics	11. Unnecessary Overhead (including training)
2. Waiting or Delay	7. Discrepancies	
3. Conveyance	8. Power and Energy	12. Inappropriate Layout
4. Inappropriate Processing	9. Human Potential	
5. Unnecessary Inventory	10. Excess Pack(ag)ing and Other Environmental Pollution	

fulfilment process which include the seven original Value Stream Mapping Tools together with three additions; the complete toolkit for this second level mapping (the 'big picture' cartoon being the first level) is illustrated in Table 9.

The first seven of these are well documented elsewhere [8] but it was felt lacked an ability to take on a more exact costing of existing wastes (and hence the potential for improvement) as well as not easily representing the human interaction stages of the value stream. Therefore three new tools are presented at this level. The first of these is the Value Analysis Time Profile (VATP). The method is partly derived from the Cost-Time Profiles used by Westinghouse [14,34] but develops this company-specific approach further to include an analysis of the relative waste and value content of the total cost of a product over time. This new tool as illustrated in Figure 4 is a time

Table 9 The second level process mapping tools for order fulfilment

1. Process Activity Mapping	5. Demand Amplification Mapping	8. Value Analysis Time Profile
2. Supply Chain Response Matrix	6. Decision Point Analysis	9. Overall Supply Chain Effectiveness Mapping
3. Production Variety Funnel	7. Physical Structure Mapping	10. Supply Chain Relationship Mapping
4. Quality Filter Mapping		

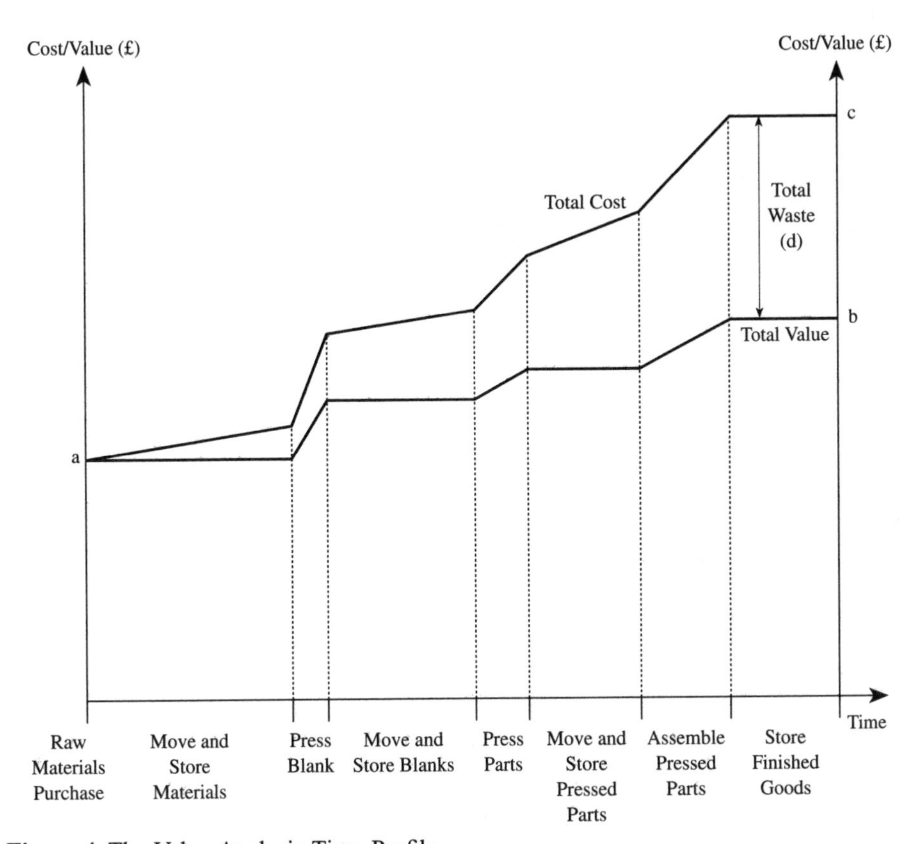

Figure 4 The Value Analysis Time Profile.

based value analysis tool which allows for the plot of both the total cost and value of the product as it moves along the supply chain under consideration.

In Figure 4, a hypothetical pressings industry case, the raw material and components are bought in at value 'a' (here assuming that there is no waste upstream of the firm concerned: an area that would be checked if the supplier integration process was also being mapped). During the various value adding steps its value is raised (for instance during the value adding parts of the blanking, pressing and assembly activity) from 'a' to total value 'b' over the period in which it is in the company. However, during this same period its cost has risen further from point 'a' to 'c'. The cost (which may bear little relation to the selling price) that has been added reflects both value added and the waste. In this example waste is added during the various non value adding steps of movement, storage and the setting up of the various processing machines. The total waste in this case is the distance 'd'. The tool therefore allows the researcher to get a very good idea of where waste is being added over time and helps pinpoint (if for instance linked to Pareto analysis) where major improvement efforts might effectively be focused. In addition it is also helpful in showing where time may be reduced for the various activities charted on the horizontal axis, providing a simple to understand time-compression tool.

Overall Supply Chain Effectiveness Mapping [35] is a tool derived from traditional shop floor Overall Equipment Effectiveness [36]. However, this precursor is generally used for the effectiveness of a particular machine or work area rather than the effectiveness of a complete supply chain. The new tool is designed to provide an total effectiveness measure for each area or section of the supply chain. The basic effectiveness measure is shown in Table 10.

Using these modifications the tool may therefore be applied right along a supply chain and summarized as in Figure 5. The use of this tool helps to identify problem areas within the supply chain, suggests sources of internal variation and emphasises the logical deployment of maintenance, quality, logistical and asset maintenance work. It also helps suggest where inappropriate stocking regimes are in place and provides a very useful bottom line effectiveness measure both within and between organizations in the supply chain.

Table 10 The Basic Effectiveness Measure

Availability for Use × Ongoing Performance Level × Quality Performance

Thus if the measure is applied to a particular bottleneck machine in a factory the effectiveness of this machine might in percentage terms be expressed as:

$$\frac{\text{Availability} - \text{Planned Downtime}}{\text{Available Time}} \times \frac{\text{Processed Amount} \times \text{Actual Cycle Time}}{\text{Operation Time}} \times \frac{\text{Total Production} - \text{Rejects}}{\text{Total Production}}$$

$$80\% \times 80\% \times 95\% = 60.8\%$$

Whilst this measure is very useful within a shop floor environment, it needs some modification within a more general supply chain setting. Here, the measure may be more broadly defined, for instance in goods inwards to be:

Components Available for Use × On Time Delivery Performance × Quality of Incoming Goods

In a similar way the measure may be applied to outgoing performance of a company:

Customer Availability of Supplied Goods × Delivery Performance × Quality of Shipped Goods

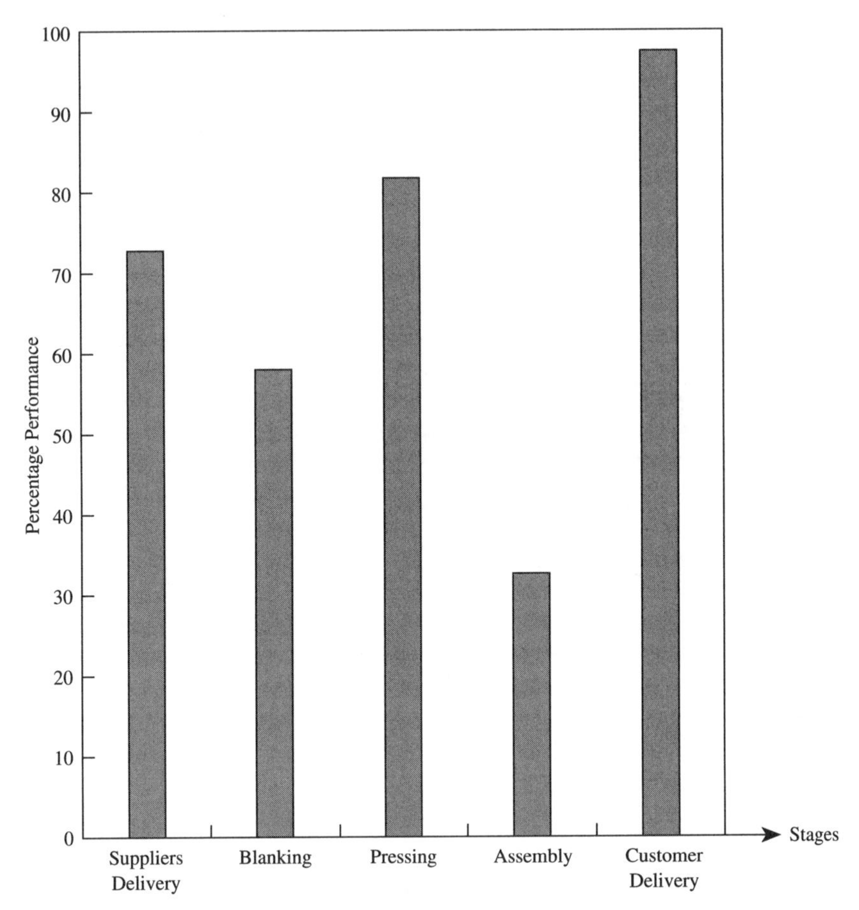

Figure 5 The use of Overall Supply Chain Effectiveness Mapping.

The last of the new tools is the Supply Chain Relationship Mapping. This map simply charts the major interactions and relationships between the different departments or sub-areas within the process being mapped. It is a widely used method but in this case the approach adopted by Alber and Walker is being followed [3]. Figure 6 gives a simple example of the use of this approach for the order fulfilment process. In this case, details are shown of the internal part of the process with the relevant links to suppliers and customers. In many cases the method is extended to include the relevant members or departments within other companies within the Value Stream. The method is quick to undertake and provides a useful insight into the existing relationships forming a firm basis for understanding which of these are working well, whether there are too many or too few relationships and perhaps who is likely to champion change or obstruct it.

UNDERTAKING SECOND LEVEL MAPPING FOR ORDER FULFILMENT

As the number of wastes and second level maps has been increased within Value Stream Mapping it is clearly necessary to extend the choice of mapping methods

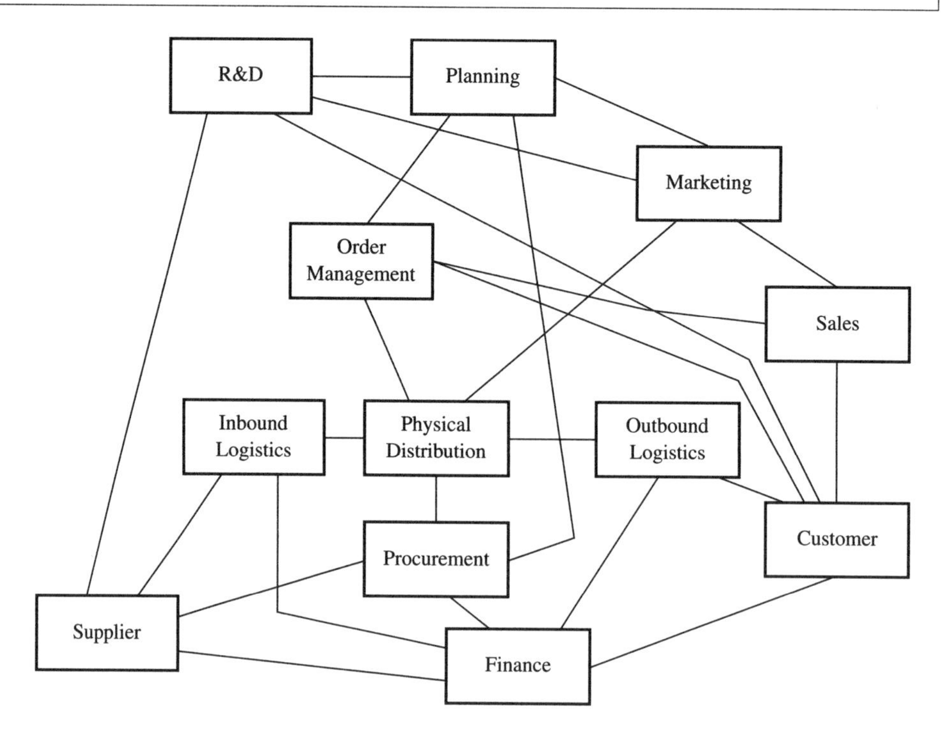

Figure 6 Supply chain relationship mapping.

from the original selection matrix used in Value Stream Mapping (Table 1). This is done by the use of an extended matrix shown here as Table 11.

The basis of choosing appropriate tools is the same as in Table 1. This involves getting the individual members of the specific process under consideration to weight the different twelve wastes on the basis of an average five points per waste with a maximum of ten and minimum of zero depending on their view of how important these different wastes are within the total process under consideration. These scores are then aggregated and divided by the total number of respondees to give an average score. This score is then used to weight the different wastes in Table 11 broadly along the lines used within the Value Stream Analysis Tool [37]. In other words, where there is a high correlation a score of nine is awarded, three for a medium correlation, one for a low one and zero where there is none. These scores are then multiplied by the weighting for each waste to give a waste/importance of correlation score for each cell. Subsequent to this the cells are summed by column to gives a total value score for the usefulness of each mapping approach in order to investigate the particular wastes under consideration. In general between three and six tools should be employed with higher scoring tools clearly given preference.

After the tools have been chosen the thirteenth stage of Value Stream Management is then to undertake the relevant mapping activity with the chosen products. Depending on the number and complexity of the maps undertaken this activity will last from anything from a day to several weeks. At the end of this mapping stage the researchers will in many cases have a clear idea of where the wastes lie. However, in almost every case where such mapping has been undertaken by the LEAP team, there

Table 11 Selection matrix for second level process mapping tools for order fulfilment

Wastes	*Process Activity Mapping*	*Supply Chain Response Matrix*	*Production Variety Funnel*	*Quality Filter Mapping*	*Demand Amplification Mapping*	*Decision Point Analysis*	*Physical Structure Mapping*	*Value Analysis Time Profile*	*Overall Supply Chain Effectiveness Mapping*	*Supply Chain Relationship Mapping*
Over production	L	M		L	M	M		H	L	
Waiting	H	H	L		M	M		M	H	L
Transportation	H						L	M		M
Inappropriate Processing	H		M	L		L	M	L	L	M
Unnecessary Inventory	M	H	M		H	M	L	M		
Unnecessary Motions	H	L					L			H
Defects	L			H				L	H	
Power and Energy	L						L	L		
Human Potential	M					L			L	M
Environmental Pollution	L			H		L	L	L	L	
Unnecessary Overhead	M	L	L		L	L	L	H	M	M
Inappropriate Design	M	M	L	L	M	H	H	M	M	M

Notes: H = High correlation and usefulness
M = Medium correlation and usefulness
L = Low correlational and usefulness

has been a need for further more detailed analysis. This may be because it has been established that machine downtime is considerable and appears to be causing considerable waste; however no accurate data is available as to what this waste is and how it is being caused. As a result, in this case, further downtime analysis is called for. Although it is beyond the scope of this paper to detail each of the individual further analysis tools that may be applied, a check list of the analysis tools that can be used within the order fulfilment process is provided in Table 12. The fourteenth and fifteenth stages of Value Stream Management is to establish what further analysis is required and to undertake this third level activity.

After the further analysis stage is complete it is then necessary to develop a customised list of the types of improvement methods that might be applied. Again following the order fulfilment process, a checklist of the implementation toolkit that may be applied is given in Table 13. At this point clearly it is not possible to apply all of these tools and indeed they should not be applied all at once. However, as a result of the mapping and further detailed analysis together with the addition of the subjective data collected throughout the Value Stream Management it is possible to identify the particular tools that should be applied. Figure 7 illustrates one example from the LEAP programme where 20 of the tools were chosen for use. However, in order to help the researcher to judge, and the company involved to understand, which tools should be used first and which later in the process the 'top hat' method is applied. This top hat is really a simple diagram that investigates the interaction of two particular improvement methods and seeks to decide whether the first should be employed before the second, at the same time or after the second. This detailed analysis as shown in Figure 7 will help provide the backbone of a successful timed change programme applied in the most efficient and effective order.

After the plan has been put in place it is necessary to develop a set of motivational control metrics to measure the progress of the plan. These should be chosen from within the general scorecard area that the senior level steering board has established for the successful application of the process under consideration. Thus, for the order fulfilment area the senior level steering board may have established that a reduction in inventory and reduction in throughput time with lower quality failures may be required. Thus the order fulfilment process in this case might construct a small number (ten or less) of metrics that will not only measure general progress but will also motivate the team members of the order fulfilment

Table 12 The third level process analysis tools for order fulfilment

1. Customer Demand Profiling	11. Delay Analysis
2. Capacity Planning for Synchronisation	12. Motion Study
3. Video Mapping	13. Time Study
4. Downtime Analysis	14. Crew Work Analysis
5. Set Up Reduction Analysis	15. Line Balancing
6. Performance Measure Review	16. Materials Handling Analysis
7. Concern, Cause, Countermeasure Analysis	17. Layout Analysis
8. Parallel Interaction & Chaos Analysis	18. Office Work Process Analysis
9. Cycle Time Reduction Analysis	19. Cost of Quality
10. Inventory Cost Drivers and Value Stream Based Costing	20. Force Field Analysis

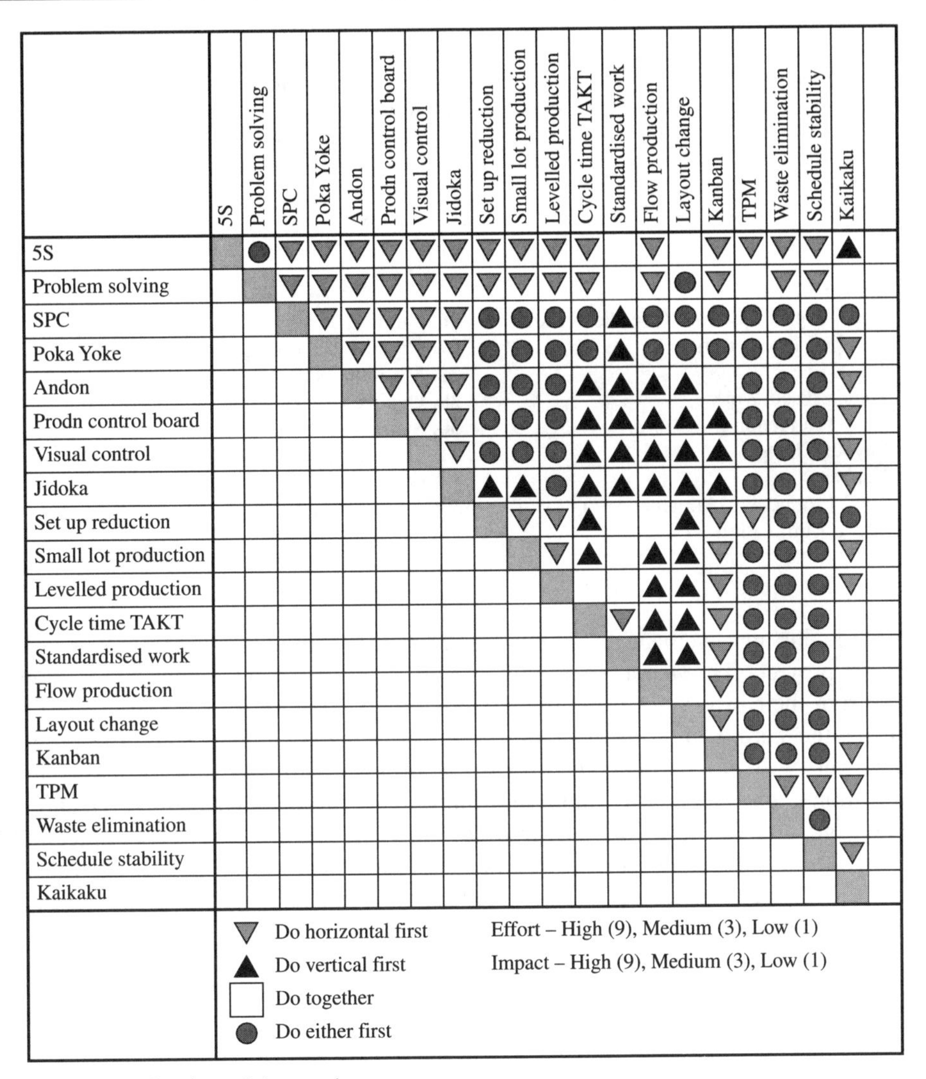

Figure 7 Application of the top hat.

process. For individual improvement projects within the change programme it may be necessary to use the same motivational metrics or even a more dynamic set of real time measures that directly measure the success of one project such as the application of 5S in a particular area of a warehouse. The specific method for applying motivational and dynamic measures is described in more detail in Dimancescu *et al.* [28].

After general control metrics have been put in place for the process under consideration, it is necessary to assemble an implementation or cross-functional action team (Figure 3) in order to enact the change. The team should be as broadly based as possible with the necessary skill range and led by a team leader. This team should be educated in lean thinking and the particular improvement method they are applying if all of the team are not aware of the method. In addition they should target

Table 13 The fourth level implementation toolkit

1. 5 Step Housekeeping (5S)
2. Problem Solving (7 Old & New Tools)
3. Statistical Process Control
4. Poka Yoke
5. Andon Boards
6. Production Control Boards
7. Visual Control
8. Jidoka
9. Set Up Reduction
10. Small Lot Production (One Piece Flow)
11. Levelled Production
12. Takt Time
13. Standardised Work
14. Flow Production
15. Layout Change
16. Kanban
17. Total Productive Maintenance (TPM)
18. Waste Elimination
19. Schedule Stability
20. Kaikaku
21. Skills Matrix (I, L, U)
22. Internal Communications and Relationship Management
23. 5 Whys
24. 5W1H (When, Why, What, Where, Who, How)
25. Failure Tree Analysis
26. Failure Mode and Effect (FMEA)
27. Hiyari KYT (Danger management)
28. Morning Market
29. Quality Circles
30. Suggestion Schemes
31. Quality Function Deployment (QFD)
32. QS 9000
33. Team Preparation

the particular improvement that is possible from the team within the timespan of the change project. The team should then undertake the improvement programme measuring their progress against the dynamic control metrics.

Over the period of implementation of the various change projects, the process champion will also constantly measure progress against the process level motivational change metrics applying appropriate project management, technical knowledge and resources as required to meet the set process level targets. At appropriate timespans the results of this progress-to-plan should be shared with the senior level steering board or by exception if there are specific roadblocks or constraints to progress.

CONCLUSIONS

This chapter has sought to provide a new methodology for the application of lean thinking within the value stream. As such it has sought to develop the existing Value Stream Mapping approach into a more strategic and holistic method. The new method described here is called Value Stream Management.

In developing this management approach the earlier Value Stream Mapping has been critiqued and has been found to have a number of constraints or problems as outlined in Table 3. As the new Value Stream Management approach has now been described, it is possible to comment on how successful this new approach will be in overcoming the previous problems. An assessment of this is given in Table 14, which demonstrates which of the various steps of Value Stream Management are likely to overcome the problems noted in Table 3. Without rehearsing each particular problem again, Table 14 shows that each of the previous problems will be largely overcome by the new method. The only exception to this is B4, namely that Value Stream Mapping was too time consuming. Indeed, due to the more strategic and wider ranging role of the new approach, Value Stream Management may be seen to perhaps be even more time consuming.

Table 14 Correlation of problems existing in value stream mapping and stage of value stream management in which solutions to the problems are applied

Stage for Solution Problem(?t)t(?b)	*1*	*2*	*3*	*4*	*5*	*6*	*7*	*8*	*9*	*10*	*11*	*12*	*13*	*14*	*15*	*16*	*17*	*18*	*19*	*20*
A1 Other Wastes											●									
A2 Other Tools						●						●		●						
A3 Value Stream Overlap														●						
A4 Chocolate Box Make-Up														●	●	●				
B1 Poor Lean Understanding					●					●								●		
B2 No Formal Education					●					●								●		
B3 Subjective Data Loss																●				
B4 Too Time Consuming																				
C1 Supply Chain Position					●					●						●				
C2 Poor Link to Strategy	●	●	●																	
C3 Other Processes Missed		●	●	●							●									
C4 Culture/Relationships				●				●	●											

However, this is not believed to be a serious problem as the Value Stream Management approach primarily focuses on the Plan stage of the traditional Deming 'Plan-Do-Check-Act' cycle [14,38]. As such it firstly encourages individual companies and supply chains to more carefully plan their change process and by so doing make sure it is more effectively focused and implemented. In addition it secondly helps companies put greater focus on change and improvement rather than maintenance of the status quo; in so doing helping to produce a more dynamic and forward focused enterprise. However, it is still true to say that the method is time consuming. As a result further research within the LEAP programme will be undertaken to see if the additional benefits of Value Stream Management may be applied within a simpler or at least quicker to use framework.

This chapter has therefore developed and described the new Value Stream Management approach which it is hoped will provide the basis for a coherent change methodology for the widespread application of lean thinking.

REFERENCES

[1] Womack, J. and Jones, D., 'From Lean Production to the Lean Enterprise', *Harvard Business Review*, March–April 1994, pp. 93–103.

[2] Christopher, Martin, Logistics and Supply Chain Management: Strategies for Reducing Costs and Improving Services, London: Pitman Publishing,1992.

[3] Alber, Karen L and William T. Walker, *Supply Chain Management: Practitioner Notes*, Falls Church, VA, USA: APICS Educational & Research Foundation, October, 1997.

[4] New, Colin, 'The Use of Throughput Efficiency as a Key Performance Measure for the New Manufacturing Era', *The International Journal of Logistics Management*, Volume 4, Number 2, 1993, pp. 95–104.

[5] Forza, C., A. Vinelli, and R. Filippini, 'Telecommunication Services For Quick Response in the Textile-Apparel Industry', *Proceedings of the 1st International Symposium on Logistics, The University of Nottingham*, 1993, pp. 119–126.

[6] Beesley, Adrian, 'A Need for Time Based Process Mapping and its Application in Procurement', *Proceedings of the 3rd Annual IPSERA Conference, University of Glamorgan, 1994*, pp. 41–56.

[7] Jessop, David and Owen Jones, 'Value Stream Process Modelling: A Methodology For Creating Competitive Advantage', *Proceedings of the 4th Annual IPSERA Conference, University of Birmingham*, 1995.

[8] Hines, Peter and Nick Rich, 'The Seven Value Stream Mapping Tools', *International Journal of Operations & Production Management*, Volume 17, Number 1, 1997, pp. 46–64.

[9] Simons, David, Peter Hines and Nick Rich, 'Managing Change in the Value Stream: An Engineering Case Study,' *Proceedings of the 3rd International Symposium on Logistics*, pp. 531–536, Padua, Italy: University of Padua, 1997.

[10] Hines, Peter, Nick Rich and Ann Esain 'Creating a Lean Supplier Network: A Distribution Industry Case', *Proceedings of the Logistics Research Network Conference*, Huddersfield, UK: University of Huddersfield, September 1997.

[11] Francis, Mark 'The Next Lean Dimension: Lean Information and its Role in Supply Chain Management' *Proceedings of the Logistics Research Network Conference*, Huddersfield, UK: University of Huddersfield, September 1997.

[12] Jones, Owen 'Optimizing the Supply Chain: Information or Inventory, *Proceedings of the Logistics Research Network Conference*, Huddersfield, UK: University of Huddersfield, September 1997.

[13] Shingo, Shigeo, A Study of the Toyota Production System from an Industrial Engineering Viewpoint, Cambridge, MA: Productivity Press, 1989.

[14] Bicheno, John, *The Quality 50*, Buckingham, UK: Picsie Books, 1994.

[15] Japan Management Association, *Kanban: Just-in-Time at Toyota*, Cambridge, MA: Productivity Press, 1985.

[16] Ishiwata, J., *Productivity Through Process Analysis*, Cambridge, MA: Productivity Press, 1991.

[17] Practical Management Research Group, *Seven Tools For Industrial Engineering*, Tokyo, Japan: PHP Institute, 1994.

[18] New, Colin, 'The Production Funnel: A New Tool for Operations Analysis', *Management Decision*, Volume 12, Number 3, 1974, pp. 167–178.

[19] Forrester, J., 'Industrial Dynamics: A Major Breakthrough For Decision Makers', *Harvard Business Review*, July–August 1958, pp. 37–66.

[20] Burbidge, J., 'Automated Production Control with a Simulation Capability', Proceedings of the IFIP Conference Working Group 5–7, Copenhagen, 1984.

[21] Hoekstra, S. and S. Romme, (eds), Towards Integral Logistics Structure – Developing Customer Oriented Goods Flows, New York: McGraw-Hill, 1992.

[22] Crowley, James, 'An Application of Pipeline Mapping in the Food Sector', *Proceedings of the Logistics Research Network Conference*, Huddersfield, UK: University of Huddersfield, September 1997.

[23] Cox, Andrew and Peter Hines, 'Advanced Supply Management in Theory and Practice', In: Cox, Andrew and Peter Hines (eds), *Advanced Supply Management: The Best Practice Debate*, pp. 1–20, Boston, UK: Earlsgate Press, 1997.

[24] Hall, Richard, 'New Directions in Purchasing – The Search for the Efficient Boundary of the Firm', *Proceedings of the Logistics Research Network 'New Directions in Logistics' Workshop*, Warwick: University of Warwick, September, 1996.

[25] Nonaka, Ikujiro, 'A Dynamic Theory of Organizational Knowledge Creation', Organization Science, Volume 5, Number 1, 1994, pp. 14–37.

[26] Delbridge, Rick and Ian Kirkpatrick, 'Theory and Practice of Participant Observation', In: Wass, Victoria and Peter Wells (eds), *Principles and Practice in Business and Management Research*, pp. 35–62, Aldershot, UK: Dartmouth Publishing, 1994.

[27] Wass, Victoria and Peter Wells, 'Research Methods in Action: An Introduction', In: Wass, Victoria and Peter Wells (eds), *Principles and Practice in Business and Management Research*, pp. 1–34, Aldershot, UK: Dartmouth Publishing, 1994.

[28] Dimancescu, Dan, Peter Hines and Nick Rich, The Lean Enterprise: Designing and Managing Strategic Processes for Customer-Winning Performance, New York: AMACOM, 1997.

[29] Kurogane, Kenji, (ed.), *Cross-Functional Management: Principles and Practical Applications*, Tokyo, Japan: Asian Productivity Organization, 1993.

[30] Womack, James, Daniel Jones and Daniel Roos, *The Machine That Changed The World*, New York: Rawson Associates, 1990.

[31] Womack, James and Daniel Jones, *Lean Thinking: Banish Waste and Create Wealth in your Corporation*, New York: Simon & Schuster, 1996.

[32] Hines, Peter, Creating World Class Suppliers: Unlocking Mutual Competitive Advantage, London: Pitman Publishing, 1994.

[33] Lockyer, Keith, Critical Path Analysis and Other Project Network Techniques, London: Pitman Publishing, 1984.

[34] Fooks, Jack, *Profiles for Performance*, Addison Wesley, 1993.

[35] Rich, Nick and Peter Hines, 'Evaluating the Effectiveness of the Value Stream: The Overall Supply Chain Effectiveness Tool', *Lean Enterprise Research Centre Occasional Paper*, Cardiff: Cardiff Business School, 1998.

[36] Nakajima, S, *TPM: An Introduction to TPM*, Cambridge MA: Productivity Press, 1988.

[37] Hines, Peter, Nick Rich and Malaika Hittmeyer, 'Competing Against Ignorance: Advantage Through Knowledge', *International Journal of Physical Distribution and Logistics Management*, Volume 28, Number 1, 1998, pp. 18–43.

[38] Imai, M., Kaizen*: The Key to Japan's Competitive Success,* New York: McGraw-Hill, 1986.

ACKNOWLEDGEMENTS

The authors would like to thank the other staff members at the Lean Enterprise Research Centre who have contributed in words, deeds or actions to the development of this paper. In particularly we would like to thank Mark Francis, Owen Jones, Donna Samuel, David Simons, Ann Esain and Daniel Jones. The work described here was primarily carried out as part of the Lean Processing Programme (LEAP).

5 Benchmarking Toyota's supply chain: Japan vs. U.K.

Peter Hines

HOW TOYOTA SUPPLIERS ARE DEVELOPED THROUGHOUT THE VALUE STREAM

The advantage gained by Japanese car manufacturers over their western competitors has been well documented.[1–6] These advantages have been felt both at the vehicle assembler level,[1,6] and at the component manufacturer level.[2–5] At the assembler level, Womack, Jones and Roos report quality gaps of 2 to 1, productivity gaps of 1.82 to 1, and inventory levels ten times higher outside Japan.[1] At the direct or first tier component supplier level, similar gaps have been demonstrated by Andersen.[3,4] In paticular the Andersen sponsored research team found 100 to 1 quality gaps, 2 to 1 productivity gaps and 7 to 1 inventory gaps when comparing the abilities of U.K. and Japanese component makers within four product category areas.[3] In addition, previous research suggests that the ability of Toyota and their suppliers within Japan may be higher than that of some of their domestic competitors due to their use of the Toyota Production System (TPS).[1,5]

The Toyota Production System is, simply put, a method of shortening the time it takes to convert customer orders into vehicle deliveries. In order to achieve this the entire sequence from order to delivery is arranged in a single, continuous flow with continuous efforts made in terms of shortening the sequence and making it flow more smoothly. The result of this is a far higher level of productivity, better quality and a major reduction in wasted time, money and effort, or, in short, better products made more cost effectively.[7]

However, although the previous research is useful, it does not answer all the key questions regarding Toyota's success in Japan. First, it fails to extend further along the value adding supply chain involved in the production of the final car. Thus, although data is available for Japanese car assemblers and first tier (or direct component suppliers) it is not readily available for lower-tier organizations (or firms that indirectly supply components to Toyota, where a third-tier firm supplies a second-tier firm who in turn supplies a first tier-organization that sends products to Toyota) or raw material manufacturers. Linked to this point is that the majority of data does not differentiate between firms primarily supplying Toyota and those primarily supplying other Japanese car makers. If both these points can be answered then it may be

This paper was first published in *Long Range Planning*, 31(6), 911–918, 1998, Elsevier Science Ltd.

possible to suggest where and why the Toyota Production System is superior to other approaches. In addition, if TPS is better than other approaches we can ask whether it is transferable to other countries.

METHODOLOGY

This research programme involved the benchmarking, verification visit, semi-structured interview and modelling of a group of 8 first-tier and 13 second-tier Toyota suppliers in Japan together with a similar grouping in the UK. The benchmarking questionnaires were sent to the participating companies approximately one month before a visit was made. The firms were requested to return the completed questionnaires before the visit date. This was the case in all but one company. Each questionnaire was then individually interrogated to ensure that errors had not occurred during completion. This was primarily achieved by entry onto a comprehensive spreadsheet with pre-determined check questions and calculations. This is believed to have identified over 90 per cent of any suspect or missing data at this pre-visit stage.

Subsequent to this analysis a visit was made to each company site for the purposes of verification and qualitative semi-structured interview. During these half-day visits, remaining missing or suspect data was checked and verified. The verification also included discussion of each data set and a tour of the factory shop floor. In addition, a semi-structured interview (1.5 to 2.5 hours in length) was undertaken to ascertain how the results demonstrated on the questionnaire were achieved. In particular, time was spent understanding to what degree the Toyota Production System (TPS) was employed, when it had been employed, how it had been learnt and how it had been disseminated to suppliers. Due to the range and detail of the techniques employed it is believed that the resulting evidence and data displays a high degree of rigour and integrity.

To supplement the above approach, structured interviews were carried out with relevant trade associations, academics and researchers in Japan and the UK. This methodology was chosen to give adequate research triangulation between the approaches used to provide as far as possible a realistic appraisal of the existing situation.

MAIN FINDINGS

The Japanese first-tier suppliers appeared to perform best in virtually every measure employed whether it was concerned with process results, internal excellence measurement or supply chain integration. With the exception of the new product development process, the Japanese second-tier firms showed themselves to be broadly superior to the UK first and second tiers. This will be demonstrated in the following section using various value stream mapping tools before a discussion of the underlying factors is undertaken.[8]

Supply chain responsiveness

As can be seen from Figure 1 *the responsiveness of the Japanese supply chain far exceeds the UK chain.* The figure plots the cumulative inventory and lead time in both countries from the point of delivery of raw materials to second tier firms to the point of delivery to the assembler. In the Japanese case this process takes 5 working days using a pull kanban system rather than the more traditional western push system involving 40 working days of lead time. In a similar way the total inventory required in Japan for this portion of the supply chain (21 working days) is far lower than in the UK (127 working days).

Process Abilities

As it is believed that key process deliverables are more important to customers than departmental excellence (Dimancescu *et al.,* 1997) data was collected for both geographical areas focusing on key process deliverables. This was added to earlier data (Womack *et al.,* 1990) to provide information at assembler, 1st, 2nd and even 3rd tier levels. Table 1 demonstrates that the gaps in quality, productivity (cost) and delivery performance are considerable at every tier. However, it is interesting to note that the quality and delivery performance gaps peak at the 1st tier whilst the productivity gap is widest at the 2nd tier.

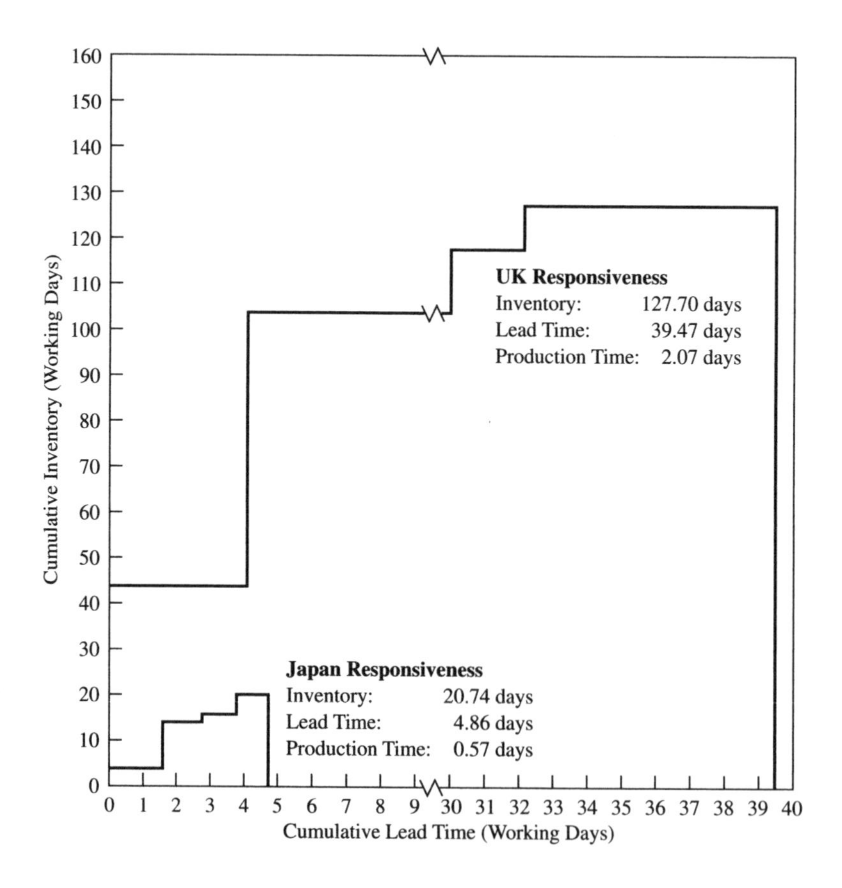

Figure 1 Supply chain responsiveness: UK vs Japanese automotive industry.

Table 1 Comparative Process Control Abilities in the Japanese and UK Automotive Industries

	Assembler[1]	*1st Tier*	*2nd Tier*	*3rd Tier*
Quality				
Customer delivery defect rate	2.00	244.50	12.01	2.57
Productivity				
Value added per qualified employee*[2]	1.82	2.84	4.35	N/A
Delivery				
-- Inventory level	10.00	14.34	4.33	N/A
– Late delivery	N/A	283.82	13.20	1.71

*[1] Based on data from Womack, Jones & Roos (1990) Japanese vs Western assemblers.
*[2] Qualified employee includes direct shop floor operators and supervisors / team leaders who spend the majority of their time as direct labour.

Value added in the supply chain

We compared the value adding profiles of the two country's automotive industries. Figure 2 segregates the raw material component of the final product outside the pyramid structure. This has been done because the principles, dynamics and style of relationships operating in the raw material value stream are very different from those operating within the parts and components value stream. Figure 2 demonstrates that there are a number of key similarities and differences between the two data sets. The first similarity is that the value added by assemblers between the two regions is very similar due to the emulation by UK based firms of earlier Japanese preferences for outsourcing.

In contrast, the value or more correctly *the cost added by parts suppliers is considerably higher in the UK (45.7% compared with 34.7% in Japan) and the cost added by raw material firms (31.3% compared with 43.1%) is considerably higher in Japan.*

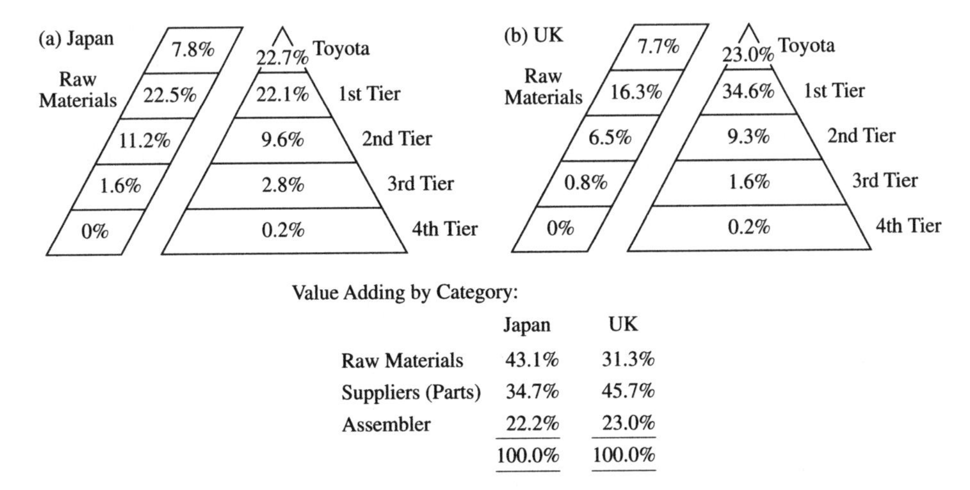

Value Adding by Category:

	Japan	UK
Raw Materials	43.1%	31.3%
Suppliers (Parts)	34.7%	45.7%
Assembler	22.2%	23.0%
	100.0%	100.0%

Figure 2 Value added by Toyota suppliers in Japan and the UK.

Variability in demand

The last key area for discussion is the demand variability within the different supply chains as shown in Figure 3. The variability is based on the difference between forecast orders one month before delivery and the actual quantity required. Although variability increases from the assembler back down the supply chain in Japan, the change is only from 2.2% up to 4.2% from the assembler's purchases, through the different decision points, to the 2nd tier firm's purchases reflecting the near exact pull of product from third tier firms.

In contrast in the UK variability from assemblers starts at 12.2%, which is dampened by the first tier by high inventory levels but greatly amplified at 2nd and 3rd tier levels. When a cost is apportioned to this variability (based on the cube of the variability) then the cost of demand amplification can be seen to be disproportionately higher in the UK case than the negligible cost to the Toyota supply chain in Japan (Figure 3).[10] The poor and declining productivity levels in the UK with lower component tiers discussed above can now be partly understood.

DISCUSSION

Although a full discussion of the data requires more space than this paper allows, a few key points can be made. The first concerns the weighted productivity gap in the total supply chain between Japan and the UK. This is shown in Table 2 and is found by finding the total productivity gap weighted by the value added at each tier. It can be seen from these crude calculations that *the Japanese supply chain, in total, shows a productivity level double the UK*. This gap is verified when attention is directed to the very low relative prices of new Toyota cars in Japan relative to other general products.

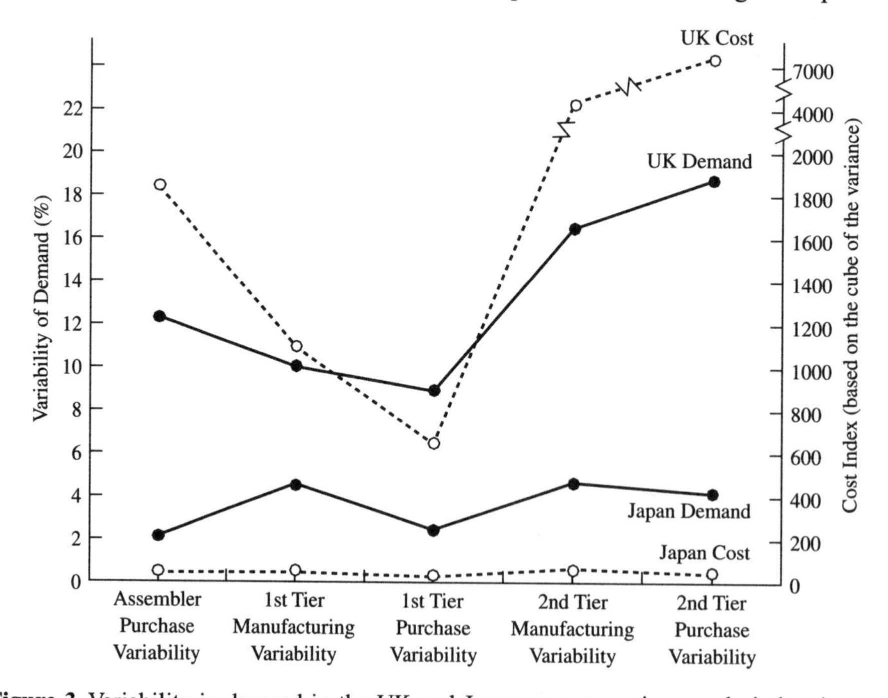

Figure 3 Variability in demand in the UK and Japanese automotive supply industries.

However, the very high productivity levels in the Japanese assembler and component firms are not matched by similar gaps in the Japanese raw material firms. These firms have followed a TQC approach that has succeeded in yielding high levels of quality. However, such firms have not emulated the Toyota Production System and largely failed to integrate their key processes with those of Toyota or their other component-making customers through activities within customers' *kyoryoku kai* (or Supplier Association) for example. This would have helped to remove inter- or intra-company waste as happened in the rest of the Toyota supply chain. The challenge then in Toyota's Japanese supply chain is to do this. The challenge in the UK is to make dramatic improvements in productivity, quality and delivery performance at the assembler and each component level in order to close the gaps with Japan. In addition if the raw materials makers can be integrated into the supplier network in the UK then there may be a source of potential competitive advantage in that country even over Toyota's approach in Japan.

A third major point for discussion is the actual mechanisms used within the Japanese Toyota supply chain that clearly differentiate it from its UK counterpart. Based on personal observation and detail discussion with the firms involved it would appear four key elements have been brought together in Japan that are absent or at least only partially realised in the UK.[11] These are the use of:

1. Policy Deployment (*hoshin kanri*) for internal strategic focusing and alignment. This approach provides a clear step-by-step planning, implementation and review process for managing change. It is an holistic systems-based approach for deploying targeted management policies and priorities down the management hierarchy in order to allow for effective planning of the means of implementation. (A fuller description of the approach can be found in Dimancescu *et al.*).[9]
2. Cross Functional Management (e.g. quality, cost and delivery) in order to actualise the policies developed. The Japanese approach to cross-functional management is guided by policy deployment which provides the focus, direction and speed of change within specific processes. For many Japanese firms the key cross-

Table 2 The competitive advantage of Toyota (Japan) vs the Toyota (UK) supply chains

	Japan Value Added (%)	*Competitive Gap*	*UK Value Added (%)*	*Indexed Competitive Gap Apportionment (%)*
Assembler	22.2	1.82*[1]	40.4	18.0
1st Tier	22.1	2.84	62.8	40.2
2nd Tier	9.6	4.35	41.8	31.8
3rd Tier	2.8	4.35*[2]	12.2	9.3
4th Tier	0.2	4.35	0.9	0.7
Raw Materials	43.1	1.00*[3]	43.1	0.0
Total	100	2.01*[4]	201.2	100

*[1] Womack, Jones & Roos (1990)
*[2] Assumed same as 2nd Tier
*[3] Based on various industry experts' viewpoints
*[4] Weighted Competitive Gap

functional processes are quality, cost and delivery, although other processes such as design, supplier integration and sales acquisition may also be managed in this way.[9]

3. Toyota Production System (TPS) to yield a standard management approach in the supply chain. TPS is a method developed by Toyota to deliver more effectively the products which their customers require, in a more timely manner than traditional management approaches.[7]
4. Supplier integration particularly focuses around the use of the *kyoryoku kai* or supplier association, which integrates each tier with the one above and below and allows for an external version of policy deployment, cross functional management and inter-company learning and development.[5]

Role of the first tier firms

The first-tier firms play a key role in Toyota's Japanese supply chain. There are six areas in which these firms act as the pivotal part of the system.

1. Quality Buffer. If a longitudinal cross-section of the supplier system is taken (Figure 4) it can be clearly seen that not only are the first tier firms most adept at controlling their own defects but significantly act as a buffer for their customers by controlling the quality of their suppliers. They thus act as a quality filter. This is particularly the case in Japan as can be seen in Figure 4. This means that Toyota can produce excellent quality products even though its second, third and lower-tier suppliers are not always so excellent in their quality performance.

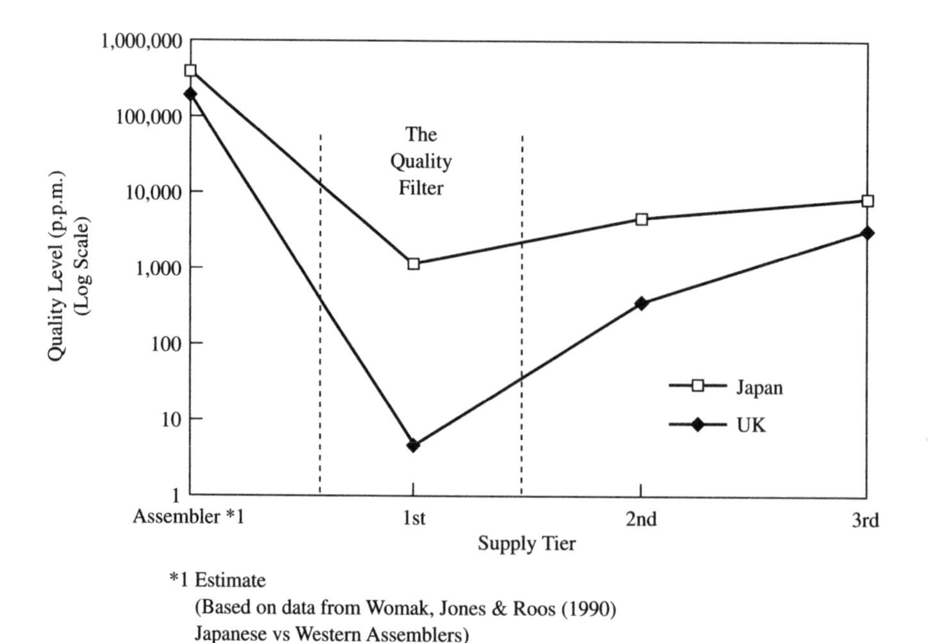

Figure 4 First-tier firms act as a quality filter.

2. Gaining (productivity) competitive advantage. As can be seen from Table 2 the competitive gaps that exist in productivity give Toyota in Japan a keen advantage in all areas except their raw material supply. However, the advantage is not uniform in the supplier network and is greater at first tier than at assembler level (and may be even greater than the figures suggest due to western assemblers closing the gap on Toyota in recent years). When these gaps are indexed according to the value adding at each level it becomes clear that the advantage that Toyota gains in Japan is largely a result of their own productivity (18%), that of the first tier (40%) and that of the second tier (32%). However, the largest advantage lies at the first-tier level.
3. System developer. Due to the focusing that the first tier firms employ in the coordination and development of their suppliers in Japan, Toyota gains significant advantage from their 2nd tier suppliers, many of whom they do not even know the names of. Thus, in addition to the advantage that the 1st tier firms exhibit themselves, they are also instrumental in unlocking another 32% directly from their suppliers (and indirectly another 10% from 3rd- and 4th-tier firms). Thus the 1st tier firms have, through their own and Toyota's work, developed a system which means that Toyota can lever their internal competitive advantage (largely gained by the prolonged and rigorous use of TPS) by at least a factor of five within their supplier network with the first tier acting as the key architects of this advantage through the use of methods such as the various *kyoryoku kai*.
4. Purchaser. The first-tier suppliers are also the focal point of raw material purchasing as they, with the exception of the direct purchases by Toyota, directly buy not only their own raw material requirements but also the majority of raw materials on behalf of their (direct and indirect) suppliers. Such materials are supplied to the second- and lower-tier firms on a Just In Time basis but are in general supported by a stockholding in subsidiary companies of the 1st tier component manufacturers. Although no verifiable data was collected as to the performance of these firms, anecdotal evidence suggests that these stockholders achieve stock turns of around 12 per year, putting them in line with the best performers in the sector in the UK.

   It would not be true to say that all raw material purchases made by second-tier firms are made in this way. However, even when first-tier firms are not actually buying on behalf of their suppliers they tend to exert a large or even total influence over which raw material suppliers are used by their suppliers and indeed even the prices to be paid.
5. Designer. The majority of the detailed design work for components is not done by Toyota themselves but by their first-tier suppliers. When comparing the design engineering hours for components made by parts suppliers in Japan and the UK, a low time input is given by Toyota in Japan (7% compared with 26% in the UK), a higher proportion at the first tier (88% compared with 69%) and a broadly similar figure at the 2nd tier (5% compared with 4%). The first-tier firms are thus key in both new product development as well as off-line research and development. In this respect Toyota has moved to being a concept designer and detailed design facilitators, keeping only a small percentage of detailed design to themselves.
6. Problem solvers. The effective use of engineering and quality staff in supplier development by the Japanese first tier firms helps them to solve their own suppliers' problems in a very effective manner. However, it is important to note

that not only is their very rapid usage of stock useful in itself for rapid cash turnover, but it critically helps identify the most important problems to be solved either in the first-tier themselves or with their own supplier networks. Thus the 1st tier acts as a facilitator of focused change not only for themselves but also for their direct and indirect supplier.

The role of the very small firm

Within the Toyota supplier network in Japan there is only a small reliance on the micro firms that authors such as Piore and Sabel (1984, *inter alia*) have suggested as being a major area of competitive advantage in such supplier networks. However, the key to Toyota's success cannot lie within such very small firms at the 3rd and 4th tiers as they are only responsible for only 3% of the value adding processes, although due to their high productivity this yields a 10% total productivity advantage to the total Toyota supplier network.

The application of the Toyota supplier system outside Japan

In Japan, Toyota should investigate how TPS can be spread to, and competitive advantage gained from the raw material industry and indeed even looking overseas for sources of steel and plastic. In addition, Toyota management might also consider how they could assist the first tier firms gain more competitive advantage from the second- and lower-tier firms. Although the Japanese second tier is in most cases superior to the UK first tier there are still large gaps between the two tiers in Japan. A recent Toyota Annual report suggests that appropriate actions in this area are already under way.

The problem for Toyota in the UK is how to increase the productivity of the UK supply chain. The answer must lie in intensive coordination and development of the first tier suppliers and major efforts to encourage these firms to spread this message on down the hierarchy. Recent evidence from the UK and the USA suggests that this is what they are doing through the application of the *kyoryoku kai* or Supplier Association approach.[5]

RESEARCH LIMITATIONS

In spite of the care taken with the survey, the research suffers from two important limitations which should be borne in mind by the reader in drawing conclusions from the work.

1. The first-tier firms in the UK and Japan can only be considered to be broadly similar. Although a strictly comparable paired research method (as employed by Andersen, 1992, 1994) was not employed due to research access and the differing research purposes involved, it is not believed that this greatly compromises the results. The reason being that although individual differences were present between firms, when these are aggregated the range and depths of product and process complexities were broadly similar in the two first tier sets of data. However, some margin for error should be allowed for in interpretation.

2. The selection criteria for suppliers may suggest a slight skew in favour of selecting a 'better' group of Japanese suppliers than British suppliers among their respective peers. The reason for this is that the British first-tier firms are the seven Toyota suppliers in Wales (a region in the UK) whereas the Japanese first-tier sample are the board members of the Tokai Kyoho Kai. As a result it may be anticipated that these latter firms are at least of average ability and expertise whereas the British set are perhaps only average within Toyota's supplier base in the UK. However, anecdotal evidence provided by another UK based assembler has suggested that the seven UK suppliers are above average within their own supplier benchmarking scheme.

As a result of these two limitations strict use of the data is not to be recommended although general conclusions may still be made.

CONCLUSION

Questions were raised at the beginning of this article as to whether the Toyota Production System, as employed in Japan by Toyota, is superior to other approaches; where does the advantage lie in the supply chain and why? Evidence is presented in the article to show that the key to this success lies all along the supply chain but is particularly marked at the first tier or direct component manufacturer level. Comparative research findings suggest that at this supply level the maximum comparative gap is to be found with the more traditional western management approach particularly in quality and delivery performance.

The first-tier component makers play a key role in achieving the success of the total Toyota supplier system. They play a number of central coordinating roles including quality buffer, gaining competitive advantage, systems developer, purchaser, designer as well as problem solver. These first-tier firms ensure that they produce high quality products, delivered on time and they also develop the expertise and abilities of lower-tier organizations.

However, to regard the system as a direct result of the first tier's efforts would be to ignore the fundamental role in Japan of Toyota themselves. Indeed it would appear that what has brought about the present situation is Toyota's ability in developing TPS and in integrating the policies and practises of their suppliers with their own, i.e. the extension of internal policy deployment through their Supplier Association into the supplier network and the active coordination and development of suppliers, directly, and indirectly through the first tier suppliers, through the widespread application of the Toyota Production System.

It would be easy to conclude that Toyota had little potential for improvement in their Japanese supplier network, but this would not be true. Significant improvements may be forecast in Japan if attention can be directed at the raw material suppliers. In addition, through the localisation and development of TPS with European-based suppliers to Toyota, a similar high performing supplier network may be developed for the UK plant. Whether this proves to be the case is an area for future research.

REFERENCES

1 Womack J., Jones D., Roos D., (1990), *The Machine That Changed The World*, New York: Rawson Associates.

2 Nishiguchi T., (1994), *Strategic Industrial Sourcing: The Japanese Advantage*, New York: Oxford University Press.
3 Andersen Consulting, (1992), *The Lean Enterprise Benchmarking Project*, London: Andersen Consulting.
4 Andersen Consulting, (1994), *Worldwide Manufacturing Competitiveness Study*, London: Andersen Consulting.
5 Hines, P. (1994), *Creating World Class Suppliers: Unlocking Mutual Competitive Advantage*, London: Pitman Publishing.
6 Womack J., Jones D. (1996), *Lean Thinking*, New York: Simon and Schuster.
7 Toyota Motor Corporation (1992) The Toyota Production System. Toyota City: Toyota Motor Corporation.
8 Hines P., Rich N., (1997), "The Seven Value Stream Mapping Tools", *International Journal of Operations & Production Management*, Vol. 17, No. 1, pp. 46–64.
9 Dimancescu D., Hines P., & Rich N., (1997), *Creating the Lean Enterprise: Design & Managing Strategic Processes For Customer Winning Performance, New York:* AMACOM.
10 Stalk G., Hout T., (1990), *Competing Against Time; How Time Based Competition is Reshaping Global Markets*, New York: Free Press.
11 Rich N., Hines P.,(1997), "The Three Pillars of Strategic Alignment: A Model for the Achievement of Internal and External Synergy", *ICPR 97 Conference*, Osaka, August.
12 Piore M., Sabel M., (1984), *The Second Industrial Divide: Possibilities For Prosperity*, New York: Basic Books Inc.
13 Toyota Motor Corporation, (1994), *How We Saved $1.5 Billion (Annual Report),* Toyota City: Toyota Motor Corporation.

Section 3

Analysing Supply Chain Performance and Identifying Waste

Waste elimination – a supply chain perspective 6

David Brunt and Chris Butterworth

INTRODUCTION

To create a competitive, world-class organization we realize we cannot stand still. We must continuously improve all aspects of our business to cope with existing and new competitors and ever-increasing customer expectations. Toyota's Taiichi Ohno realized that improvement must never stop. He identified that only activities that the customer is willing to pay for are 'value added' and that when we actually analyse the activities that we do in a company, large proportions of them add no value through the eyes of the customer.

The understanding of what waste is and how to remove it is fundamental to the creation of a lean enterprise or supply chain; however, many managers and companies often do not understand or realize the importance of the concept. Frequently we hear that managers and workers are too busy to allow people time to plan and implement improvement activities, but as customers most of us would not want to pay for the activities they currently do.

Value chain analysis has been adopted as a method of achieving an understanding of strategic capability by identifying how the activities of the organization underpin its competitive advantage. The technique is used to shed light on the profitability of separate steps in complex manufacturing processes, in order to determine where cost and/or value creation can be improved (1). While analysis of the individual links in the value chain can be of use, it is argued that the full benefits will only be realized when integration of this chain occurs (2). This linking of value added activity forms the value stream that creates, sells and services a family of products. In understanding different value streams, Hines and Rich (Chapter 3) contend that 'it is necessary to map these intercompany (activities between companies) and intracompany (activities inside a company) value adding processes'. The main reason for this is to help researchers or practitioners identify waste with the aim of finding ways to remove or reduce it.

The paper is divided into four parts. First, the subject of waste removal will be discussed. This will lead to a description of the methodology of waste elimination in the LEAP programme. It is recognized that the waste methodology is part of a Value Stream Management method, a strategic and operational approach designed to help a company or complete supply chain begin the journey of becoming lean. In the third part of the paper, a new classification of waste relating to a supply chain perspective will be discussed and new wastes drawn from the research highlighted. Finally a summary and conclusions will be drawn from the work.

UNDERSTANDING WASTE REMOVAL

In his book *Today and Tomorrow,* Henry Ford reveals his philosophy of industry (3). With regard to the nature of waste in industry, Ford suggests that 'it is use – not conservation – that interests us. We want to use material to its utmost in order that the time of men may not be lost.' Therefore if something has labour expended upon it, and is subsequently wasted ('we do not put it to its full value'), then the time and energy of men are wasted. Therefore Ford suggests that 'we will use material more carefully if we think of it as labour'. The fundamental driver behind eliminating waste is 'true efficiency', which Ford said is simply a matter of doing work using the best methods known, not the worst.

This view raises the question of efficiency and whether it is a function of quantity and speed. Ohno (4) suggests that 'efficiency is never a function of quantity and speed . . . the Toyota production system . . . has always suppressed overproduction, producing in response to the needs of the marketplace'. The Toyota production system was born out of the need to develop a system for manufacturing small numbers of many different kinds of automobiles (5). This approach is in direct contrast to the view supported by Ford, which results in the practice of producing large numbers of similar vehicles. Thus to eliminate waste in the production process, Ohno classified waste into seven categories.

The wastes as classified by Ohno are productivity, rather than quality related (6). However, quality and productivity are closely linked. The contention is that improved productivity leads to leaner operations which make quality problems more visible. Furthermore, improved quality improves productivity by removing wasteful practices such as rework and inspection. Ohno regarded overproduction as the most serious waste as it is the root of so many problems. For example, it discourages the smooth flow of goods or services and is likely to inhibit quality and productivity. Overproduction can also cause excessive lead-time and storage times. This can result in quality problems not being detected early and products deteriorating. In addition, work-in-progress stocks may become excessive, which can result in operations becoming dislocated.

The wastes outlined in the Toyota production system are shown in Table 1 along with a number of features common when these wastes are present in the production environment.

In addition, when identifying the steps required to create, order and produce a specific product it is possible to categorize each step into three areas: firstly, those that actually create value as perceived by the customer; secondly, those that create no value but are currently required to produce the product; and thirdly, those activities that do not create value as perceived by the customer (8,9).

Finally, stepping aside from the Toyota Production System view of waste elimination, Canon (the Japanese manufacturer of cameras, copying machines and computers) identifies nine wastes in its production system. These are shown in Table 2.

It can be seen that there is a large amount of overlap between Ohno's categorization of waste and that of Canon. For example, the Toyota wastes of unnecessary inventory (work-in-progress), defects (rejection), waiting (facilities) and motion (motion) display large similarities with the Canon headings in brackets. However, the fundamental difference appears to be the emphasis of overproduction within the Toyota production system. In addition, the subjects of transportation and inappropriate processing are not covered directly in the Canon system.

The nine Canon wastes, while still in a production setting, encompass a wider business view as the issues of design, new product run-up, expenses (investment) and talent (personnel issues) are explicitly categorized. However, it could be argued that this might reduce the ability of practitioners to identify and eliminate waste in their work environments, as Canon's wastes are less focused than those exhibited in the Toyota Production System.

This point becomes particularly relevant when considering that:

- waste elimination is relevant across complex supply chains and in environments other than production;
- a number of new factors affecting competitiveness are now in existence.

Table 1 The seven wastes of the Toyota Production System

Waste Category	*Nature of Waste*
Overproduction	Smooth flow of goods difficult Piles of WIP Target and achievement unclear Excessive lead-time and storage times
Waiting	Operators waiting Operators slower than line Operators watching equipment and operation
Transportation	Stacking and unstacking of components Conveyors Many busy forklifts Widely spaced equipment
Inappropriate processing	Variation between operators methods Variation between standard and actual operation Processes that are not statistically capable
Inventory	Prescribed storage volume exceeded Deteriorating material Old dates on material Stocks of containers for WIP Sophisticated stores system
Motion	Components and controls outside easy reach Double handling Layout not standardized Widely spaced equipment Operators bending
Defects	Poor material yield Work in scrap bin High inspection levels Difficult assembly Large rework area High customer complaints Irregularity of work

In order to understand waste and remove it, a number of techniques have been developed. For example, Kobayashi (10) suggests the use of 'treasure maps' to provide an easy way to see where waste exists. The treasure map categorizes waste into three levels of severity: activities that are extremely wasteful (gold) through to activities with progressively smaller degrees of wastefulness (silver and copper). Activities that appear to contain very little waste are labelled flatlands, which may have hidden waste after the treasure mines (gold, silver and copper) have been depleted. Kobayashi suggests that once the waste has been highlighted, a friendly contest to raise the actual work ratio in each workplace can commence and improvement plans can be developed to reduce the waste.

LEAP METHODOLOGY FOR WASTE IDENTIFICATION AND REMOVAL

To remove waste from a supply chain, it is necessary to understand the waste in the particular companies along that chain. Therefore the starting point for the waste exercise involves identifying the generic wastes that exist. This is done by selecting a number of personnel from across the value stream and asking them to complete a 'Waste Workshop' as shown in Figure 1. Figure 2a shows examples of the areas from which personnel may be selected. In a manufacturing environment these can include logistics, purchasing, production, maintenance, personnel, stores and sales. The selected personnel are given a written summary of each of the wastes found in the Toyota production system. They are asked to read these and assign a score to each of the wastes they feel are relevant to the company.

Following the 'Waste Workshop', a structured interview is used. There are a number of reasons for this. Firstly, it provides the practitioner or researcher with a

Table 2 Canon's Nine Waste Categories. Source: Imai (1986:249)

Waste category	*Nature of waste*	*Type of economization*
Work-in-progress	Stocking items not immediately needed	Inventory improvement
Rejection	Producing defective products	Fewer rejects
Facilities	Having idle machinery and break-downs, taking too long for set-up	Increase in capacity utilization ratio
Expenses	Over-investing for required output	Curtailment of expenses
Indirect labour	Excess personnel due to bad indirect labour system	Efficient job assignment
Design	Producing products with more functions than necessary	Cost reduction
Talent	Employing people for jobs that can be mechanized or assigned to less skilled people	Labour saving or labour minimization
Motion	Not working according to work standard	Improvement of standard work
New-product run-up	Making a slow start in stabilizing the production of a new product	Faster shift to full-time production

background to the organization. This background is useful when using mapping tools with the organization, when analyzing and interpreting any data collected in the organization and also when educating or implementing improvement activities, as cultural issues, special causes to problems and one-off events may be highlighted. Secondly, it allows the researcher to meet relevant people throughout the company. This is again of use later when the value stream is mapped. Thirdly, it gives people an opportunity to tell you what is wrong. While this is a soft issue, the effects of which are difficult to quantify, it is often useful as the early involvement of personnel to the project can assist with the generation of a sense of ownership of the problems and potential solutions. Fourthly, the interviews provide a check that people understand the wastes. This occurs as a result of direct questioning of why particular scores were given and from examples personnel give to illustrate particular wastes in their work environment. A noted weakness of the first part of the methodology is the possibility of respondents' not understanding the wastes and their relationship with the work environment. As a result the interview process can be used to check this understanding.

In addition to questioning personnel about the waste in their organization, the interview is also used to highlight the positive aspects of the firm. This can be done by using strengths, weaknesses, opportunities and threats as a framework that provides a balance in the interview between the benefits and concerns in the organization.

Figure 2a shows the Waste Mapping Control Grid. The scores from each of the personnel are input into the grid and the average score for each of the wastes calculated. As discussed in the introduction, the waste methodology is part of a Value Stream Management methodology. These average scores are used to assess which value stream mapping tools will be used to collect further information on the business (as shown in Figure 2b). Thus the collection of subjective data in the waste exercise is converted so that we can use tools which give us objective data to quantify the issues in the organization and check the initial views of personnel in the waste workshop.

NEW WASTES AND NEW APPLICATIONS

The original Value Stream Mapping methodology described in Chapter 3 was applied primarily within a shop-floor order fulfilment environment and relied on the seven wastes of the Toyota Production System. However, in revisiting the methodology and using it in a supply chain context, seven new wastes are identified and explored. In

- Read each of the sections on the seven wastes
- Give a score to each waste that you believe is applicable to your organization
- You have 35 marks to allocate to the seven wastes
- The maximum score for any of the wastes is 10 marks
- The minimum score is 0 marks.

  Therefore if you believe each waste has equal importance in your organization, score 5 marks each = 35 marks.

Figure 1 The Waste Workshop.

addition, new sectors in which to apply waste analysis are identified. The seven new wastes are illustrated in Table 3. While these wastes are frequently encountered within the production environment, they were not fully recognized as waste in the early work.

The first of the new wastes is that of wasted power and energy. An example of this is leaving machines switched on when not in operation. This is found to be equally applicable in a shop-floor or office environment. This waste is thought to be particularly relevant as service costs of electricity, gas and water increase with an increase in the use of mechanization on shop floors and computers in offices. The second waste is that of human potential. Examples occur where people are being under-utilized in their daily routines or where their full potential is not realized as they are not consulted about potential improvements in their working environments. The third waste identified is environmental pollution. This is particularly relevant given the regulatory controls concerning topics such as chemical discharges, packaging, noise levels and health and safety issues to which companies are required to adhere. Fourthly, the waste of unnecessary overhead is identified. Examples include over-investing for the required output, having too many indirect personnel or too many forklift trucks. In addition, carrying out training that is unnecessary or not implemented correctly falls into this category. The fifth waste is inappropriate design. This may be in terms of the product (its functionality, features or weight) or the manufacturing process. The last two wastes are those of departmental culture and inappropriate information where departmental politics and over-complicated systems slow up the organization.

It is suggested that a degree of translation is required to make the wastes applicable along the supply chain in environments other than component and assembly production manufacturing. This view has been the subject of previous work at the Lean Enterprise Research Centre, an example being the introduction by Jones of the Toyota production system philosophy to a warehouse environment (11).

The LEAP team developed a list of seven generic work environments (described in Figure 3) in which the seven new wastes can be applied with adequate renaming of the wastes. It is recognized that some of the wastes will be understandable in a number of environments without the need for translation. An example of a waste that

Table 3 The seven new wastes within component assembly manufacturing

Waste category	*Nature of waste*
Wasted power and energy	Machines and lights switched on when not in operation. Leaks on machines, facilities
Wasted human potential	Employing people for repetitive, non-value added activities. No development/succession plans
Environmental pollution	High usage of disposable materials such as packaging. Lack of recycling policy
Unnecessary overhead	Inappropriate training. Over investment for the required output
Inappropriate design	Lack of design for manufacture. Late design changes
Departmental culture	Functional silos. Conflicting performance measures
Inappropriate information	Overcomplicated IT solutions. Demand amplification

													Average	Rank
Waste	1	2	3	4	5	6	7	8	9	10	11	12	13	14
	Logistics Manager	Production Manager	#2 Press Shop Production Controller	Raw Material Scheduler	Senior Buyer	Goods Inwards	Goods Inward Steel Stores	Press Shop Supervisor	Supervisor	Despatch Controller	Kaizen Manager	Operations Manager		
1. Overproduction	2	10	7	5	8	0	4	3	7	0	2	2	4.17	5
2. Waiting	3	5	7	10	5	0	4	10	6	5	7	5	5.58	3
3. Transportation	10	5	4	5	7	10	7	5	8	5	8	8	6.83	1
4. Inappropriate processing	3	5	3	0	3	0	4	3	2	10	3	2	3.17	7
5. Unnecessary Inventory	4	5	7	10	8	10	9	4	2	10	3	5	6.42	2
6. Unnecessary Motion	10	0	0	0	2	6	0	0	5	5	8	8	3.67	6
7. Defects	3	5	7	5	2	9	7	10	5	0	4	5	5.17	4
Total	35	35	35	35	35	35	35	35	35	35	35	35	35.00	
1st Waste	3	1	1	2	1	3	3	2	1	4	2	3		
2nd Waste	6		2	5	3	5	5	7	3	5	3	6		
3rd Waste			5		5	7	7				6			

Figure 2 (a) The Waste Mapping Control Grid.

	Process Activity Mapping	Supply Chain Response Matrix	Production Variety Funnel	Quality Filter Mapping	Demand Amplification Mapping	Decision Point Analysis	Value Analysis Time Profile	Physical Structure Mapping
1. Overproduction	4.17	12.50	0.00	4.17	12.50	12.50	37.50	0.00
2. Waiting	50.25	50.25	5.58	0.00	16.75	16.75	50.25	0.00
3. Transportation	61.50	0.00	0.00	0.00	0.00	0.00	0.00	6.83
4. Inappropriate processing	28.50	0.00	9.50	3.17	0.00	3.17	9.50	0.00
5. Unnecessary Inventory	19.25	57.75	19.25	0.00	57.75	19.25	57.75	6.42
6. Unnecessary Motion	33.00	3.67	0.00	0.00	0.00	0.00	3.67	0.00
7. Defects	5.17	0.00	0.00	46.50	0.00	0.00	0.00	0.00
Totals	201.83	124.17	34.33	53.83	87.00	51.67	158.67	13.25
Rank	1	3	7	5	4	6	2	8

Figure 2 (b) Score sheet ranking the tools.

is applicable to the environments of both component and assembly production manufacturing and warehousing and distribution may be unnecessary motion. However, the concept of overproduction in manufacturing is difficult to explain in environments such as office and information systems and warehousing and distribution. Thus the terms 'producing too soon' or 'faster than necessary pace' may be more applicable respectively.

It is recognized that some of the wastes will be understandable in a number of environments without the need for translation. An example of a waste that is applicable to the environments of manufacturing, warehousing and distribution may be unnecessary motion.

CONCLUSIONS

This chapter has attempted to provide a methodology for understanding waste in order to find appropriate mechanisms for removing it along the value stream. Practical steps have been outlined which, if followed, can lead to the identification of both intracompany and intercompany waste. It has been possible to extend the use of waste identification and elimination techniques by identifying seven new wastes which encompass both an individual company along the value stream and the supply chain setting, the factors affecting competitiveness and a number of different work environments.

The original Value Stream Mapping methodology developed from Hines and Rich was applied primarily within a shop-floor order fulfilment environment and relied on the Seven Wastes of the Toyota Production System. However, in revisiting the methodology we have subjected it to the principles of continuous improvement in order to improve the productivity of the process and the quality of the product. As a result of using the methodology in a supply chain context, a number of new wastes are identified and explored and new sectors in which to apply waste analysis are identified; where necessary, the wastes are retitled to make them applicable to the setting. In a sense this process of improving and enhancing the waste analysis methodology exemplifies the contention in the lean literature that 'the more waste that is removed, the more waste it is possible to see' and that 'the more you find out about lean management and lean thinking, the more you realize you need to know'.

- Component and Assembly Production Manufacturing
- Complete Supply chain
- Office and Information Systems
- Process Manufacturing
- Warehousing and Distribution
- New Product Development
- Service and Retail

Figure 3 The seven work environments.

REFERENCES

(1) Johnson, G. and Scholes, K. (1993) *Exploring Corporate Strategy*. New York: Prentice Hall.
(2) Womack, J. P. and Jones, D. T. (1994) From lean production to the lean ernterprise, *Harvard Business Review*, 72(2), 93–103.
(3) Ford, H. (1926) *Today and Tomorrow*. New York: Doubleday.
(4) Ohno, T. (1988) *Toyota Production System, Beyond large-scale production*. Portland: Productivity Press.
(5) Imai, M. (1986) *Kaizen, The key to Japan's Competitive success*. New York: McGraw-Hill.
(6) Bicheno, J. (1994) *The Quality 50, A guide to gurus, tools, wastes, techniques and systems*. Buckingham: Picsie.
(7) Suzaki, K. (1987) *The New Manufacturing Challenge, Techniques for continuous improvement*, New York: The Free Press.
(8) Monden, Y. (1993), *Toyota Production System: An integrated approach to just-in-time*. Georgia: Industrial Engineering and Management Press.
(9) Womack, J. P. and Jones, D. T. (1996) *Lean Thinking, Banish waste and create wealth in your corporation*. New York: Simon and Schuster.
(10) Kobayashi, I. (1994) *Twenty keys to workplace improvement*. Portland: Productivity Press.
(11) Jones, D. T. (1995), Applying Toyota principles to distribution, Supply Chain Development Programme I, Workshop No. 8 Workbook, Britvic Soft Drinks Ltd, Lutterworth.

7 Creating big picture maps of key processes: company activities viewed as part of the value stream

David Brunt

INTRODUCTION

The LEAP team's search for a methodology to analyse supply chain performance and identify waste has indeed been a challenge. When faced with analysing the steel supply chain – from steel manufacture, through distribution to manufacture of automotive components – it quickly becomes apparent that the physical flows of products are complicated by the huge variety of products made by the firms involved. In addition, complex information flows exist between each individual firm, inside each firm and at different levels between each firm.

Interviewing employees to ask them where waste exists in their companies and supply chain yielded a raft of suggestions for improvement. In most cases, employees at the 'coal face' are experts in what they do and can become frustrated by what they see as waste in their organizations. However, being faced with a long list of problems (often at a very detailed level in the organization) can lead to confusion about what improvements need to take place and exactly what are the root causes of the symptoms employees see on a day to day basis.

Faced with this apparent complexity, the research team looked to the tools and techniques of Taiichi Ohno's Toyota Production System for inspiration. In the book *Kanban Just-in-Time at Toyota* (1), the authors suggest that 'we have repeatedly stated that the most important aim of the Toyota Production System is the thorough elimination of waste. Yet it is difficult to recognize what waste really consists of. In contrast, it is not too difficult to determine the methods or ways of eliminating waste. Thus if we can make it obvious to everybody what waste is, we have taken the first positive step towards its elimination.' In the Toyota system, the workplace is made into a showcase that can be understood by everyone at a glance. In terms of quality, it means to make defects immediately apparent. In terms of quantity, it means that progress or delay, measured against the plan, is made immediately apparent. Ohno suggested that when this is done 'problems can be discovered immediately, and everyone can initiate improvement plans'.

Visual Control at the workplace makes it possible to let machines operate automatically when conditions are normal, and lets workers engage in abnormality control when abnormalities occur. Thus if Visual Control is an important concept within Toyota's system, we need to make the performance of the supply chain visual in order to explain what occurs and communicate potential improvements to create a vision of how the supply chain could operate.

A critical success factor to making supply chain performance visual is an understanding of process management. However, quite often it is difficult to see how any of the activities people carry out on a day-to-day basis can be categorized into a particular process. Often organizational structures of departments and functions (both within and between firms) seek to mask and complicate what currently happens to deliver a product or service to market.

This chapter is divided into four parts. First, the background to process analysis is discussed. Secondly, the tools used to analyse processes will be examined and their strengths and weaknesses discussed. The third part outlines the way in which these tools can be used together to illustrate the big picture of the supply chain, its performance and how improvements can be made which have a benefit to the whole system and not just its individual parts. Finally, conclusions will be drawn from the work.

BACKGROUND TO PROCESS ANALYSIS

Employees can usually describe their company's organizational chart. Depending on the size of the organization, the traditional company can be organized into divisions, departments and functions. In some firms the structure is possibly a left-over from Taylor's scientific management school of thought which promoted job specialization in pursuit of efficiency or from the command and control regimes developed as organizations became unwieldy due to their size. The result is what Dimancescu (2) calls a 'vertical' management style (Figure 1) where 'individual tasks are often completed in one department or group and then handed off to the next without full appreciation of the whole process'.

Increasingly, the competitive pressures faced by organizations have resulted in them turning the organization on its side so that the key activities carried out are customer-facing. In addition, the company starts to see itself as a whole entity, thus leading management to manage key processes rather than discrete activities.

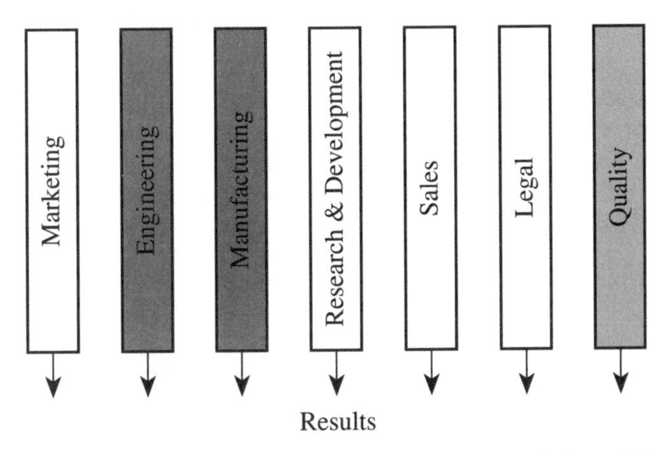

Figure 1 Vertical/functional management (from Dimancescu (1992, p. 13)).

Dimancescu *et al.* (3) suggest that 'many of the fundamentals of process management were pioneered in Japan in the 1960's, but the concepts did not gain serious corporate attention in the United States until the late 1980's, fuelled by a wave of interest in process engineering'. From a process re-engineering viewpoint, Champy (4) suggests that every business has core processes, for example new product development, customer service and order fulfilment. In *Reengineering the Corporation*, Hammer and Champy contend that focusing on the fundamental redesign of these core operational processes offers a company the best leverage for raising performance.

WHAT IS A PROCESS?

A number of authors have sought to define the term process. Oakland (5) in his book on Total Quality Management, defines the term process as 'the transformation of a set of inputs, which can include actions, methods and operations, into outputs that satisfy customer needs and expectations, in the form of products, information, services or – generally – results'. Similarly, Harrington (6) defines a process as 'an activity or interrelated series of activities that takes an input, adds value to it, and produces an output. A process can be as small as a single activity or it can include many activities and sub-processes. Processes are usually divided into activities.'

Keen and Knapp (7) suggest that there are two interpretations of business process among the process movements. The first sees process as a workflow, a series of activities aimed at producing something of value. Both the definitions above fall into this category. The second interpretation sees a process as the coordination of work, whereby a set of skills and routines is exploited to create a capability that cannot be easily managed by others. For example Prahalad and Hamel (8) suggest that 'the real sources of advantage are to be found in management's ability to consolidate corporate wide technologies and production skills into competencies that empower individual businesses to adapt quickly to changing opportunities'. Similarly, Senge (9) cites Arie de Geus, then head of planning for Royal Dutch/Shell, who remarked that 'the ability to learn faster than your competitors may be the only sustainable competitive advantage'.

At a practical level these two interpretations of business process can be combined. Thus Keen and Knapp suggest that business process implies:

- organization of work to achieve a result;
- multiple steps and coordination of people;
- an element of design or implementation that renders a business process as distinctive a competitive asset as research and development or product development, a 'firm specific asset', 'core competence' or 'dynamic capability';
- management as the enabler and sustainer of process change.

Finally, when analysing the definition of process it is important to understand the scope at which analysis can take place. For example, it is possible to analyse a process as narrow as one occurring in a department or assess a process at a company level. However, when undertaking lean thinking, Womack and Jones (10) suggest that we 'must go beyond the firm, the standard unit of score-keeping in business across the world, to look at the whole: the entire set of activities entailed in creating and producing a specific product, from concept through detailed design to actual availability, from the initial sale through order entry and production scheduling to delivery, and from raw materials produced far away and out of sight right into the hand of the customer'.

PROCESS ANALYSIS TOOLS

Process flow charts

Using the premise that 'you cannot measure improvement if you don't know where you started from', process flow charts are a useful step in the recording of events and activities. One of the seven tools of quality, originally assembled by Ishikawa for use with quality circles, process flow charts (illustrated in Figure 2) use standard symbols to aid understanding of separate activities that make up the process. For example, the starting point is usually a circle, each processing step a rectangle and the end point of the process is shown using an oval. The process flow chart is particularly useful for displaying the decisions points present in many processes. These decision points are usually illustrated using a diamond shape.

The brown paper process

Brown paper charting is very similar to the process flow charting techniques, but differs in its method in that (as the name suggests) brown paper is used to show a pictorial overview of the process being analysed. In order to map many processes in an organization, a cross-functional team is required so that the relevant information

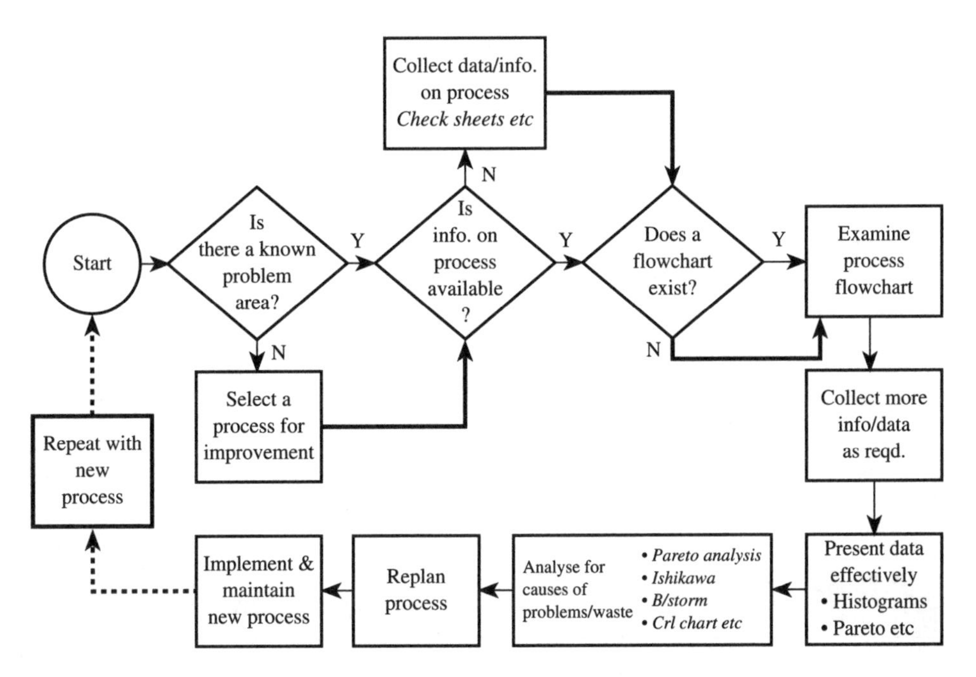

Figure 2 Process flow charting – the strategy for process improvement (from Oakland, 1995, p. 199).

can be collected. With the team in place, firstly a draft flow of the process is made with 'Post-it' notes being used to develop an understanding of the main activities. Secondly, once the brown paper has been drawn, the team needs to get agreement on the flow from all relevant actors along the chain. As a third step, the team can then illustrate the separate activities using copies of paperwork used at different steps; for example, a printout of the computer screen used to order materials or a quality report. In addition, photographs can be used to illustrate activities. The team needs to focus on what really happens and not what is supposed to happen, so that all the relevant loops and bottlenecks become apparent on the chart. The fourth step of the brown paper technique is to look for potential improvements to the process. The team does this by writing potential suggestions on more 'Post-it' notes (using a different colour to those used previously). Fifthly, having brainstormed potential improvements, the brown paper can then be 'incubated' for a short time so that other members of staff can also look at the process and ask questions or suggest improvement ideas. When the incubation period is completed, the questions and suggestions can be analysed and an improvement plan drawn up.

Figure 3 shows an example of a brown paper from a company in the LEAP programme. The process starts with the receipt of sales orders from companies at another tier in the LEAP value stream. On receipt of the order a job card is generated, time allocated for the order with the production department and transport informed of the shipping instructions. Two copies of the schedule are created, one of which goes with the job card. When material is present on site, and the schedule calls for the product to be produced, material is brought to the line and waits while the machine is set up to produce the product. The product is then made, packed and stored. Transport loads the material to a truck, raises the necessary advice notes and the product is delivered to the customer. The darker Post-it notes show the physical flow of the product and the lighter Post-it notes show the information flow required to produce the physical flow.

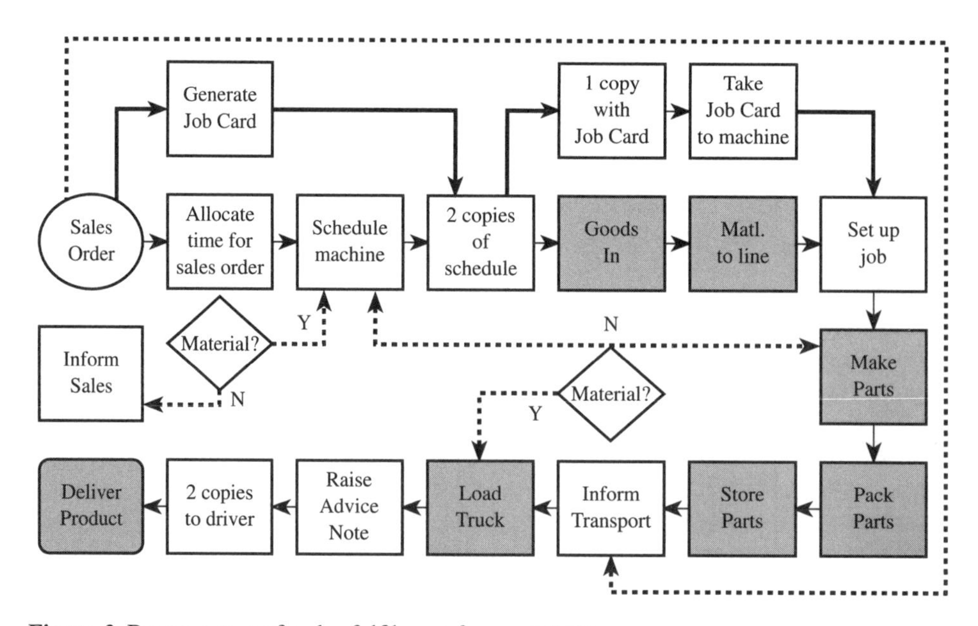

Figure 3 Brown paper of order fulfilment from a LEAP company.

When used at a high level, the brown paper technique is very powerful as it immediately facilitates the team to ask questions about the process undergoing analysis. In the example above, the team were able to visually display the after-effects of problems associated with material availability and quantity. In addition, at a more detailed level the complexity of the information flows between sales and transport became apparent. In terms of the physical flow, the time spent between getting material and setting up the job to be processed was highlighted. The brown paper process also highlights where other tools can be used to quantify issues with process performance, for example areas along the process where tally sheets, Pareto analysis and problem solving would be beneficial.

There are, however, several concerns with a straightforward process flow brown paper. The first of these is that while relevant details surrounding the process such as computer printouts and paperwork can be put onto the brown paper, quite often these do not quantify key aspects of performance for the process or what actually matters to the customer. Examples of this would be quality, delivery or cost information. Secondly, when the steps of the process are defined at such a high level there is a danger that the team can over simplify the process, which can lead to a loss of important detail.

The brown paper technique can also be used to display time through a process (Figure 4). This is particularly useful when looking at leadtime and where delays exist for a particular product or service. The analysis is set up in the same way as the exercise described above but instead two axes are used and a flow line or critical path added. On the y-axis the process stages are added. For example, send order, receive

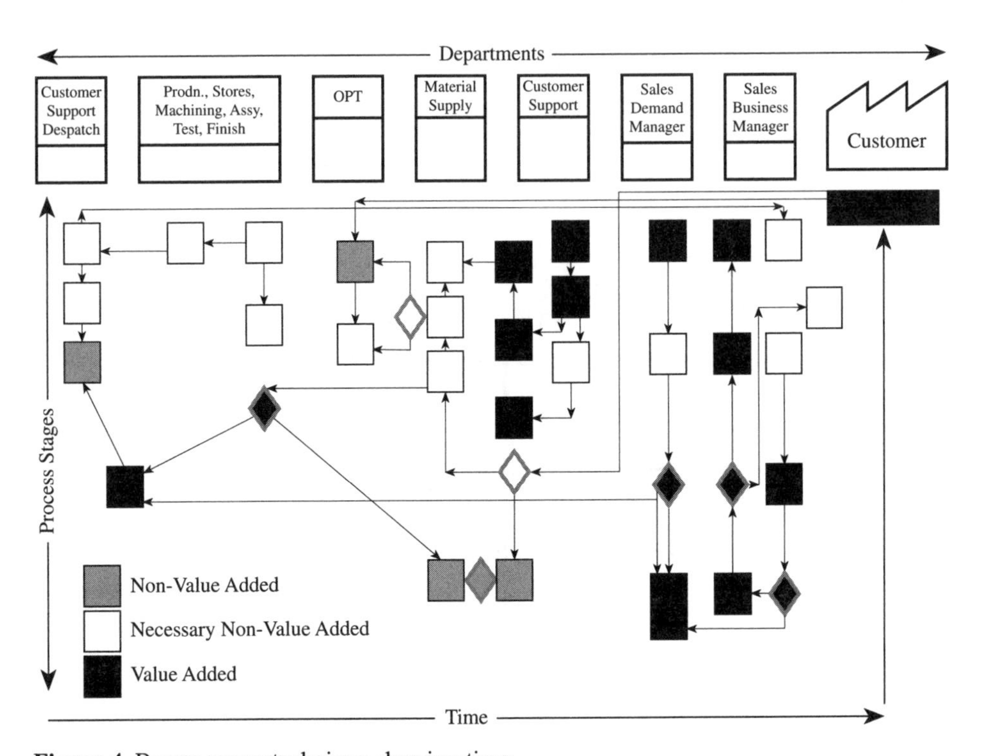

Figure 4 Brown paper technique showing time.

new order, develop production plan through to receive delivery. On the x-axis the departments involved in the process are illustrated, for example sales, purchasing, manufacturing and despatch. This analysis helps managers to gauge which of the key processes need to be understood in more detail as it can clearly show the bottlenecks, time delays and rework loops present in current practice.

ILLUSTRATING THE 'BIG PICTURE' OF THE SUPPLY CHAIN AND ITS PERFORMANCE

While the brown paper shown in Figure 3 shows the order fulfilment process for one company in the LEAP project, the process can be carried out across a whole value stream to show where the largest opportunities for waste removal occur. Hines and Rich (11) contend that waste may be intracompany (inside a particular firm) or intercompany (between different firms along a value stream). Therefore carrying out a brown paper analysis in a single company would attempt to identify waste at the intracompany level, whereas joining brown papers together to create a picture of the whole value stream would identify intercompany waste as well as further waste within each of the individual firms.

This point can be further explained by discussing the process view of an organization or value stream. Camp (12) suggests that processes have inputs, process steps, outputs, feedback and results. Figure 5 shows this generalized process. For each of the LEAP companies, inputs from suppliers are fed into the process. For example, steel making at the steel manufacturer; slitting, blanking or resizing at the steel service centre; and blanking, pressing, assembly, painting and finishing at the first-tier component companies. Once actors along the value stream have carried out their process, an output is delivered to the customer.

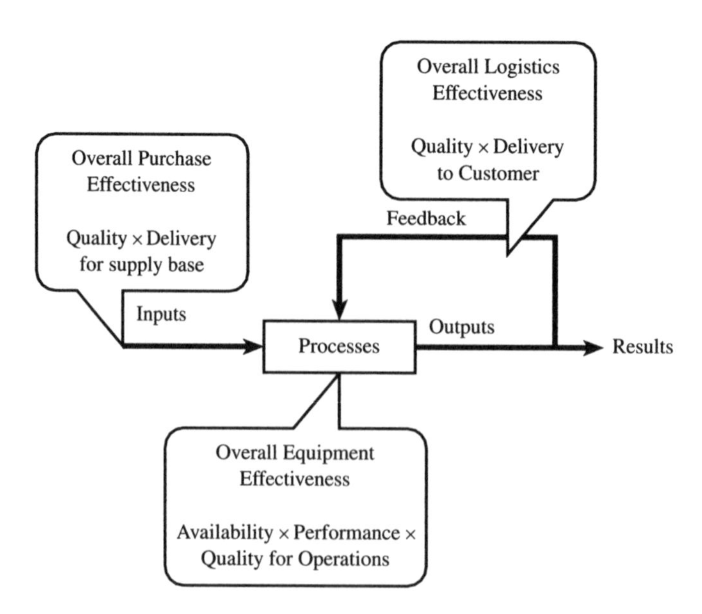

Figure 5 Generalized process view of an organization.

Rich and Francis (13) suggest that 'the greatest of performance measurement problems . . . is the ability to create a series of common measures to evaluate the "health" of the supply chain and to use these measures to co-ordinate improvements within departments, between departments and throughout the supply base'. They propose that the commonality of measures and the alignment of these measures with the strategic goals of the business are key factors. In the context of the LEAP supply chain, exposure to such measures already exists in the form of first-tier suppliers being measured by vehicle manufacturers in terms of Quality, Delivery and Cost (QDC) targets. However, the use of these QDC measures between first and second tiers and second and third tiers was not comprehensive. In addition (and arguably most importantly), where QDC measures did exist, they had not been combined to show the performance of the value stream from raw materials (steel producer through to car manufacturer). By using the brown paper process in conjunction with the quality and delivery measurements highlighted in Figure 5 it is possible not only to show the steps contained in the key processes for the LEAP project, but to highlight the overall performance of each of these processes. Hence it is possible to indicate where future improvement potential exists.

An example of the technique is illustrated in Figure 6 (although the actual performance figures are for illustrative purposes only). For a first-tier component assembler, performance to the vehicle manufacturers appears high. First-tier firms' delivery and quality performance is often close to 100 per cent when supplying the mainstream vehicle makers. At first sight, internal performance also appears impressive. However, if quality and delivery are measured (in terms of availability 80 per cent, performance 80 per cent and quality 95 per cent) then the overall effectiveness of the process – the ability to supply right first time, on time – within the walls of the first-tier firm would be 60.8 per cent. In reality this performance loss is hidden by excess inventory and excess time (i.e. working overtime) in order for the

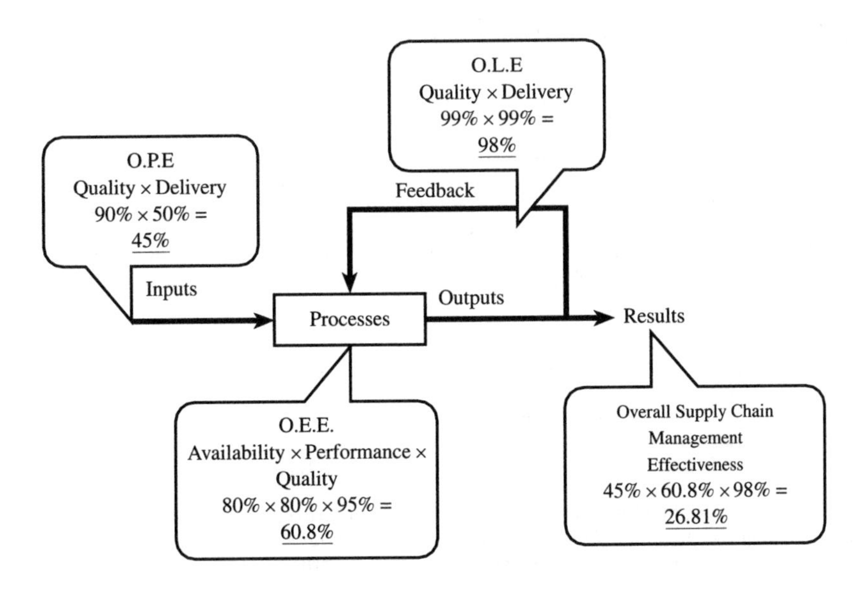

Figure 6 Performance of an individual firm in terms of Quality and Delivery.

system to deliver. In addition the performance of suppliers to these firms can be taken into account. Here we have assumed a quality performance of 90 per cent and a delivery performance of 50 per cent, resulting in a supplier performance of 45 per cent. If all these figures are linked together to form a performance measurement represented within the brown paper process, then the overall performance within the first tier firm would be 26.81 per cent (45% × 60.8% × 98%.) When the same calculation is carried out along the value stream (to include the steel company and steel service centre) then the overall performance of the order fulfilment process could be as low as a theoretical 1.92 per cent (26.8% × 26.8% × 26.8%.)

SUMMARY AND CONCLUSIONS

This chapter has attempted to describe the background to process analysis before outlining the tools used to analyse them. Using information gathered during the LEAP programme, the strengths and weaknesses of the different tools has been highlighted before showing the way in which these tools can be used together to illustrate the big picture of the supply chain. In addition, the tool has been linked to the measurement of supply chain performance in order to illustrate where the largest emphasis in terms of improvements needs to occur in order to have a benefit to the whole and not just individual parts of the system.

In doing this there is an important learning point captured by Norman Bodek (14) in the introduction to Kobayashi's book *Twenty keys to workplace improvement*. He suggests that 'Companies that lead the world in their markets do so by improving more than one thing at a time, and by doing it over the long term. They recognize the importance of synergy between different improvement efforts and the need for commitment at all levels of the company to achieve total, systemwide upliftment.'

REFERENCES

(1) Japan Management Association (ed.) (1986) *Kanban Just-in-Time at Toyota*. Oregon: Productivity Press.
(2) Dimancescu, D. (1992). *The Seamless Enterprise, Making Cross Functional management work*. New York: HarperCollins.
(3) Dimancescu, D., Hines, P. and Rich, N. (1997). *The Lean Enterprise, Designing and Managing strategic processes for Customer-winning performance*. New York: AMACOM.
(4) Champy, J. (1995) *Reengineering Management: The Mandate for new Leadership*. London: HarperCollins.
(5) Oakland, J.S. (1995) *Total Quality Management*. Oxford: Butterworth-Heinemann.
(6) Harrington, H.J., Hoffherr, G.D. and Reid, R.P. (1999) *Area Activity Analysis, Aligning work activities and measurements to enhance business performance*. New York: McGraw-Hill.
(7) Keen, P.G.W. and Knapp, E.M. (1996) *Business Processes*. Boston: Harvard Business School Press.
(8) Prahalad, C.K. and Hamel, G. (1990) The core competence of the corporation. *Harvard Business Review*, May–June.
(9) Senge, P. (1990) *The Fifth Discipline, The art and practice of the Learning Organisation*. London: Century Business.
(10) Womack, J. P. and Jones, D. T. (1996), *Lean Thinking, Banish waste and create wealth in your corporation*. New York: Simon and Schuster.
(11) Hines, P. and Rich, N. (1997) The seven value stream mapping tools. *International Journal of Operations and Production Management*, 17(1), 46–64.
(12) Camp, R.C. (1995) *Business Process Benchmarking: Finding & Implementing Best Practices*. Wisconsin: ASQC.
(13) Rich, N. and Francis, M. (1998) Overall supply chain performance measurement: Focussing improvements and stimulating change. *Proceedings of Logistics Research Network Conference*, pp. 442–455.
(14) Kobayashi, I. (1994) *Twenty keys to workplace improvement*. Portland: Productivity Press.

The Value Analysis Time Profile – an approach to value stream costing

8

David Brunt, Peter Hines and James Sullivan

INTRODUCTION

Work within the Lean Processing Programme has focused on how previous attempts to map the value stream can be improved to develop an holistic methodology which addresses both the strategic and operational areas of complete supply chains. While developing this suite of tools, it has become apparent that quantifying the benefits of value stream improvement by measuring the cost benefits both communicates the potential of such improvement and focuses senior management on the most important areas where waste (and hence cost) can be driven out of the value stream for a given product or service (1).

By assigning costs to both non value added and value added activities required to create, produce and deliver a product (2), the chapter shows how waste is highlighted either within companies or along supply chains using a graphical technique called the 'Value Analysis Time Profile.' This tool is now one of a suite of tools used in the LEAP programme, drawn from a variety of origins such as industrial engineering, operations management, systems dynamics and time compression/logistics.

The chapter is divided into four parts. Firstly, a review of the use of waste removal with reference to costing is carried out. Secondly, the steps used to produce the 'Value Analysis Time Profile' are described. In the third part of the chapter, a case study from within the LEAP project (automotive supply chain) is used to illustrate the methodology and to highlight the role of this tool as a simulation device to quantify the impact of improvement activities. Fourthly, a discussion and conclusions are drawn from the work.

COSTING AND WASTE REMOVAL

Global competition and the changing pace of business have led to increased pressure on the key actors in the LEAP automotive supply chain. Quality and delivery performance (the ability to deliver 'right first time, on time') became prerequisites for suppliers to the mainstream auto-makers. In addition, these competitive pressures faced by the industry required all the firms along the value stream to find ways of reducing costs year on year (3, 4).

To successfully adopt this cost reduction approach, traditional business practice needs to be questioned. Thus a company can no longer seek a profit on a 'cost plus' basis where:

Cost + Profit = Selling Price

In contrast, the market determines the selling price. In the automotive industry this has become increasingly visible. For example, in Europe the industry's exemption from Competition Law is reviewed in 2002 (the 14-year-old block exemption in effect allows carmakers to distribute new cars through dealers). Currency fluctuations and differing tax rates for new cars mean that importing vehicles is currently favourable. The ability to compare prices using mechanisms such as the Internet and sites such as Autobytel has increased transparency and increased competition. Therefore if the market dictates the price, cost reduction can bring more profit and more business if the equation described above becomes:

Selling Price − Cost = Profit

Looking at the key processes the business needs to manage facilitates this management approach. Wickens (5) suggests that 'future profitability depends upon present actions and we have, therefore to be concerned with *the quality of inputs as well as the performance of outputs*'. In support of this process view, Deming (6) argued that by concentrating on financial measures you are forced to look at short-term performance. Therefore it can be argued that by concentrating on the fundamentals of the business (by managing its key processes) then, providing pricing is competitive, the financial performance will look after itself.

However, while this theory appears logical, there are a number of inhibitors to its implementation. Firstly, the time frames in which businesses are managed may cause leaders to look after short-term pressures (for example, from financial institutions and shareholders) rather than focus on longer-term waste elimination techniques. Elliot (7) suggests that all the new programmes, initiatives and change management processes of information-age companies are being implemented in an environment governed by quarterly and annual financial reports. The financial reporting process is derived from an accounting model developed centuries ago for an environment of arm's length transactions between independent entities (not for the management of a value stream.) Information-age companies are still using this financial accounting model as they attempt to build internal assets and capabilities, and to forge linkages and strategic alliances with external parties.

Secondly, the current measures used within a business may also reinforce short-term behaviour or actions that improve departmental rather than business optimisation. Kaplan and Norton (8) suggest that 'the collision between the irresistible force to build long-range competitive capabilities and the immovable object of the historical-cost financial accounting model has created a new synthesis: the Balanced Scorecard'. The Balanced Scorecard attempts to bridge the gap between traditional financial measures that track past events and measures which can help guide firms to create future value through investment in customers, suppliers employees, processes, technology, and innovation.

Thirdly, the career objectives of management may cause leaders to look short-term. Increasingly, aspiring professionals are told of the importance of career planning and the need to move and develop into new roles. Without careful management this could lead to short-term gains being crafted at the expense of a longer-term strategy.

However, despite these potential constraints, the power of looking at the whole, rather than parts of the system and quantifying the benefits of value stream improvement can help focus senior management on the most important improvements that need to take place. This process is paralleled between buyer and supplier firms in terms of the evaluation of total cost. Merli (9) suggests that 'suppliers must be chosen and later evaluated in their operations on the basis of not only price but of the total cost they require the client company to bear'. Factors to be considered in total cost evaluations are shown in Table 1.

Hines and Rich (10) suggest that 'the use of waste removal to drive competitive advantage inside organizations was pioneered by Toyota's chief engineer, Taiichi Ohno, and sensei Shigeo Shingo and is oriented fundamentally to productivity rather than quality. The reason for this is that improved productivity leads to leaner operations which help expose further waste and quality problems in the system.'

Womack and Jones (11) contend that operations can be categorized into three areas. Firstly, those that actually create value as perceived by the customer; secondly, those that create no value but are currently required by the product development, order filling or production systems and so cannot be presently eliminated; and thirdly, those actions that do not create value as perceived by the customer and so can be eliminated immediately.

Bicheno (12) has noted that Westinghouse made extensive use of cost time profile charts. The company used the profiles in a hierarchical fashion. Therefore the profile for each sub-process or product can be combined to form a profile for a whole section which, in turn, can be combined into a profile for a complete plant or division. In the

Table 1 An example of total operational cost evaluation (price + quality performance + logistical performance) from Merli (1991)

Quality costs	• Inspection and testing • Reserve supplies • Waste and rework • Litigation management • Technical assistance, guarantees, complaints • Loss of image • Other induced costs (both for production and technical assistance)
Costs related to delivery reliability	• Reserve stock • Production interruptions • Delivery delays • Loss of sales
Response time costs (supply lead-time)	• Need for planning and scheduling • Reserve supplies for forecast changes
Supply lot costs	• Average stock for a particular code • Risk of obsolescence
Costs linked to lack of improvement	• Lack of increase in contribution margins • Lack of reduction in poor quality costs
Technological obsolescence cost	• Costs due to late updating • Value of late opportunities

Westinghouse case, total costs are used so that it is necessary to multiply the unit cost profiles by the average number of units in process. All processes must be considered; value adding as well as support activities and overheads (Fooks (13)).

THE VALUE ANALYSIS TIME PROFILE

The Value Analysis Time Profile method seeks to plot the accumulated cost against time for a given product or process. The starting point for the generation of the map is the construction of one of the other mapping tools used in Value Stream Mapping: the Process Activity Map. Originally an industrial engineering tool, this map sets out the sequence of flow by recording all the individual steps that take place. Each step is categorized as one of the following: operation (O); transport (T); inspection (I); delay (D); or storage (S). The magnitude of each step is calculated in terms of time taken (minutes), distance travelled (metres) and the number of people involved in each step of the process. In Chapter 3, Hines and Rich summarize five stages when using this tool:

1. the study of the flow of processes;
2. the identification of waste;
3. the consideration of whether the process can be rearranged in a more efficient sequence;
4. the consideration of a better flow pattern, involving different flow layout or transportation routing; and
5. the consideration of whether everything that is being done at each stage is absolutely necessary and if superfluous tasks can be removed.

A theoretical example of a small part of a Process Activity Map is shown in Table 2.

Once the Process Activity Map has been constructed, financial data is collected which enables a cost to be put against each of the steps involved in the process. An example of some of this data is shown in Table 3.

To illustrate the use of the tool, Figure 1 shows a hypothetical pressings industry case (Hines *et al.* (14)) in which the raw material and components are bought in at value ‘a’. During the various value adding steps its value is raised (for instance, during the value adding parts of blanking, pressing and assembly activity) from ‘a’ to total value ‘b’ over the period, during which it is in the company. However, during this same period its cost has risen further from point ‘a’ to point ‘c’. The cost (which may bear little relation to the selling price) that has been added reflects both value adding and waste. In this example waste is added in the steps of transportation, storage and the setting up of the various processing machines used. The total waste in this case is the distance ‘d’.

Table 2 The Process Activity Map. Summary of Product (Main Flow)

Step Name	*Flow*	*Area*	*Distance*	*Duration*	*People*	*Op. VA*	*Op. NVA*	*T*	*I*	*D*	*S*
Lorry arrives at factory	T	Outside Building 2	15	0.00	2			1			
Unload truck	T		0	20.00	0			1			
Store in location	S		27	960.00	1						1

When interpreting the graph, it is useful to think of the area under the line as the time that money is tied up. Bicheno (15) suggests that the aim is 'to reduce the area under the graph by reducing time and/or cost'. Therefore the tool allows researchers or practitioners to obtain an understanding of where waste is being added over time and helps pinpoint where major improvement efforts might most effectively be focused. In addition, it shows where time may be reduced for the various activities charted on the horizontal axis, providing a simple time compression tool.

CASE STUDY

To illustrate the methodology and to highlight the role of this tool as a simulation device to quantify the impact of improvement activities, a case study from the LEAP project is used. The company from which the case study is drawn produces pressed components and assemblies as a first-tier supplier to the automotive industry. The Value Stream Mapping methodology was used on a number of components produced by the organization. These were either components which exhibited high value to the firm, were high volume or were parts produced for a strategic customer.

Table 3 Key data required to complete the Value Analysis Time Profile

Data	*Definition/Measure*	*Process*	*cost p.a.*	*Utilization %*	*% time devoted to producing component*	*Average production batch size*
Capital interest rate (% p.a.)						
Stock hold interest rate (% p.a.)	including capital cost, storage, stock keeping	Paint		60.00	2.00	1.00
Direct labour cost (£/hour)	including NI, pension, but excluding o/h	Assembly		65.00	100.00	1.00
Forklift cost (£/p.a.)		Test		85.00	5.00	1.00
Floor space	(per m2/p.a.)	Press 1		50.00	40.00	5000.00
Raw material cost	£/unit	Press 2		50.00	40.00	5000.00
Raw material cost	£/unit	Blank 1		55.00	2.00	10000.00
Average batch size	from supplier					
Total bought out finished cost (per component)	£/unit	Details for each process are required in terms of each item of major equipment used in each of these areas.				
pieces per year						
Sales price	£/unit delivered to customer					
Payment period to supplier	in days					
Payment period from customer	in days					
Cost of consumables	£/p.a.					
Work time/shifts	Hrs/week					

By using a number of the mapping tools the following key points for the business were generated:

- Quality, cost and delivery measurements highlighted on-time delivery to the customer as a key concern. This was shown in terms of large finished goods stock required by customer.
- Large amounts of waste were evident in the pressing and blanking operation where non-value added set-up times were large. The large set-up times also impacted on the batch sizes and frequency of parts produced and hence the stocks of components between blanking and pressing and pressing and assembly operations.
- Reliability of machines and tooling were a further complication in the blanking and pressing areas.
- The result of these concerns was that overtime was required to produce the output required for on-time delivery to the customer.

The mapping data enabled both researchers and management in the organization to focus on the areas of the company in which the largest benefits would be gained from improvement. In this example a number of ‘quick hits’ were carried out by the firm, which facilitated an improved flow of products through the operation. Once these had been completed, the more difficult task of improving the pressing operation was targeted. The area was analysed over a two-day period in more detail than that required for the initial Value Stream Mapping activity. People from the production and planning areas were formed into two teams each facilitated by a member of the LEAP team. The first team focused on the physical flow of materials and work on the shop floor such as setting the tools in the presses. The second team analysed the information flows, including the production planning mechanisms used.

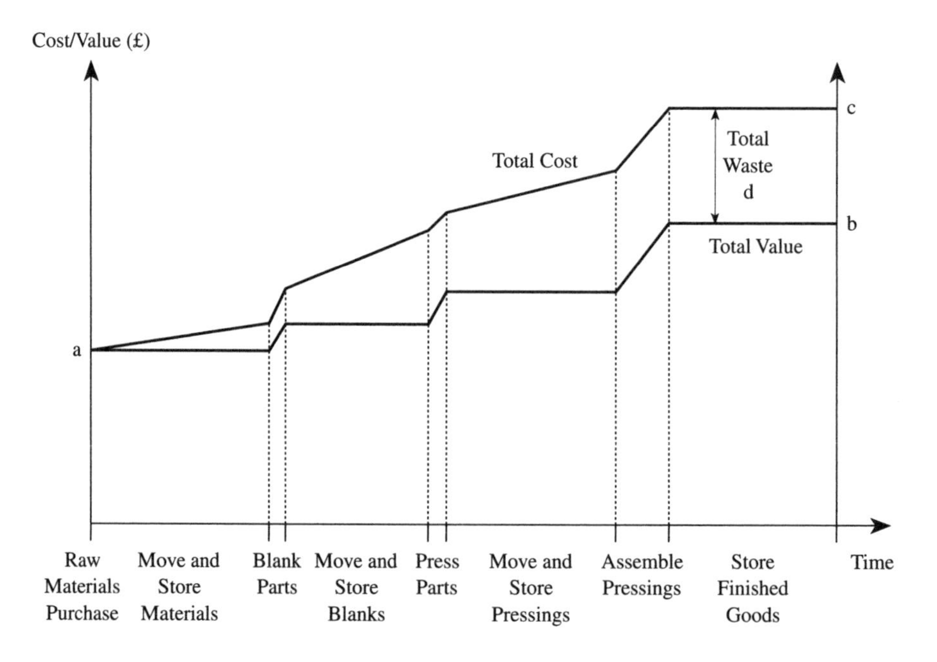

Figure 1 The Value Analysis Time Profile.

Analysis of the physical flow of materials and work on the shop floor revealed a number of possible improvements. Examples included reducing the time required to set up the presses by reducing the external time required for set-up (Shingo (16)) and allocating tools to specific machines. In addition, reducing the double handling of material before it arrived at the line and also reducing the number of points at which material was stored by locating the material at the press that was to use it were identified. This second point required close communication with the team looking at information flow, as they were involved in the planning of production and materials and interaction with the supply base. It quickly became evident that if tools and materials were to be successfully allocated to a given press then the planning processes would be affected and that the issue of reliability of tools and presses would require management attention.

Analysis of the information flows and, in particular, the production planning system attempted to understand the implications of producing the output required to meet on-time delivery to the customer with minimal overtime. A capacity planning model (Butterworth (17)) for the press line was created which was used as a scenario manager, taking account of customer demand, overall equipment efficiency (including set-up time), machine speed (blow rate), blank weight, coil sizes and material-handling requirements such as packaging type. The reduction of overtime on the bottleneck press was targeted by smoothing the workload over the presses and moving parts to different presses. In an attempt to achieve flow, the order load on the presses was considered with reference to runners, repeaters and strangers (Parnaby (18)). Harrison (19) suggests a definition as follows:

- runners (products or key features produced every day, every week);
- repeaters (products or key features produced regularly but at longer time intervals);
- strangers (products or key features produced irregularly).

Finally, a link was set up to the team focused on the physical flow of materials and work on the shop floor by assessing the set-up time thought to be readily achievable to produce a minimum batch size without failure to deliver to the customer and minimizing the necessary overtime required. The shop-floor team had identified considerable set-up reduction potential (60 per cent), which would be achievable within a longer-term project improving maintenance activity and establishing standard operations. (Suzaki (20)).

Having analysed the situation in terms of both physical and information flows, the two teams assembled together to formulate a plan to integrate their findings. The capacity plan showed a series of scenarios: a 90-minute set-up time, a 60-minute set-up time and a set-up time required to produce every part every day. The scenarios were then modelled using the Value Analysis Time Profile so that potential savings could be quantified. The scenarios were presented to management by the two teams in terms of what was achievable in the short term (60-minute set-up) and medium term (to produce every part every day). In addition, an action plan was produced which highlighted the sequence of activities required to get to the medium-term goal.

Practical examples of the next steps include tools and materials placed close to machines and blanks put in set locations on the shop floor to minimize the wastes of transportation and motion. The capacity planning model gave the physical flow team a clear target for the set-up time required to produce every runner every day with no overtime required. As a result the team targeted their set-ups, standardizing the process and producing tool boards to locate spanners as close to the machine as possible.

DISCUSSION AND CONCLUSIONS

Using the Value Analysis Time Profile in conjunction with the Value Stream Mapping methodology provides a number of benefits for researchers or practitioners attempting to make a change either at an intra or inter-firm level. Firstly, quantifying potential savings when making improvements communicates to employees at all levels of an organization the potential improvement activities can bring. This is particularly important when investments need to be justified as part of the change project. Secondly, the Value Analysis Time Profile communicates in the language of management the benefits a 'Lean' approach can have on the business. Thirdly, the tool can be used for projects between buyer and supplier firms so that potential savings can be identified between the participants. Fourthly, the approach reduces the potential for undertaking what Jim Womack (21) calls 'kamikaze kaizen' in which firms carry out continuous improvement activities which do not necessarily have an impact on the performance of the business.

However, while there are a number of benefits to the use of the tool, there are also a number of constraints which need to be addressed. The first of these concerns the data required to produce the chart. A number of Value Analysis Time Profile charts have been attempted both in the LEAP project and in other work carried out by the Lean Enterprise Research Centre. While it is usually possible to gather the information, there have been instances where firms do not break their costs down into distinct areas which we can compare with the Process Activity Map. In addition, the methods used to allocate overheads can be an issue when calculating the individual steps. The second constraint concerns completing the chart. This is time consuming and is, at present, a largely manual operation, which can result in errors occurring in the data. The third constraint is also linked to the complicated nature of completing the Value Analysis Time Profile. This means that often researchers complete the chart rather than management in the business. This has led to management being sceptical about the results of the analysis until it is communicated in detail.

To conclude, the chapter has provided a review of the use of waste removal with reference to costing in order to explain the steps used to produce the 'Value Analysis Time Profile.' A case study from within the LEAP project (automotive supply chain) has been described to illustrate the methodology and to highlight the role of this tool as a simulation device to quantify the impact of improvement activities. In future, research can be directed to the use of the tool both within firms and along supply chains in order to focus improvement efforts where most impact can be gained.

REFERENCES

(1) Johnson, G. and Scholes, K. (1993) *Exploring Corporate Strategy.* New York: Prentice Hall.
(2) Womack, J. P. and Jones, D. T. (1994) From lean production to the lean enterprise. *Harvard Business Review,* 72(2), 93–103.

(3) Christopher, M. (1992) *Logistics and Supply Chain Management: Strategies for reducing costs and improving services*. London: Pitman.
(4) Alber, K. L. and Walker, W. T. (1997) *Supply Chain Management: Practitioner Notes*. Falls Church, VA: APICS Educational & Research Foundation, October.
(5) Wickens, P. D. (1995) *The Ascendant Organisation*. London: MacMillan.
(6) Deming, W. E. (1982) *Out of the Crisis*. Cambridge: University Press.
(7) Elliott, R. K. (1992) The third wave breaks on the shores of accounting. *Accounting Horizons*, 62–65.
(8) Kaplan, R. S. and Norton, D. P. (1996) *The Balanced Scorecard*. Boston: Harvard Business School Press.
(9) Merli, G. (1991) *Co-makership: The new supply strategy for Manufacturers*. Cambridge: Productivity Press.
(10) Hines, P. and Rich, N. (1997) The seven value stream mapping tools. *International Journal of Operations and Production Management*, 17(1).
(11) Womack, J. P. and Jones, D. T. (1996) *Lean Thinking, Banish waste and create wealth in your corporation*. New York: Simon and Schuster.
(12) Bicheno, J. (1994) *The Quality 50: A Guide to Gurus, Tools, Wastes, Techniques and Systems*. Buckingham: Picsie Books.
(13) Fooks, J. (1993) *Profiles for Performance*. Addison-Wesley.
(14) Hines, P., Rich, N., Bicheno, J., Brunt, D., Taylor, D., Butterworth, C. and Sullivan, J. (1998) Value Stream Management. *International Journal of Logistics Management,* 9(1).
(15) Bicheno, J. (1998) *The Lean Toobox*. Buckingham: Picsie Books.
(16) Shingo, S. (1985) *A Revolution in Manufacturing: The SMED System*. Cambridge: Productivity Press.
(17) Butterworth, C. (2000) Forthcoming paper on capacity planning.
(18) Parnaby, J. (1988) A systems approach to the implementation of JIT methodologies in Lucas Industries. *International Journal of Production Research,* 26(3).
(19) Harrison, A. (1992) *Just-in-time Manufacturing in Perspective*. London: Prentice Hall.
(20) Suzaki, K. (1987) *The New Manufacturing Challenge, Techniques for continuous improvement*. New York: The Free Press.
(21) Womack, J. P. (1998) Lean Thinking: How far have we got? Lean Summit, Nottingham, UK.

9 Dynamic distortions in supply chains: a cause and effect analysis

Matthias Holweg

INTRODUCTION

The initial value stream mapping exercises in the LEAP programme highlighted the low performance of the analysed value streams by showing high inventory levels throughout all tiers of the system. Furthermore, the demand patterns along the system were found to be distorted and highly amplified.

The value mapping tools, however, as defined by Hines and Rich 1997, are more of an investigative nature and the outcomes of the mapping exercise did not provide any analytical evidence of the causes for distortion of the demand and material flows in the supply chains. Therefore, following on the outcomes of the mapping, the 'Scheduling Project' was initiated, as it was felt that more detailed understanding of the supply chain dynamics was needed to determine the root causes for the distortion and amplification. Furthermore, it was intended to define potential improvement areas to counteract the effects.

To seek the root causes of the behaviour in the supply chain, both the demand and supply flows and the decision points in the system had to be analysed. Therefore, the internal scheduling and ordering procedures at the different companies were analysed and supplemented with data on inventory levels, batch sizes and lead times. The supply chain dynamics were studied by quantifying the demand and supply patterns for a range of parts over all levels of the value stream over a significant period of time. And, although not taking part in the programme, the vehicle manufacturers as the origin of demand for the analysed supply chains were considered.

The research approach applied to this research problem is derived from the system dynamics logic, and in the course of this chapter considerable synergy between theory and practice will be pointed out. After identifying the root causes for disturbances in the system, the last section will propose improvement actions on how this supply chain could be improved on both a short- and long-term perspective. Also, a tentative concept will be introduced on how the total supply chain could be scheduled more efficiently.

THE BACKGROUND – SYSTEM DYNAMICS

Concerning the research on supply chain behaviour and characteristics, the fundaments were laid by Jay W. Forrester (1958, 1961) in the early 1960s with his seminal work on 'Industrial Dynamics'. 'Industrial Dynamics' is defined as the study of the information-feedback characteristics of industrial activity to show how organizational structure, amplification (in policies), and time delays (in decisions and actions) interact to influence the success of the enterprise. It treats the interactions between the flows of information, money, orders, materials, personnel, and capital equipment in a company, an industry, or a national economy.

Using a quantitative and experimental approach to model business environments, combined with the utilization of the first available computer technology, Forrester simulated several information-feedback systems to analyse industrial and economic entities. The models used were closed-loop information-feedback systems, based on an abstract, dynamic, non-linear and transient (not 'steady-state' = characteristics change over time) mathematical model. These models cannot be seen as comprehensive, as an 'all-inclusive' model would not be feasible due to the complexity of the involved variables in the real world. Instead, several different models might be used for different classes of questions to model a particular real-world system. Furthermore, Forrester argues that mathematical equations cannot describe a business problem accurately due to the great number of system and exogenous variables involved. Therefore, the mathematical solution found in the simulations is not the optimum solution for the managerial problem, but it gives evidence as to the potential behaviour of the system, in case of policy changes for example. This idea finally brought out the differentiation between OPT® and 'Synchronous Manufacturing', as Umble and Srikanth (1990) believe that a mathematical model cannot achieve an optimal solution, as too many exogenous variables interact.

Forrester only modelled very limited systems in terms of integrated variables, a 'one product' distribution chain model in an isolated Factory-Distribution-Retailer-Customer chain, as the state of computer technology at the time did not give the possibility to model more sophisticated systems with more complex equation system calculations. But even with his limited systems he could prove that, for example, an increased demand level at final customer level causes an amplified demand wave once it reaches the manufacturing level. This effect is known as the 'Forrester Effect' or 'Demand Amplification' (Figure 1).

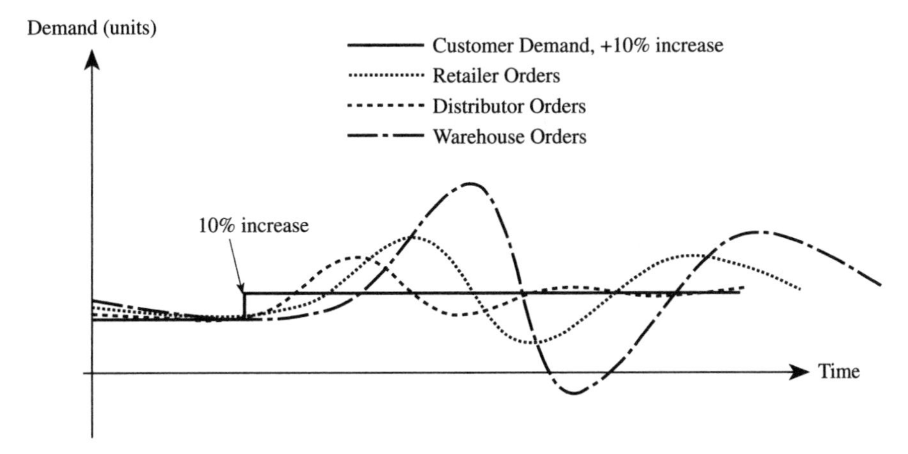

Figure 1 The Forrester Effect.

Forrester identifies the main causes as:

- **No demand visibility** along the network, so changes in demand are not transmitted to give the 'trigger' to adjust the production level to the new demand level. Therefore, when this information finally reaches the production level, the demand 'backlog' or overproduction already is the difference between the increased / decreased and the old demand level, accumulated over time. The longer this information lead time is, the greater the backlog or overproduction will be. Additionally, in case of increased demand, all inventory levels along the chain will be adjusted to a new increased demand level, causing extra order volume without real demand and worsening the production peaks.
- **Information distortion** along the chain, as Reorder-Points (ROP) or other decision points cut out the real demand pattern and issue their 'own' demand pattern to the upstream levels, according to their inventory levels.
- **'Playing around' or frequent adjustments to inventory levels**, as a change in inventory policies creates either additional orders without demand to fill up the storage, or fewer orders than the actual demand is, to lower current inventory levels. Either case creates erratic order patterns for the upstream processes.

A common management strategy to cope with the 'increased' or amplified demand would be to increase capacity, to meet the new order volumes. Forrester as well can prove with his models that such an increase in production capacity at the manufacturer can even lead to higher peaks at production level, as a limited production capacity hinders the system from further oscillation.

John Burbidge, a fellow researcher at MIT, focused on a different aspect of amplification in the system. Whereas the 'Forrester Effect' is rather a 'systems' related source of distortion, Burbidge (e.g. in Towill, 1996) identified 'reordering' as another source of amplification (Figure 2). The effect is caused by a lack of synchronization of reordering ('multi-phasing'), and most often is not recognized, yet causes considerable swings in demand and hence in inventory levels, potentially causing additional orders, if safety stock levels are breached. Burbidge never explicitly published his work on 'multi-phasing', yet became well known for his 'five

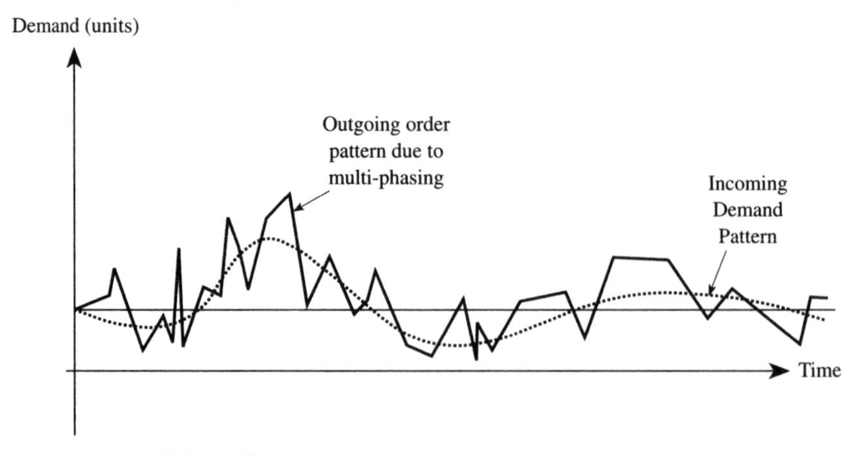

Figure 2 The Burbidge Effect.

golden rules' on how to overcome both types of amplification and 'avoid bankcruptcy' (Burbidge, 1983).

With the recent focus on supply chain management, Forrester's concept of distorted information flow or demand amplification in supply chains was taken up in the last decade, by companies as well as by academia. For example, Procter & Gamble (Saporito, 1994) found that the diaper orders issued by the distributors have a degree of variability that cannot be explained by consumer demand fluctuations alone. It was found that the orders placed by the retailers have much greater variations than customer demand. The information distortion caused wrong assumptions in planning and decision making and was subject of an internal project.

In an inventory management experimental context, Sterman (1989a,b) reports evidence of the demand amplification in the 'MIT Beer Distribution Game'. The experiment involves a supply chain with four players who make independent inventory decisions without consultation with other chain members, relying only on orders from the neighbouring player as the sole source of communication. As root cause for the demand amplification Sterman names human behaviour, such as misconceptions about inventory and demand information.

In recent publications about supply chain behaviour, the demand amplification phenomenon is sometimes referred to as the 'Whip-lash Effect' (Johnson, 1997) or 'Bullwhip Effect', as Lee *et al.* (1997a, 1997b) call it.

Unlike Forrester and Sterman, Lee *et al.* believe that the bullwhip effect is a consequence of the player's rational behaviour within the supply chain's infrastructure. This important distinction implies that companies wanting to control the demand amplification have to focus on modifying the chain infrastructure and related processes rather than the decision maker's behaviour. They developed simple mathematical models of supply chains that capture essential aspects of the institutional structure and optimizing behaviours of members. Mathematical models are employed to explain the outcome of rational decision making, as opposed to deriving an optimal decision rule for managers. The following causes for the bullwhip effect are given:

- **Demand signal processing** refers to the situation where demand is non-stationary and the past demand information is used to update the forecasts. Distortion of demand information arises when the retailer issues orders based on this updated demand forecast. As a result, the manufacturer loses sight of the true demand in the marketplace. The production schedule is – based on the distorted signals – inevitably inefficient. The distortion effect gets amplified as the number of intermediaries in the channel increase.
- **'The rationing game'** refers to the strategic ordering behaviour of buyers when supply shortage is anticipated. Information distortion can arise as a consequence of strategic decisions by the retailer who assesses the possibility of being placed on allocation by the manufacturer. The order data has little or even negative informational value to the manufacturer, and he or she needs to exercise great care in interpreting the order signals for inventory / capacity planning.
- **Order batching** – when the fixed order cost is non-zero, ordering in every period would be uneconomical, and batching of orders would occur. Batching of orders is a consequence of two factors: the periodic review process and the processing cost of a purchase transaction. Demand distortion due to the periodic review

process can be alleviated by providing the manufacturer with access to sell-through data and / or inventory data at retail level. The manufacturer uses this information to create a production schedule that is determined by sales as opposed to orders.

- **Price variations** refer to the non-constant purchase prices of the product. One way to control the bullwhip effect due to price fluctuations is to reduce the frequency as well as depth of manufacturer's price promotions (i.e. wholesale discounts).

Demand signal processing and order batching are interrelated, since they are driven by each member trying to optimize internal operations of inventory management. The rationing game and price variations are also related to each other, since they both reflect the member's reaction to the market dynamics. Furthermore, Lee *et al.* show that balanced and 'perfectly synchronized' retailer ordering can be achieved. Under that scenario, and only then, the variability of demand experienced by the supplier and the retailers is identical, and the bullwhip effect disappears.

To overcome the demand amplification effect, Forrester suggested a modification in the behavioural practices, whereas Sterman recommended better individual education and communication amongst all partners in the supply chain.

Lee *et al.* argue that to overcome the bullwhip effect the institutional and inter-organizational structure and related processes have to be attacked. Three major areas are identified as critical: information sharing, channel integration and operational efficiency. With information sharing, demand information at a downstream site is transmitted upstream in a short period of time. Channel alignment is the coordination of pricing, transportation, inventory planning and ownership between the upstream and downstream sites in a supply chain. Operational efficiency refers to activities that improve performance, such as reduced costs and lead time. Just-in-time replenishment, small batch policies and frequent re-supply are seen as effective measures to mitigate the effect.

The literature on demand amplification is summarized in Table 1.

The following section will describe how the system dynamics research was applied to investigate the research problem.

THE RESEARCH APPROACH

The research approach applied in the scheduling project basically follows the system dynamics methodology by defining supply chains as a system of two basic flows, demand and supply, and decision and stocking points, at which the flows are altered.

A decision point is defined as the point at which the original or incoming flow of demand information is altered, for example the customer places an order for 20 units. The sales department acknowledges the order, but forecasts that the customer will need another 10 units shortly, and passes an order for 30 units to production control. Production control takes the orders and passes on raw material requirements to purchasing for 27 units, as 3 units are still in the finished goods stores. Purchasing, though, orders material for 50 units, as this represents the minimum order quantity, etc. It is important to notice that there are usually several decision points within the company, both for the information flow as for the material flow. For the material flow these decision points are stocking and processing points, as quality losses, conversion or rework can alter the original flow.

Table 1 Demand amplification – literature overview

Key author	*Root causes for demand amplification*	*Contributing factors*	*Proposed countermeasures*
Forrester (1961) 'Demand Amplification'	• No demand visibility • Information distortion • Inventory level adjustments	• Reorder points • Excess delays	• Time compression • Removal of unnecessary echelons in the system
Burbidge (1983) 'Multi-phasing'	• Multi-phased ordering	• Unsynchronized order flow • Poor information, uncertainties	• Ordering policies adjustments
Sterman (1989a,b) 'Beer Game'	• Human misperceptions • Decision-making processes	• No visibility of end demand	• Improved communication in the supply chain • Improved education ('awareness')
Lee *et al.* (1997a,b) 'Bullwhip Effect'	• Demand signalling	• No visibility of end demand • Multiple forecasts • Long lead-times	• Information sharing, i.e. demand visibility • Channel integration, i.e. coordination of transportation, inventories and pricing • Operation efficiency, i.e. JIT deliveries
	• Order batching	• High order cost • 'Full truck load' economies • Random or correlated ordering	
	• Fluctuating prices	• High-low pricing • Delivery and purchase asynchronized	
	• Shortage or rationing game	• Proportional rationing scheme • Ignorance of supply conditions • Unrestricted orders and free return policy	

Outside the companies, the flows *per se* are not altered, reflecting the period while the order is sent via mail or the time the lorry needs to deliver goods to the customer's production site.

Supply chains are a complex network of information and material flows. For every single product in the chain an individual demand and supply pattern exists, and some products even have different routings along the chain.

To understand these demand and supply dynamics, quantitative research was undertaken for a range of parts in the steel supply chain. A multiple three-tier supply chain (see Figure 3) was chosen, and for the analysed link both the information flow (forecasts, schedules and call-offs) and the material flow (deliveries according to the orders) were analysed.

The features shown in Table 2 were identified as the key measures that needed to be considered.

Owing to the complexity of the information flow in the supply chain, there are two dimensions to be considered: **stability** and **consistency**.

Table 2 Applying system dynamics to the research problem

System dynamics	*Supply chain equivalent*	*Relevant features analysed*
Information Flow	All demand information, i.e. • Forecasts • Schedules • Call-offs (DCI) • Late amendments	• Information stability or variability • Information consistency • Time horizon covered • Status, i.e. forecast or firm order • Detail of information
Material Flow	• All physical deliveries	• Delivery reliability • Quality
Decision Points	• Manufacturing Planning and Control Systems (MPC) • Stocking points	• Production batches • Throughput times • Stock levels • Safety stock policies
	• Ordering and purchasing systems	• Ordering policies • Minimum order quantities • Purchasing discounts

Stability refers to the behaviour of the demand over time, i.e. how much the demand changes from day to day. For example, the demand might be 1000 on Monday, 2000 on Tuesday, 500 on Wednesday, etc. The stability can be measured in deviation from the average and applies to forecast and firm orders. However, as there are both forecast and firm orders, a second dimension needs to be considered – consistency of the demand information.

Consistency in this case refers to the deviation of forecast to actual demand. For example, on 1 December 1999 a delivery for 1 January 2000 might be scheduled as 400 units, yet the actual call-off arriving on 30 December 1999 only states 200 units for delivery on 1 January 2000. Consistency of demand is important to plan long-term decisions such as capacity planning and in some cases even raw material purchases, which would directly affect the dynamics of the system.

Furthermore, the time horizons and detail given by the different types of information need to be considered, as a schedule, for example, could provide three months' forecast on a monthly basis or in weekly requirements, or the first month in weeks, the rest in months, etc. Also, it needs to be considered at what stage the demand information is commercially binding, i.e. turns from a forecast into a firm order, as only then are the actual requirements known to the suppliers.

In conclusion, the demand flow in the supply chain is complex, and needs to be analysed in two dimensions, consistency and stability, furthermore taking detail, time horizons and commercial status into account. This complexity is often not appreciated, yet is essential to understand the implications of the impact on the system. The scheduling project tried to take this fact into account, although it was difficult in some cases to determine the status of orders. Some vehicle manufacturers, for example, provide a cost coverage guarantee for the raw material purchase of their suppliers if the actual demand differed from the amount of raw material purchased. The vehicle manufacturer would in this case still pay for raw material purchased by the supplier. This guarantee is usually granted four weeks ahead of delivery, yet the firm order for the components is not issued until the week before delivery. Consider the demand information that is sent upstream within the chain in this case, giving false signals to all preceding supplier tiers.

In summary, the research approach to investigate the root causes for demand distortion and amplification in the LEAP supply chains consisted of the following two major elements:

- Analysis of **scheduling and order procedures** along all tiers of the system, as the scheduling and ordering systems are the major decision points in the system. Additionally, data on inventory levels, batch sizes and lead times was collected to provide a full picture. Inventory levels are particularly of interest, as inventory increases the total lead time for the part to pass through the supply chain and is used to cover up quality and process reliability problems. The phenomenon is known as the 'Rock-boat-analogy' (Vollmann *et al.*, 1992; Monden, 1998).
  This also involved the analysis of the effects of discounting policies on the ordering patterns, and an evaluation of MRP-based scheduling systems.
- A **detailed analysis of the supply chain dynamics** by quantifying the demand and supply patterns for a range of parts over all levels of the value stream over a significant period of time. Seventeen parts were chosen and their demand and supply patterns through three tiers of the system were monitored for a period of three months. Additionally, delivery performances were recorded and evaluated. Furthermore, the quality of demand information submitted by the vehicle manufacturers was analysed, although they did not formally take part in the programme. The vehicle manufacturers are the origin of the demand for the supply network analysed, therefore they are a significant factor to be considered when analysing the information quality further upstream, i.e. defining which variability was already induced by the OEMs and does not originate from the supply chain.

Owing to constraint resources, the scheduling project was limited to a selected number of LEAP companies, as shown in Figure 3. The vehicle manufacturers (or customers to the first tier suppliers) include all major European manufacturers and Japanese transplants in Europe.

The next two sections will show selected outcomes from these research areas, before concluding in the final section.

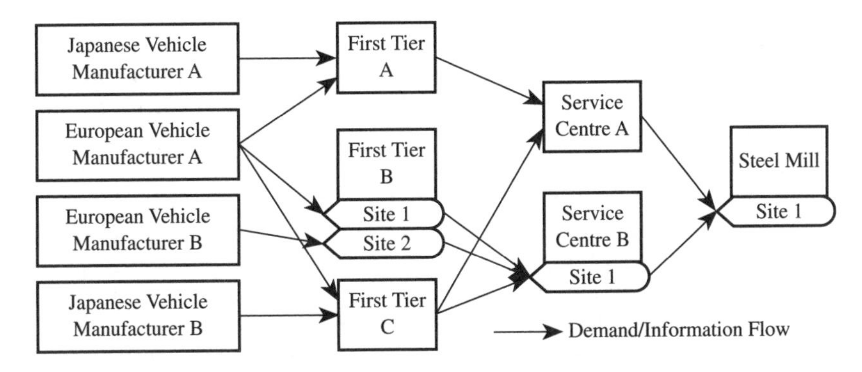

Figure 3 The Scheduling Project – value streams involved.

ANALYSING THE DECISION POINTS

This section will report on selected evidence found by analysing the decision points in the system, i.e. scheduling systems and practices, as well as ordering and batching policies. Also, as MRP systems were criticized significantly for causing demand distortion, a separate section will discuss the impacts of MRP-based systems as perceived in this study.

The research findings cannot be discussed comprehensively here, therefore only representative examples are shown. The complete analysis and data set can be found in Holweg (1999).

First-tier scheduling approaches

At first-tier level, three basic types of scheduling approaches were found. The first group uses MRP-based scheduling systems, driving the process through a master scheduling (MPS) and capacity management (CRP) system. These users load the MPS according to customer orders and communicate the resulting material requirements directly to the service centres. Although daily call-offs from the vehicle manufacturers (VM) take place, the schedule is driven by forecasts from the VMs, with a finished goods buffer taking up the difference between forecast and actual demand.

The second group uses kanban-driven production control in assembly, but uses MRP for medium-term material acquisition. This group takes demands directly from some customers who have demonstrated small variance between forecast and actual call-off. In such cases there is no 'MPS'. However, for other customers who produce more erratic and unstable forecasts, these suppliers tend to produce according to a smoothed schedule based on average or anticipated demand, again buffered by FGI inventory.

The third group uses their own 'home-grown' spreadsheet scheduling system, which loads batches onto press and assembly. These systems are generally based on spreadsheet applications and contain a logic similar to MRP.

Overall, the demand patterns which the first-tier companies issue on their suppliers, the service centres, were found to be reasonably stable, with one exception showing 'nervous' MRP system settings (see Vollmann *et al.* 1992) due to frequent rescheduling.

A major reason for short-term adjustments to supplier schedules at first-tier companies are coil size variations and short deliveries from the service centres. The 'call-off' patterns from the first-tier companies show fluctuations, but generally match the schedules in a reasonable range. It was found that the degree of schedule stability correlates with the degree of variation between the call-off and the schedules: the better and more reliable the schedules, the less problematic or fluctuating the resulting 'call-offs'. Or, using the terminology proposed in the section on system dynamics, the more stable the schedules, the more the likely the demand information is to be consistent.

Comparing the scheduling approaches, it can be said that none of the analysed approaches showed significantly superior performance and characteristics. Even MRP-driven plants are able to issue stable order patterns, if the MPS is levelled out before the MRP execution.

In terms of performance, measured by inventory levels within the plant, MRP-driven companies show slightly increased work-in-progress (WIP) and finished goods inventory (FGI) levels compared to manual-driven spreadsheet scheduling and JIT / kanban pull systems, but generally all inventory profiles of the analysed first-tier companies are similar. The reason is that all press shops studied work on an economic batch quantity (EBQ) basis, even in companies applying JIT techniques in their assembly operation.

Batch sizes differ between daily press runs and batches of four weeks' requirements; some companies apply a strict '10-day press batch' policy, but on average press batches range from 5 to 10 days. In terms of overall supply chain performance, the EBQ concept is an attempt at a local optimum supported by cost accounting oriented performance measures, but neglecting the impact on the overall company and the supply chain. Furthermore, the press shops represent the most unreliable process step, and unreliability and tooling problems have to be buffered with inventory. This directly reflects in all inventory profiles. Figure 4 gives a summary of all first-tier companies' inventory profiles, measured in days of required production. The minimal and maximal lines do not represent one single company, but refer to the extremes found at the particular station in the companies. The significant increase at the press and blanking operations reflect the buffer stocks necessary due to the large batches and the process unreliability.

However, it has to be considered that the scheduling approach itself cannot be seen as an isolated factor, as several factors contribute to the overall success of the scheduling system. The business structure, i.e. the customers, the matrix of supplied plants per customer and the product range play an important role, as shown in the section on vehicle manufacturers demand pattern analysis, below. It is definitely easier to respond efficiently in terms of supply chain inventory, if the customer is providing stable and reliable information. The quality of the information provided by the VMs will be discussed in the next section.

In addition, the human factor has great importance. Good planning staff using any system probably will be more efficient and cost-effective than non-motivated staff

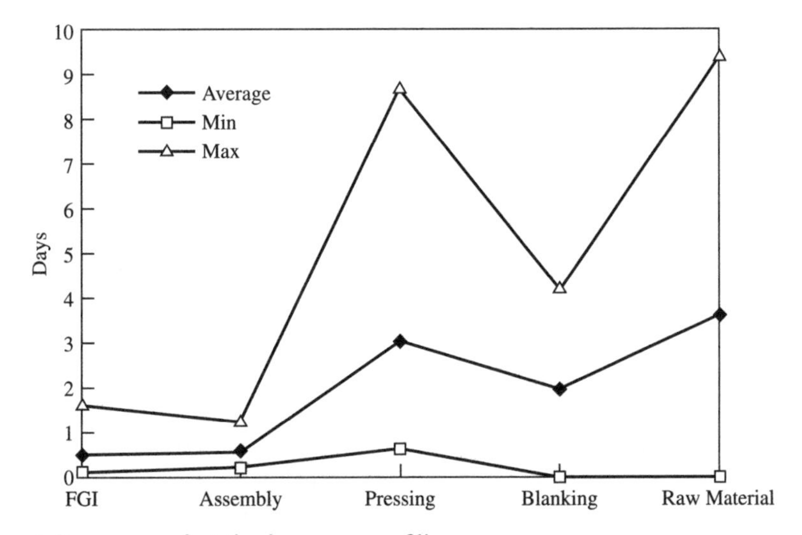

Figure 4 Summary – first-tier inventory profiling.

driving the most sophisticated system. Furthermore, the available machinery and equipment contributes. Press shop unreliability caused by tooling problems or old presses, for example, might bring up severe problems in generating valid and stable schedules, irrespective of which system is used.

MRP systems and their effect on supply chain dynamics

MRP systems are a common tool in the automotive supply base, but are often blamed for causing information distortion due to issuing of 'nervous' schedules. Distorted demand leads to demand amplification and finally to increased supply chain inventories. The point of criticism is the direct execution of the MRP-generated production schedules, which might lead to unstable supplier schedules and creates demand fluctuations for all upstream levels in the chain.

Therefore, a closer analysis was undertaken to investigate the effects of MRP systems within the research environment. The following analysis compares the supplier schedules for two similar pressings. The pressings are manufactured at two similar component manufacturers, both of which have to deal with a problematic customer who only provides unstable demand patterns, judging from past experience. The only difference between the companies is that one directly executes the MRP II generated schedules, whereas the other does not, although applying the same MRP II system.

The schedule stability will be shown by comparing the (rolling) supplier schedules over a certain time period. Owing to the schedule issuing cycles, at least three schedules (and the call-off) are available for every part at any time. The stability of these schedules of course has a direct impact on the demand pattern quality at the service centre.

The quantity actually called off will not be reflected here, as the call-offs usually are not MRP generated but manual schedule adjustments to determine the actual press sequence. The call-off quality depends mainly on the process reliability in the press shop – the more breakdowns and problems, the less stable the call-off. The section on demand and supply dynamics will examine the complete demand patterns.

Company 1 uses the MRP to generate all schedules, internal and supplier schedules. The MRP system runs every Saturday, so every Monday new supplier schedules are issued, defining the requirements for the next week based on last week's production.

Company 2 uses MRP as well to generate all schedules, production and supplier schedules, but a smoothing algorithm is applied to level production in assembly and to level the order patterns to the supplier. The MRP system is not executed directly, as customer schedules are smoothed manually in a spreadsheet before being re-entered into the MRP system to level production and supplier schedules, which are issued monthly, as opposed to weekly at Company 1.

The examples in Figures 5 and 6 show the schedules for a high- and a low-volume part at Company 1. The schedules fluctuate and show a low degree of matching. Both parts show unstable demand patterns. The reasons are the longer cycle of issuing schedules (monthly instead of weekly) and the internal 'schedule smoothing' strategy, which takes out the 'noise' and converts the outgoing schedules into relatively stable demand.

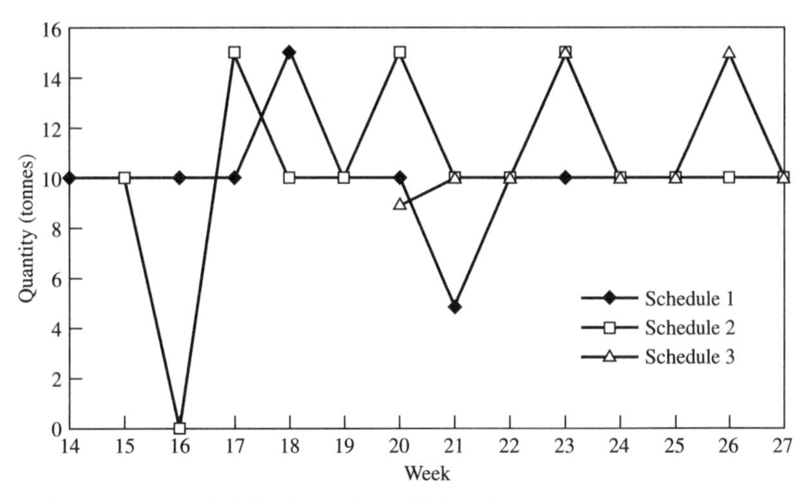

Figure 5 Company 1 – schedule fluctuations, high-volume part.

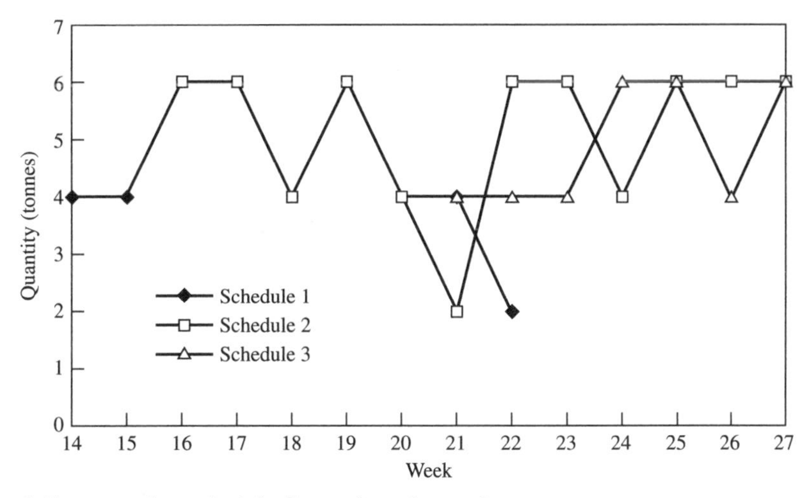

Figure 6 Company 1 – schedule fluctuations, low-volume part.

The schedules at Company 2 generally were found to be stable for most of the products. Looking, for example, at a high-volume part (Figure 7), which is a component of a subframe, the schedule shows a reasonably stable level. Looking at a low-volume part at Company 2 (Figure 8), the schedule shows less stability over time, but still match as to a high extent. The low demand in week 22 was caused by company holidays. Summing up, the schedules at Company 2 are relatively stable and match with the previous schedules.

Judging from these two cases and the demand patterns submitted to their steel service centre, it can be said that direct MRP execution used to generate internal and supplier schedules can disturb demand patterns and create unstable and fluctuating order patterns, which in consequence are a root cause for demand amplification. The reasons are 'nervous' system settings, as well as frequent rescheduling cycles and narrow-ranged system variables.

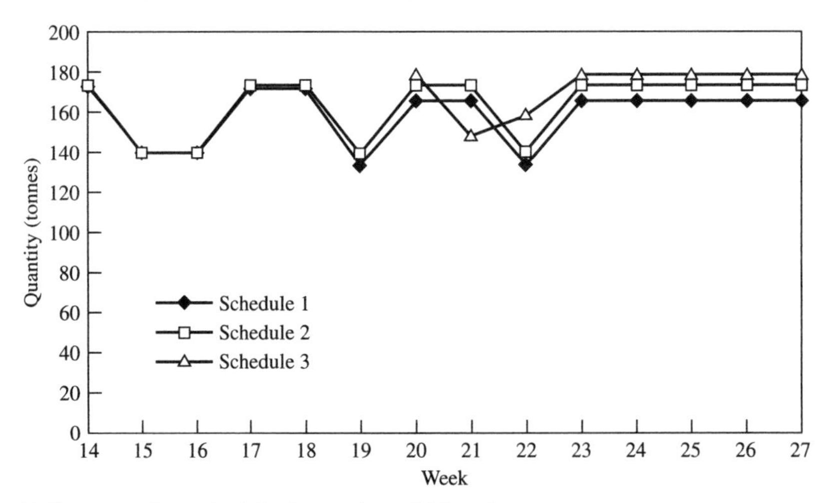

Figure 7 Company 2 – schedule fluctuations, high-volume part.

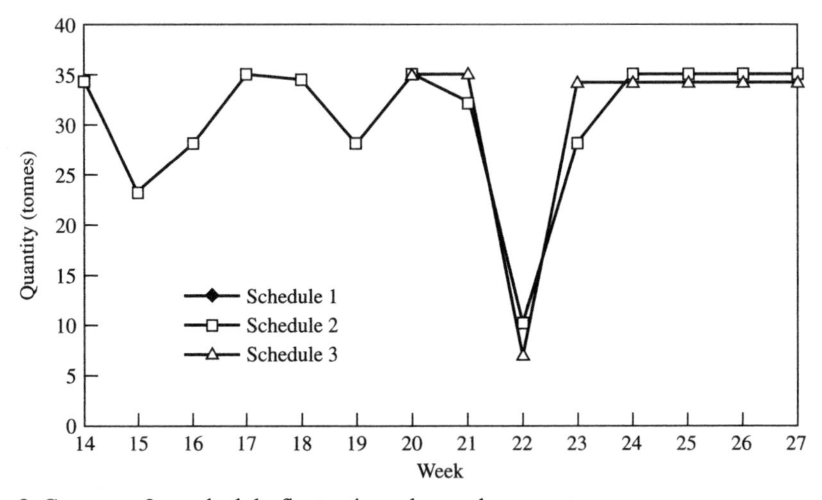

Figure 8 Company 2 – schedule fluctuations, low-volume part.

On the other hand it can be said that stable and reliable demand patterns can be transferred to the second-tier supplier, if the MRP system is not directly executed and the customer requirements are smoothed out manually before being re-entered into the MRP system. The applied production smoothing algorithm described at Company 2 for example, provides the basis for long-term stable production and supplier scheduling, but requires excess FGI to smooth out fluctuations; however, the overall benefits of applied level scheduling and a smooth production should equalize the additional cost for the excess inventory.

MRP utilization is not generally contra-productive to supply chain performance and can be a useful calculation tool, as long as it is not directly executed or set up unaware of the potential consequences for the supply chain. Direct MRP execution and 'nervous' system settings result in fluctuating supplier schedules and might cause demand amplification due to distortion of the information flow.

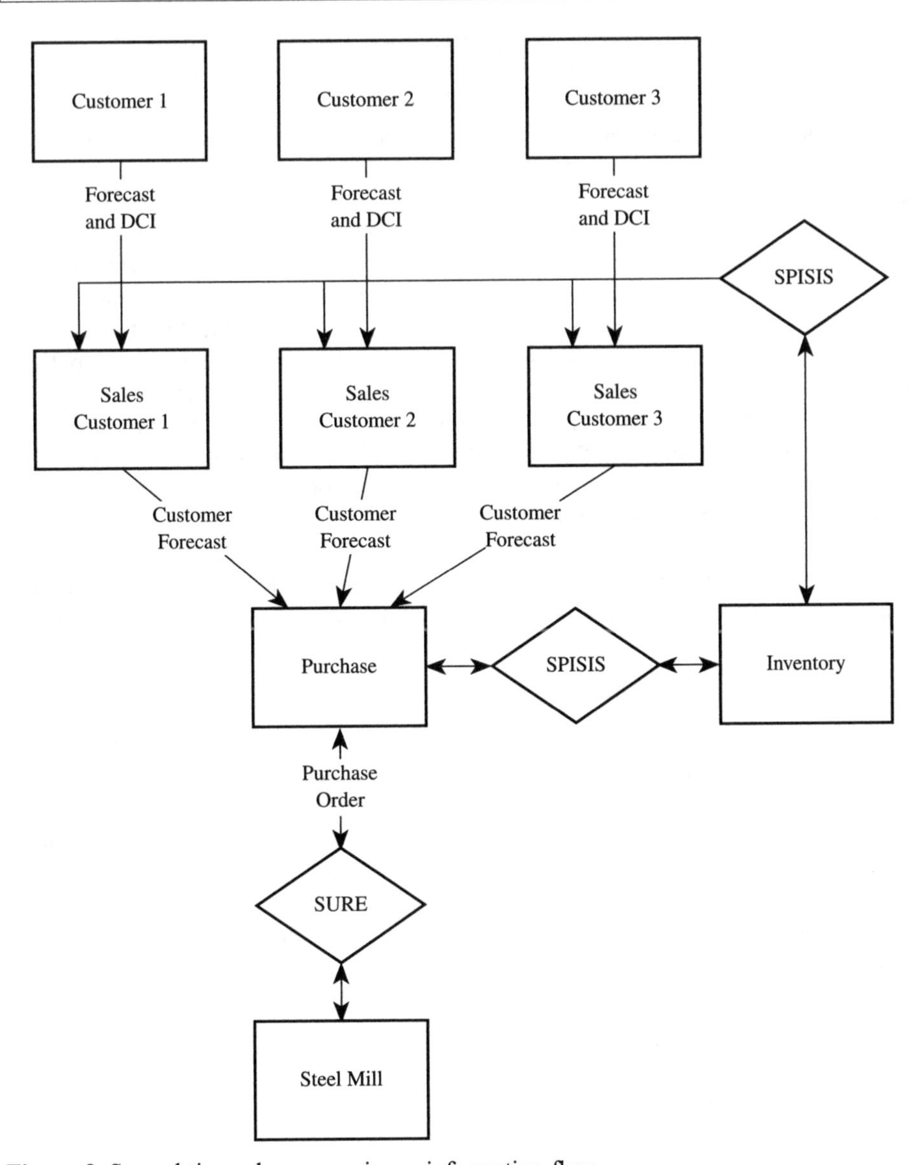

Figure 9 Second-tier order processing – information flow.

Second-tier order processing

The steel service centres use different systems to the one described above, although the tasks to be performed by the systems are basically identical:

- Customer order processing
- Inventory control
- Production scheduling of the blanking and slitting operations
- Raw material purchasing

The systems found were either based on spreadsheets or were directly derived from a pure inventory control system. The reason is clearly an historical issue, as the service centres developed from so-called 'steel stockholders', which simply used to store the material and send it out in case of an order coming in. The systems used to control these infrequent orders were mainly used to control the inventory, and not to process orders or schedule production. However, with the automotive sector becoming more and more demanding in terms of stock efficiency and delivery cycles, these systems cannot cope and now prove to be a major issue or 'legacy' for the service centres. This is the core reason why at second-tier level the phenomenon of information flow distortion becomes most apparent – even in case of very stable first-tier demand patterns. Current order processing systems are simply unsuitable for the repetitive automotive business, which requires frequent small batch production.

A further obstacle for the service centres is the internal departmental structure, which has grown alongside the 'stockholder' legacy, usually comprising of a centralized purchasing department. This in itself is another obstacle to communication within the company and with the supplier.

The information flow within a service centre is shown in Figure 9. The sales departments are customer-oriented, yet the purchasing department is physically separate and functionally centralized and in effect 'de-coupled' from the customers.

Steel purchase discount policy

In addition to the inappropriate order processing systems at the service centres, the current steel purchase discount policy operated by the steel mill gives an incentive for the service centres to batch their orders. The order batching practice is a root cause for information distortion and the often quoted lack of demand visibility at the steel mill.

The order batching can be shown in the demand and supply dynamics graphs, as shown in Figure 10, whereby a relatively stable weekly demand of the first-tier component suppliers (180 tons / week) is converted into erratic orders of 100, 400 and 500 tons.

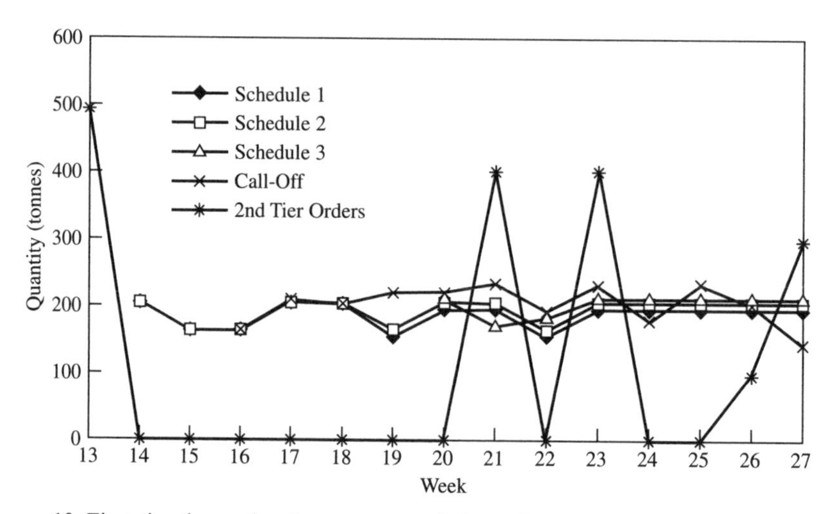

Figure 10 First-tier demand pattern vs. second-tier orders.

As Lee *et al.* (1997a) show, non-constant purchase prices of a product can be a driving factor for demand amplification within a supply chain. Price fluctuations and the frequency and depth of manufacturer's price promotions (i.e. discounts) should be reduced to a minimum to obtain true demand patterns in the chain.

This theory has been developed in the context of a distribution chain in the retailing sector, but it is as valid as relevant for the automotive steel supply chain, as the steel mill offers significant quantity discounts for steel purchases.

The intention of these purchase discounts is to achieve big batches of orders for the same grade of steel, which makes it simpler to handle and process the orders and finally to fit the orders into the caster schedule. The caster is the main constraint in the steel-making process, and therefore the objective is to achieve a maximum of output by generating 'coffin-shape' schedules by gathering as much quantity per grade as possible. Priority number one at the steel mill is volume achievement by maximizing throughput through the caster, unfortunately at the expense of the customers, as due date performance is second priority to the technical casting exigencies – if it is measured at all.

Looking at the discount policy on an abstract level, it has primary and secondary effects. The main primary positive effects are of course 'easy to handle' demand patterns in terms of generating a 'coffin-shape' caster schedule. The batched demand leads primarily to efficient caster scheduling and in consequence to better capacity utilization which is the main performance measure at the steel-making plant.

The secondary effect, however, is the distortion of the real demand patterns. Batched orders do not give the information of the real point of demand. The information, when the orders are required or 'due' at the customer, and consequently which order to prioritize in case of capacity problems, gets completely lost.

The discount policy can even be seen as a vicious circle. In order to profit from the discount, the service centres order ahead of real demand by batching their orders. The steel mill receives big order batches and falls short on some deliveries due to limited casting capacity. As a result, the service centres receive only partially fulfilled orders and fall short on delivery to their customers. In consequence, next time the service centres order even earlier to assure delivery to their own customers, taking excess inventory into account to ensure customer service. The steel mill receives the orders and gets the impression that the overall demand quantity has risen, as the real demand information is not available. Owing to this 'increased' demand, delivery performance further drops, leading to very early orders from the service centre, etc.

This circle can be temporarily broken by increased and excess capacity at the bottleneck (the caster in this case), but in the long term only transparent demand at the bottleneck operation can break the circle. The reason why increased capacity does not solve the issue is to be found in organizational behaviour. If the excess capacity were installed, the remaining backlogs would be cleared fairly soon, and the demand would drop back to normal level for some time. Yet any disturbance or uncertainty will raise demand levels again and create the next set of amplified orders, which in the long run will not be met by the available capacity, and as backlogs rise the same situation develops again.

In conclusion, the positive effect of better caster capacity utilization stands here against the negative effect of distorted supply patterns, which leads inevitably to poor delivery performance. However logical this argument might sound, in reality the costing of the latter proves to be difficult, hence the business case is currently not strong enough. Additionally, two departments are involved, sales and production, and inter-departmental conflicts do not foster these kinds of considerations. It will remain to be seen if the pressure in the market becomes too high to sustain this policy, i.e. if the continuous lack of delivery performance threatens to drive the company out of business.

Steel mill scheduling

The steel mill is a classic V-plant (Umble and Srikanth, 1990), converting a limited range of raw materials into a great variety of finished products. All of these end items are produced in essentially the same way, using highly capital-intensive and specialized equipment.

The process flow diagram (see Figure 11) shows the simplified process steps from the input of the raw materials to the different points, from where the coils are despatched to the customers. Basically, the steel can be despatched after hot rolling, pickling, cold rolling, annealing or galvanization. Apart from the type of steel, the gauge, width, grade, mechanical characteristics, surface quality, roughness, coil weight, any possible coating, and the quality determine the variables the steel can have.

In simplified terms, a mixture of coke, limestone and iron ore is used to melt iron in the blast furnace. The molten iron is transferred in a 'Kress carrier' to the steel-making plant, where it is poured into ladles. There, the iron is desulphurized, before it is transferred into the 'vessel', where scrap steel is added. Then, simply speaking, oxygen is blown in to remove the carbon, and the iron converts into steel. Minor additions of specific alloys or further degassing and desulphurizing may be undertaken in the secondary steel-making process.

Finally, the steel is poured into continuous casting machines, from where it goes directly through a mould, and with only the outer shell solidified the steel is drawn downwards through an arrangement of supporting rolls and water sprays. Automatic gas or plasma cutters cut the continuous cast into slabs.

From there, the slab goes either directly into hot rolling, where it is rolled into horizontal and vertical shape, or, as many slabs are too hot for immediate rolling, it goes via the slab stock and the slab reheating furnace to hot rolling at a later time.

After hot rolling, the wide coil either is despatched, or it is passed on to the hydrochloric pickling line. After pickling the coil, the coils are either despatched or sent to the cold rolling and finishing processes, which include cold rolling, annealing or galvanizing.

On average, there is a three-month order lead time from the steel mill. The bottleneck operation is casting, but once the slab is cast, rolling and coil processing are not capacity constraints. The lead time from casting a slab to despatching the finished coils can take up to eight weeks, as the batch annealing of cold-rolled strip steel, for example, takes up to one week itself, and the time from the slab cut to hot roll a coil is about 36 hours on average.

The steel mill uses several self-developed multi-stage scheduling systems for the production of steel coils. The planning system for coils and slabs gives only two weeks or 1000 tonnes visibility of the orders per grade. Also, the system will always wait to fill up a cast, before it is released. The long-term forecast ability of the planning system is not appropriate; it does not provide the necessary demand

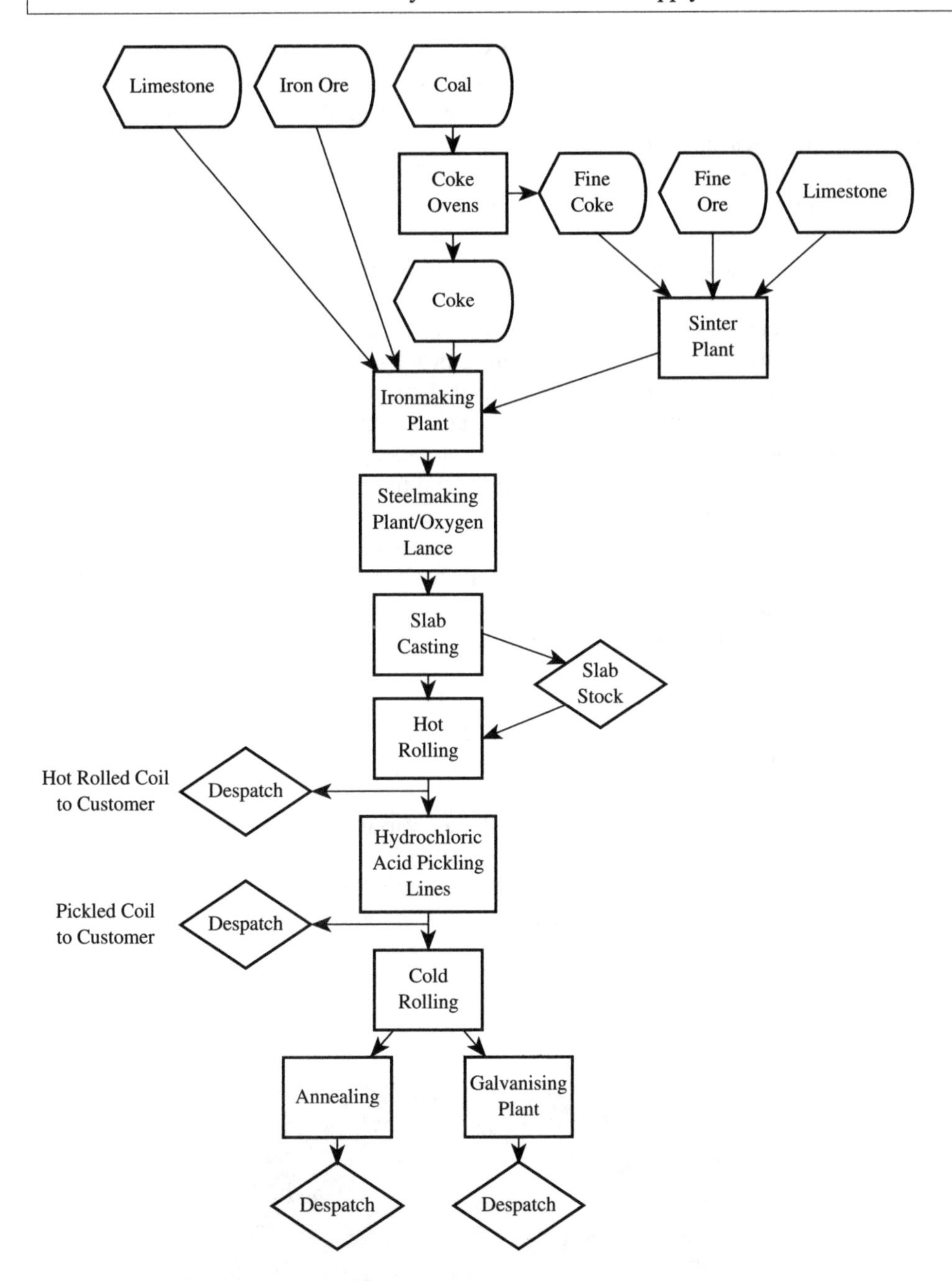

Figure 11 Steel-making process flow.

visibility and is not capable of showing repetitive high-volume demand patterns, such as found in the automotive steel supply chain. In terms of overall volume, 20 per cent of total cast steel is destined for automotive customers, of which a certain percentage is directly delivered to the body shops in the car plants, and the rest goes via the service centres to suppliers.

The scheduling problem involves the sequencing of slabs in the hot mill, which must conform to a 'coffin' shape in terms of width and thickness and are made in 'campaigns'. 'Coffin shape' means that width and grade have to be sequenced in ascending then descending order.

There are uncertainties with regard to yield and quality, and coils have to be produced in finite sizes not necessarily conforming to customer order multiples. Slabs are converted into coils which are sent on to skin pass and/or pickling. These stages have localized capacity constraints. Buffer stock, often in considerable quantities, is held between each stage and in dispatch, but nevertheless delivery shortages occur regularly, thereby encouraging customers to inflate their lead times. Coils are transported by truck or train – a scheduling problem in itself.

However, the discount policy and lack of demand visibility distorts the demand visibility and finally result in increased coil stock. This in return causes despatch bay problems due to multiple handling, surface damage, physical damage, and inventory control problems. This in turn was found to lead to erratic delivery performance and frequent rescheduling.

The coil damage problem, however, has to be seen in relation to the inventory level in the despatch area. Owing to excess inventory (>40k tons in some cases), the coils have to be stored in multiple layers, which means that access to the lower layers is only possible by moving the top layers – and multiple handling becomes inevitable, [see Figure 12].

These problems are even more exaggerated for the significant proportion of coils that require further processing (for example, pickling) which results from a loose coupling of scheduling systems and from damage that occurs in the intermediate buffer.

Further disturbance results due to coil size variations. Varying coil sizes cause delivery uncertainty, and force subsequent tiers to reschedule their production in case of over- or under-delivery. Coil size variation is a technical issue of the steel-making process, which hardly can be changed, yet affects all subsequent tiers in terms of planning unreliability.

Figure 12 Steel coil storage in a despatch bay.

ANALYSING THE SUPPLY CHAIN DYNAMICS

This section is split into four parts: the first part will report on the analysis of vehicle manufacturers' demand patterns submitted to the first-tier component suppliers, the second part will examine the steel mill delivery performance, and the third part will show exemplary evidence from the supply chain dynamics analysis. The last part will briefly describe a newly discovered effect in supply chains – 'Reverse Amplification'.

Vehicle manufacturers demand pattern analysis

The quality of demand patterns of four vehicle manufacturer customers for the analysed supply chains, as the original demand for the supply network, were analysed in terms of stability and consistency. Stability refers to the standard deviation of the orders, whereas the demand consistency refers to the match between forecast demand and the actual orders or 'call-offs'. Two of the VMs analysed are European manufacturers, the other two are Japanese transplant operations in the UK.

It was found that the patterns analysed differ significantly in terms of stability and consistency. For instance, one European VM shows highly volatile schedules, whereas the other provides relative stability within a 16-week window but low forecast quality beyond that, yet the two Japanese VMs provide reasonable and very stable demand patterns on both short and long terms.

To cope with demand fluctuations, buffers at finished goods inventory (FGI) level are applied successfully within two first-tier companies, providing the benefit of an internally smoothed production at the cost of higher FGI levels.

The example in Figure 13 shows the variation between the latest and the previous schedules issued by a Japanese VM. It can be seen that the average variation ranges between 2 and 4 per cent of the previous schedule, with the exception of part 5, which is very low volume and therefore very sensitive to mix alterations.

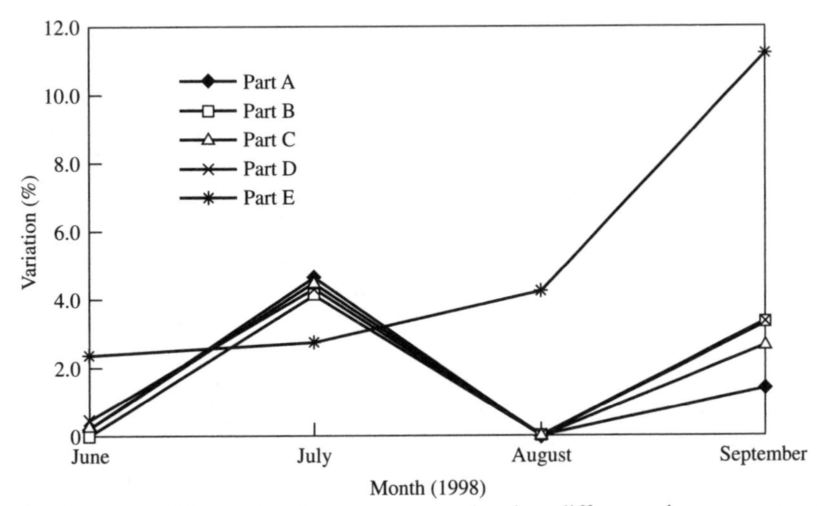

Figure 13 Japanese VM – schedule consistency, showing difference between newest and previous schedule in percent.

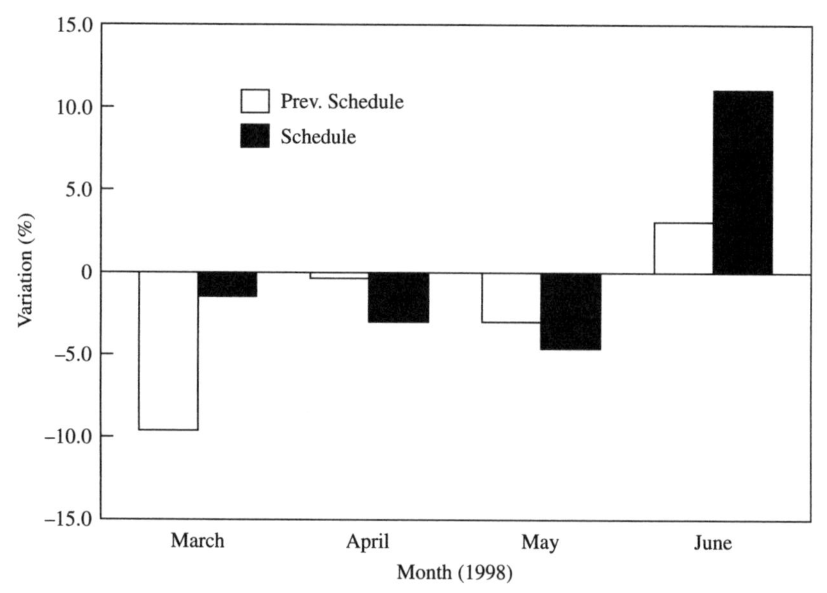

Figure 14 European VM – schedule consistency, showing actual DCI vs. the last two schedules.

The example in Figure 14 shows the schedule consistency of a European VM. What can be seen is the variation of the schedules from the actual call-off, which is not received until the week before. For example, the first schedule for May (issued in March) forecast a demand of 9 per cent below the actual; the following one (issued in April) predicted a demand of 2 per cent below actual. In June, the last schedule overestimated demand by 11 per cent, etc. What can be seen is that regardless of the time frame the forecast varies from max +/– 11 per cent from the actual demand, which has to be catered for by production availability.

Steel mill performance evaluation

The steel mill is the generic source of all material in the analysed supply chain – its delivery performance therefore is critical for the complete supply chain. Comparable to the impact that the demand patterns from the vehicle manufacturers have on the information flow, the delivery patterns from the steel mill have crucial influence on the material flow all along the supply chain. The steel mill delivery performance has been criticized extensively and has been blamed for poor overall supply chain performance. This section will therefore investigate the delivery performance of the steel mill to one major service centre, which mainly serves automotive customers.

Based on the demand data for the 49 purchase orders, this analysis will sum up the order patterns and the according delivery patterns. It has to be considered that the volume (6500 tonnes of steel) represents approximately 25 per cent of the overall yearly quantity ordered by the service centre from the steel mill.

Conducting the analysis, it was striking that in fact two dates need to be considered in relation to the delivery performance. The reason is that two dates are assigned to each order, the 'due' date which represents the real point of demand at the service centre, and the 'acknowledged' date, which is the date at which the steel mill promises the delivery for this order. All measurements at the service centre are based on the acknowledged date, given by the steel mill, but the real point of demand is the due or wanted date. Therefore, it was found to be necessary to analyse all 49 orders in terms of the difference between due and acknowledged dates.

Figure 15 shows the percentage of orders against the difference between the wanted and acknowledged week for the examined orders. It can be seen that only in 45 per cent of the orders did these dates match, and the latest found was an order acknowledged seven weeks after the wanted week. The fact that more than 50 per cent of the orders are acknowledged up to seven weeks late has serious effects on the order practices and the necessary buffer stocks to cover this practice. Possible reasons for this late acknowledgement might be:

1. **unreliable forecasts and lack of demand visibility at the steel mill,** resulting in invalid long-term planning;
2. **the internal steel mill sales and capacity planning process**, which does not allocate appropriate capacity to meet the service centre requirements;
3. **order batching practices at the service centre,** which cause erratic and amplified demand patterns.

Current business practice works within the constraint of poor correlation between wanted and acknowledged week, but as far as the Lean Processing Programme is concerned, it is an obstacle for any approach to pull scheduling or repetitive and synchronous deliveries on which a lean supply chain would be based.

But even evaluated on the basis of the acknowledged week, the deliveries from the steel mill show very poor results. Generally, the steel mill does not deliver all steel for

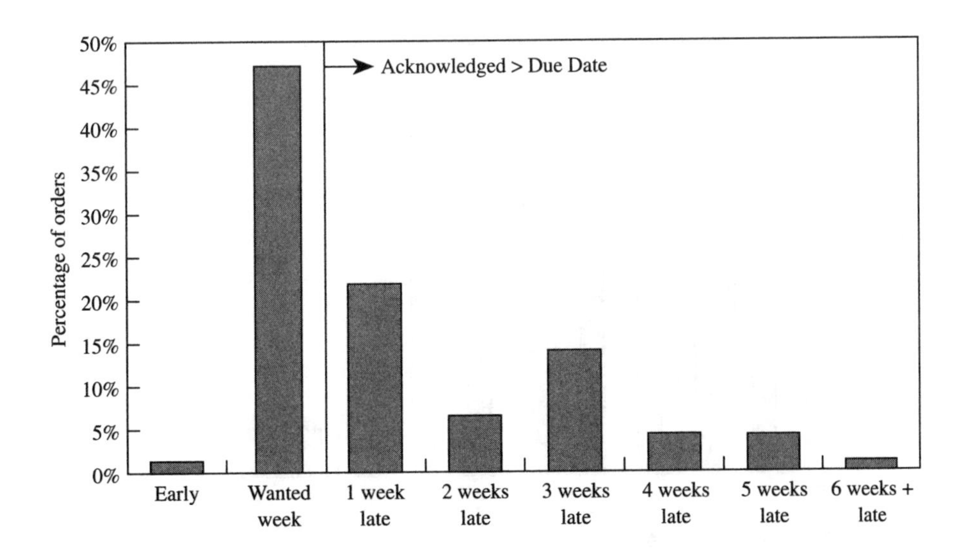

Figure 15 Steel mill deliveries to service centres – differences between Due and Acknowledged dates.

one purchase order in a batch, but spreads out the deliveries over time, as will be shown in Figure 18.

Figure 16 shows these delivery patterns, summing up all 49 orders. Every single delivery for the 49 orders has been cross-checked with the acknowledged week of the order and then has been marked as 'early delivery' (delivery before acknowledged), 'within the acknowledged week' (on-time deliveries), or 'X weeks late' (X weeks after the acknowledged week).

Summing up, the study of 49 analysed purchase orders shows that the deliveries of the orders arrive over a period of several weeks. The latest duration for an order fulfilment was 14 weeks.

The total percentage is 78 per cent, less than 100 per cent, because still outstanding orders at the point of collecting the data are not integrated. Therefore, all outstanding orders (22 per cent) will be late by definition.

The steel mill delivery performance shows that a total of 7 per cent of the ordered quantity is delivered on time, 16.5 per cent is delivered early, in the right week or up to 1 week late, and 43 per cent of the material is being delivered more than 4 weeks late.

Concerning the steel mill delivery performance it can be said that the service centre due or wanted week does not correlate with the steel mill acknowledged week. The due date is earlier in 55 per cent of the orders studied. However, even measuring against the acknowledged week the steel mill deliveries do not correlate with the promised delivery date. Only 16.5 per cent of the ordered quantity was delivered early, on time or up to one week late. The remainder was > 2 weeks late. Only 7 per cent of the overall quantity reached the service centre on time. Also, no significant differences can be found concerning the delivery performances against particular first-tier requirements, even in case of complex routings for certain grades.

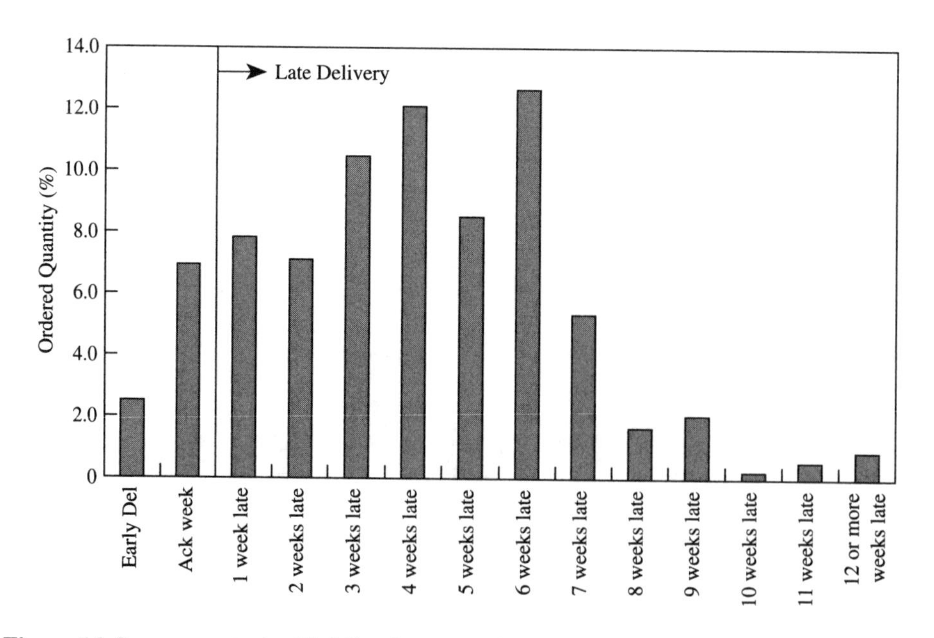

Figure 16 Summary: steel mill deliveries to service centres.

However, this fact needs be further discussed, as a delivery performance of 7 per cent on time cannot be seen isolated from order batching practices at the service centres and information distortion in the supply chain, making it very difficult for the steel mill to meet the highly amplified demand patterns.

Three-tier supply chain dynamics

To analyse the impact of the scheduling systems on the supply chain, 17 parts were mapped in terms of their dynamics of information and material flows. In particular, the orders and promised delivery dates were opposed with the actual deliveries between the three tiers. Two examples will be discussed here, shown in Figures 17 and 18, whereby erratic patterns in the information flow cause erratic delivery patterns, leaving potential delivery shortfalls which have to be covered with inventory.

The effects described above can best be seen in the demand and supply dynamics graph, shown in Figure 17. What can be seen is the first-tier forecast demand and actual demand for slit coil steel which is more or less stable around 30 tonnes per week, and the resulting service centre orders to meet this first-tier demand. Service centre orders are shown as passed onto the third-tier supplier, who confirms delivery in both cases shown one week late. The service centre or second-tier orders reflect the batching policy, as the three orders are 100 tonnes each, instead of 30 tonnes weekly. The steel mill delivery response to these batched orders (shown as underlying curve) does not match these orders, but spreads out over time, resulting in low on-time delivery performance.

Delivery shortfalls for the service centre occur when the steel mill deliveries are lower than the first-tier requirements in that period. If the service centre in this case does not hold enough stock, the first-tier company will be delivered short. The only way to overcome this effect would be to abandon the steel purchase discount policy and to deploy full demand visibility to the steel mill.

The outcome is poor delivery performance, both from the steel mill to the service centre and from the service centre to the first-tier companies. A typical service centre delivery performance to a first-tier supplier was on average 60–70 per cent (data for 1997 and 1998).

Analysing the demand and supply pattern in this study, the only conclusion that can be drawn is that the demand and supply patterns are erratic and do not correlate with any logical pattern. The first-tier call-offs show a certain stable and repetitive character which one would expect in the automotive business. These call-offs seem to be the last ‘realistic’ demand patterns; further onwards any pattern gets lost. The demand patterns which finally reach the steel mill are distorted and severely amplified, and it cannot be a surprise that the delivery performance against this amplified demand is very poor. The following section will sum up all deliveries from the steel mill analysed for this study, but in the opinion of the author this is of limited significance – for the above reason. The focus has to be turned to find the root causes for the information distortion, and these causes will have to be clearly divided from the numerous effects this amplified demand causes.

The second example (Figure 18) reinforces the observations made in the previous example. Stable first-tier demand is distorted due to batching at second-tier level, and subsequently the steel mill deliveries are erratic and bear no resemblance to the actual

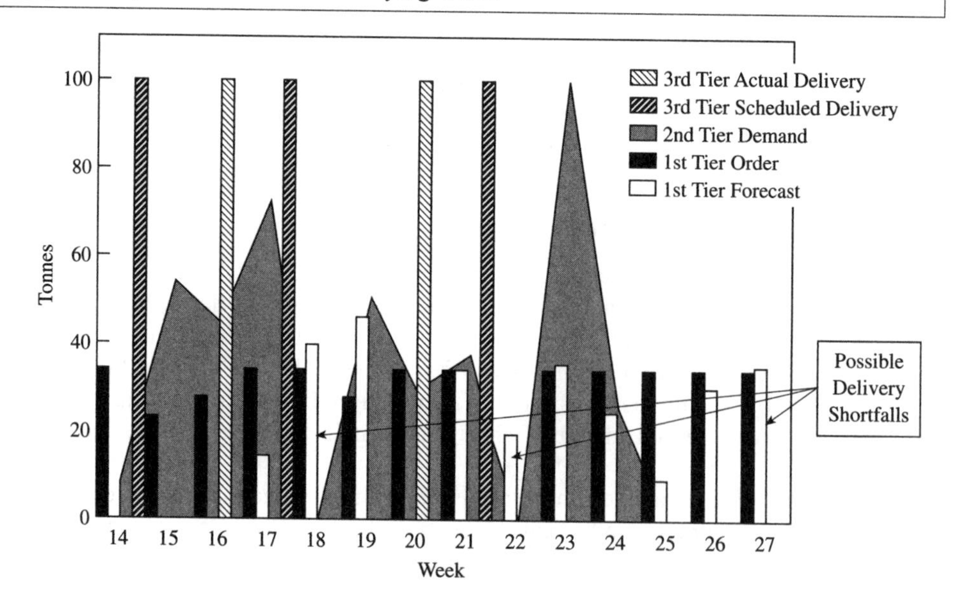

Figure 17 Three-tier demand and supply dynamics, first example.

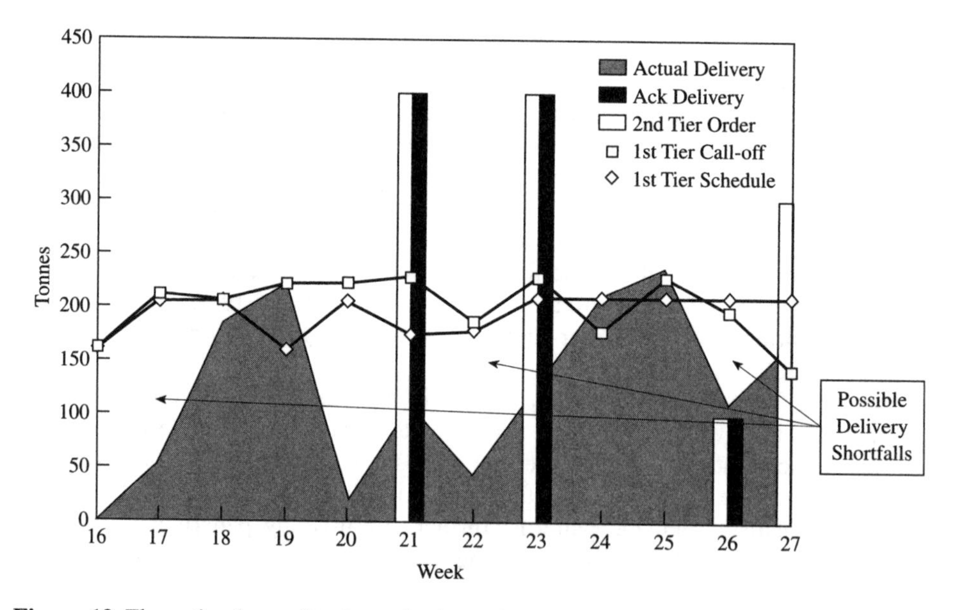

Figure 18 Three-tier demand and supply dynamics, second example.

demand. In this case, however, the acknowledged dates coincide with the due dates given by the service centre. Nevertheless, the steel mill deliveries are matching neither first-tier nor second-tier demand patterns, and inventory in the system is the only way this value stream can operate without compromising the service to the final customer, the vehicle manufacturer.

The reverse amplification effect

The 'demand amplification' effect has been previously discussed and could be shown in the LEAP research. However, a second and similar effect could be observed which is not described in the system dynamics research so far – what could be called 'supply or reverse amplification'. Reverse amplification will hereby be defined as 'amplified and distorted supply patterns caused by supply or throughput constraints'.

What happens is that in case of a supply constraint, as for example the caster operation in the steel supply chain, order backlogs build up over time. As described by Lee *et al.,* customers of this bottleneck operation start to over-order as a safety measure (the 'rationing game'); the demand amplifies and soon an order backlog builds up. As the constraint generally is operated on large batches, it will not be able to supply against these orders until the next batch of the right product is run through the bottleneck. If, however, the right product is produced, suddenly a large quantity of products becomes available and can be supplied. As the order backlog has built up, a large 'wave' or quantity of material is subsequently sent down the system, flooding the stocking points with material. The downstream tiers will then stop ordering immediately until the product is in short supply again. Then the over-ordering starts again, etc., and the circle is closed.

The steel supply chain is an obvious candidate for this effect, yet any other supply chain that has a common bottleneck operation through which several products have to flow might be affected. Any additional process unreliability, quality problems or other delays worsen this effect.

Reverse amplification would show the following symptoms in the demand and supply dynamics:

- **Amplified orders** hitting a constraint operation. High orders are a 'safety measure' by the subsequent tiers ('rationing game') to ensure supply and to build up safety stock.
- **Multiple products that need to go through one constraint operation**, hence large production batch sizes are used, further lengthening the order lead times. The throughput possibly is even further compromised by quality or process problems.
- **Large order backlogs** at the throughput constraint operation.
- **'Wave'-type supply patterns** of the bottleneck operations, way beyond the actual order quantities as response to the backlogs.

The 'reverse amplification' is a corresponding effect to the 'demand amplification', and both effects complement each other. Their existence and causal link could be reproduced and verified using a quantitative simulation approach (see Chapter 17).

In the real-life example, the 'reverse amplification' effect shows typical 'wave' supply patterns (Figure 19). In the example below, the underlying demand from the first-tier company is on average 15 tonnes / week. The service centre, however, orders in batches of 60 tonnes. The steel mill acknowledges the deliveries for week, 14, 15 and 18 (in case of weeks 14 and 15, the order on the mill has been put earlier on, but delivery is acknowledged late). The actual deliveries only start in weeks 14 and 15 with small amounts; the 'wave', though, comes in week 18, when the main part of the production batch / campaign is ready for shipment and is then supplied against the backlog, exceeding the ordered quantity for that week by 33 per cent.

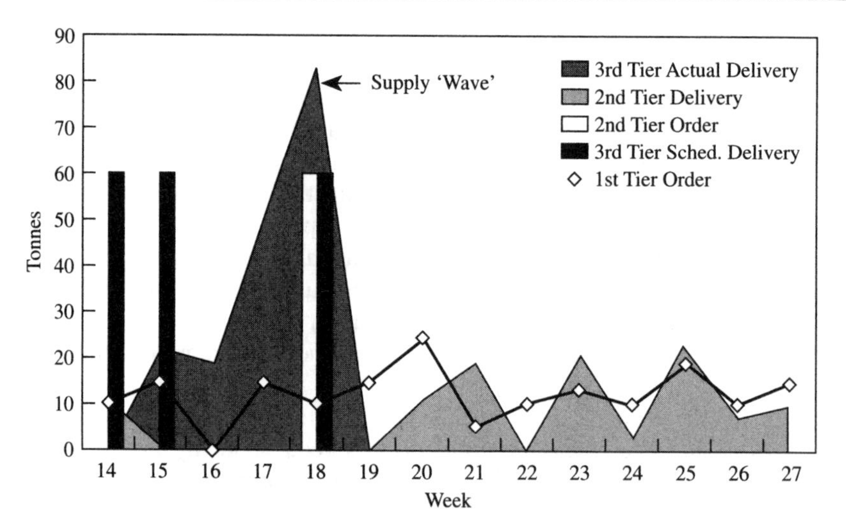

Figure 19 Reverse amplification effect.

After the 'wave' hits the service centre, no more deliveries are made until week 23, when another remainder is supplied.

CONCLUSION

The automotive steel supply chain has been described and reflected under various aspects. Sources of distortion could be pointed out all along the value stream, and it can clearly be said that only the focus on the whole supply chain enabled the discovery the wide range of root causes for the distortions occurring in the system. It became clear that any system or action in the system can only be evaluated by seeing it in conjunction with its effects on the entire supply chain, otherwise it might lead to local and restricted solutions that would obstruct the global optimum, the 'seamless' supply chain (Towill, 1997).

In this context, it could be proven that all levels of the analysed automotive steel supply chain owe certain responsibility to the endeavoured deficiencies in the system, and that the lack of performance currently experienced in the analysed supply chain is the result of these problems.

It was not intended to expose individual companies or tiers in the supply chain as scapegoat or single source of the problems experienced in the system – by interpreting the sets of data discussed here in that spirit the main conclusion of this research would be missed out.

In detail – the demand and supply dynamics analysis showed classic symptoms of amplified and distorted demand patterns over all three tiers analysed. These symptoms were:

- **excess supply chain inventory;**
- **unstable demand patterns;**
- **low forecast quality;**
- **poor delivery performance in all levels.**

In terms of scheduling, it was found that a variety of different manufacturing planning and control approaches were applied along the supply chain. It also can be said that none of the approaches showed significant superior performance in terms of delivery and schedule stability. There is some inconclusive evidence to show that MRP-driven companies show increased WIP and FGI levels compared to manual-spreadsheet production control systems and kanban-driven approaches. What appears clear is that the performance of the scheduling systems is heavily dependent upon the performance of the customer (in relation to the accuracy and stability of forecast), and upon the performance of the supplier, in terms of meeting delivery requirements.

Several root causes for demand distortion could be identified in all tiers of the system, some of which coincide with the value stream mapping findings. Root causes of the demand and supply pattern distortion include:

- **instability of vehicle manufacturers' demand pattern;**
- **batch orientation in the press shops;**
- **process unreliability;**
- **information distortion due to inappropriate scheduling systems;**
- **order batching and steel purchase discounts;**
- **coil size variations.**

It can be further shown that demand pattern stability decreases the further upstream the pattern goes, and makes it particularly difficult for the steel mill to supply according to this amplified demand. Not surprisingly, the steel mill delivery performance against these distorted and batched demand patterns is very poor. The consequence is excess safety stock in the chain to buffer this supply uncertainty, and frequent rescheduling is often necessary.

Furthermore, a new system dynamics effect, 'reverse' or 'supply' amplification, could be described, providing further explanation of the distortion found in both information and material flows across the value streams analysed in this study.

Table 3 summarizes the major root causes for distorted and amplified demand in the analysed supply chains and also points out their effects on the different tiers in the supply chain.

Summing up, the question arises as to how the whole supply chain can work within a framework of distorted information and supply patterns at all tiers in the chain, considering that the final customers, the vehicle manufacturers, do not tolerate short deliveries.

The first reason why the system still works within these parameters is that the service centres submit orders which over-anticipate future demand. The steel is simply ordered way ahead in time to obtain discounts and to cover possible steel mill delivery shortfalls. Therefore, late deliveries are not problematic, unless they fall outside the anticipated delays in supply.

The second reason why the supply chain works within these parameters is the amount of buffer stock held at all levels. Inventory is used all along the chain to cover demand and delivery uncertainty and process unreliability. The amount of inventory held in the chain can be displayed with the process activity map which shows that parts are waiting up to 97 per cent of their time on their way along the chain. Value added time is usually less than 0.1 per cent.

Table 3 Cause and effect matrix

Causes	*Description*	*Efect on* 1st tier	2nd tier	3rd tier
Vehicle manufacturer's supplier scheduling	• Demand fluctuations in VM suppliers' schedules	↑↑	↑	
Supplier scheduling at first tier level	• Unstable demand patterns due to frequent rescheduling, nervous MRP settings, etc.		↑↑	
Press shop scheduling at first tier level	• EBQ scheduling in the press shop might distort the information flow • Batch orientation alters repetitive and stable demand patterns into batched demand		↑↑	←→
Press shop unreliability at first tier level (i.e. tooling)	• Unreliable press shops cause supplier rescheduling due to short-term adjustments to press plan		↑↑	←→
Order processing at second tier	• Order batching practices • Order processing can alter the information pattern		↑	↑↑
Steel purchase discount policy	• Steel discounts for big orders severely alter the demand patterns		↑	↑↑
Steel mill delivery performance	• Poor delivery performance from the steel mill causes rescheduling at all first and second tier companies • Poor delivery performance is covered by excess safety stock at all levels of the chain	↑	↑↑	
Performance measures and company philosophy at the steel mill	• Capacity utilization is highest priority, delivery performance to customer ranks second only • This prioritization is the company philosophy and reflects in the performance measures	↑	↑↑	

Key: ←→: low / infrequent impact, ↑: medium / frequent impact, ↑↑: high / frequent impact

THE WAY FORWARD

Having pointed out the deficiencies of the current supply chain, the remainder of the paper will focus on short and long-term improvement actions. Some of the short-term actions are currently being implemented, whereas the long-term actions represent tentative proposals.

Short term – communication, visibility and reliability

In the short term, improved communication along the chain is an obvious candidate for implementation. This refers to the achievement of demand visibility, which must go hand in hand with an understanding of the system dynamics of the network. The first is being addressed in projects implementing level scheduling and demand smoothing (see Chapter 10), the second by involving participants in the network in the chain game, 'The Lean Leap Logistics Game' (see Chapter 17).

Level scheduling or demand smoothing begins with awareness of final customer demand, which is communicated to all parties in the network, with the aim of creating stable repetitive schedules. Separate projects are in progress to improve the particular issued pointed out here, as for example to improve press shop reliability, to improve steel mill performance, to improve awareness and communication along the supply chain and to encourage level scheduling.

Furthermore, process reliability is being addressed in projects relating to overall equipment effectiveness (OEE), set-up reduction (SMED), and total productive maintenance (TPM). Short-term waste removal is being addressed through a series of kaizen-blitz exercises (see e.g. Bicheno, 1999; Butterworth and Bicheno, 1999; Sullivan and Bicheno, 1999).

Long term – synchronized supply chain scheduling

The long-term vision aims at extending the level scheduling project to 'synchronized scheduling', whereby processes are laid out to create the overall effect of a 'conveyor' which pulls steel products in a steady flow towards their destination.

The synchronized supply chain scheduling concept aims to integrate a whole supply chain into one scheduling concept and to deploy demand visibility to all levels of it in order to achieve a 'zero waste' value stream. The aim is to include all different scheduling systems found in a supply chain and to integrate as many parts as possible, even those that provide less stable and repetitive demand patterns. The concept is based on the following concepts:

1. **The five lean principles**, aiming at waste elimination in the supply chain (Womack and Jones, 1996), such as inventory or overproduction, and achieving 'flow' production.
2. **The 'synchronization of demand and supply'** in order to minimize non-value adding operations. 'Synchronization' relies on the two basic principles of **demand visibility** (all demand patterns are visible for all participants in the chain) and **process visibility** (all production processes in the supply chain are timed against each other – 'synchronized') (Forrester, 1961; Lee *et al.*, 1997a).
3. **The 'quick response' approach to communicate information** along the supply chain. The final demand pattern is made known to all tiers via EDI, the Internet or other means, similar to the 'point of sale' system used in other sectors such as the food or textile industry (Hunter, 1990; Kurt Salmon Associates, 1993).
4. **The 'time buffer' concept** applied in synchronous manufacturing to buffer the system against any kind of uncertainty or process unreliability (Umble and Srikanth, 1990).

Furthermore, removing echelons from the system might even be considered. The obvious candidate would be the service centres, which could be either part of the blanking operation at first-tier suppliers, or integrated into the coil despatch operations at the steel mill. This integration would reduce the total lead time and inventory in the supply chain, and hence eliminate sources for demand distortion in the system. The benefits of echelon removals on the supply chain dynamics were initially described by Forrester (1961), but continue to be revisited (e.g. Wilkner *et al.* 1991).

As the demand patterns are known to all participants, there is no need to hold any stock for the time of no demand. In a synchronized supply chain, the final customer demand is known, and all levels work according to the final demand. The main prerequisite is, of course, stable and level demand from the end customer. In case this demand is not stable enough, the introduction of an FGI buffer at first-tier level can create this stability.

The benefits of known demand along a supply chain have been widely proven; for example, Lee *et al.* (1997a) prove that 'double forecasting' can be a driver for the 'bullwhip effect' (demand amplification). Double forecasting means that the incoming forecasts are altered at every decision point within the chain and then submitted to the preceding level, where the same process occurs and the information is altered again. In the synchronized chain, no double forecasting is happening, as all levels use the final demand to forecast their production. This approach is commonly suggested (Evans *et al.*, 1993; Lummus and Alber, 1997).

Even Forrester (1961) proved the benefits of visible demand. He showed, although in a limited system simulation, that in case of a +10 per cent increased final demand the production peak at manufacturing level can be reduced from +45 per cent to +26 per cent by transmitting the information directly from the customer to the manufacturer. Furthermore, Lee *et al.* showed for retail distribution chains that balanced and 'perfectly synchronized' retailer ordering can be achieved. Under that scenario (and only then), the variability of demand experienced by the supplier and the retailers is identical, and the 'bullwhip effect' disappears.

Synchronization improves the overall supply chain performance, as the demand visibility erases demand amplification, facilitates inventory reduction where safety stocks were necessary to cover demand and supply uncertainty, and improves the quality of forecasting and long-term planning in all levels.

Unlike in 'pull' scheduling systems, the synchronized approach does not need inventory to pull from or to replenish, as direct demand information is available for all levels of the supply chain. Therefore, all levels can schedule their production accordingly and produce / purchase the material only when it is required by the customer.

The set-up of the necessary 'demand transmission system' will not need much time or resources, as Internet, e-mail or existing EDI links can be used. What definitely is necessary is the commitment of all participating companies to share information beyond the supplier / customer horizon, i.e. over all tiers of the chain. Figure 20 shows the synchronized supply chain scheduling concept.

The origin of the demand is the final customer demand, in this case the vehicle manufacturer, whose orders and forecasts are transferred to all levels in the chain. This data is used for planning and to give the real demand picture, but still all levels in the chain will have to communicate to synchronize their processes.

In case of unstable demand patterns or even a call-off system between first tier and vehicle manufacturer, safety stock is necessary at FGI level to buffer the fluctuations and create a smooth schedule. In the current supply chain safety stocks are required throughout the chain in order to protect against uncertainty of any kind.

The synchronized value stream will introduce 'time buffers' instead of safety stock at all points, where unreliable processes might destroy the synchronized flow. Time buffers are not safety stock – safety stock is an agreed amount of inventory in front of a process step which will always be filled up to the agreed level. Safety stock covers

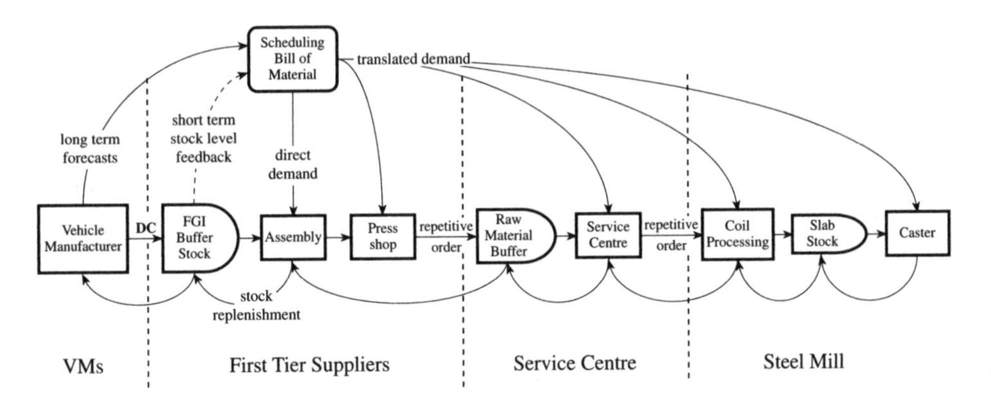

Figure 20 Synchronized supply chain scheduling concept: information and material flow.

against demand and process uncertainty, whereas time buffers will provide 'safety time' for unreliable processes. Time buffers are not stationary inventory; they are the 'right' product in the 'right' quantity and quality, just ahead of time. Time buffers will be introduced in front of all problematic processes to give more time to process the products. If more time (for unplanned set-ups, for example) is required, the time buffers give the possibility to take more time than scheduled. Once the products are processed, the buffer will not be replenished. The long-term objective, of course, is to decrease unreliability and hence the time buffers.

A special case is the steel mill, as the synchronized scheduling approach relies on a slab stock, in the same way as kanban systems would. This slab stock is not a time buffer, but safety stock which covers against casting process reliability, but will be continuously replenished. The reason is that the caster schedule is driven mainly by technical issues and is not flexible enough to respond adequately to the demand, although known and stable, without this buffer stock.

In comparison to the Quick Response or Efficient Consumer Response approach, the difficulty is that it is not sufficient to measure the quantities in 'number of parts', as this information is only relevant for first-tier assembly lines and press shops. The overall demand has to be translated for every level of the supply chain. For first-tier companies, the quantity of finished parts is the basis, for second tier the number of blanks or slit coil, for third tier the quantity of steel in wide coil is necessary. Therefore, the demand pattern has to be translated in a 'Scheduling Bill Of Material (SBOM)', which follows the generic BOM concept, but takes all the levels of the supply chain into account.

FINAL REMARKS

The automotive steel supply chain analysed in the chapter has been reflected from various different angles, and the causes and effects of distorted demand on the supply chain performance could be shown. However, the complexity of the causal interactions that drive these distortions is hardly understood and acknowledged, and even the study presented in this chapter is far from being a comprehensive analysis of the research problem.

Nevertheless, this exploratory study tried to acknowledge this complexity and to reflect the issue from various perspectives, i.e. from the point of view of all tiers in the system, whereas so far supply chain analysis mostly has only focused on single tiers or single customer–supplier links. Within this study, however, the limitations of single link approaches could be shown. The findings strongly suggest that any analysis or improvement activity in supply chains needs to be investigated from a holistic perspective, otherwise any conclusion would not only be biased (probably in favour of the tier or company analysed in the research), but also any improvement will remain an island solution and potentially cause more additional deficiencies in the system than it aimed to cure.

REFERENCES

Butterworth, C. and Bicheno, J. (1999) Kaikaku: A LEAP forward for automotive supply. *FT Automotive Components Analyst*, February, 12–13.

Bicheno, J. (1999) Kaizen and Kaikaku. *Logistics Focus*, 7(3), 12–17.

Burbidge, J.L. (1983) Five golden rules to avoid bankruptcy. *Production Engineer*, 62(10).

Evans, G., Naim, M.1 and Towill, D. (1993) Dynamic supply chain performance: assessing the impact of information systems. *Logistics Information Management*, 6(4), 15–25.

Forrester, J. (1958) Industrial Dynamics: a major breakthrough for decision makers. *Harvard Business Review*, 36(4), 37–66.

Forrester, J. (1961) *Industrial Dynamics*. New York: MIT Press and John Wiley & Sons.

Hines, P. and Rich, N. (1997) The seven Value Stream Mapping Tools. *International Journal of Operations and Production Management*, 17(1), 46–64.

Holweg, M. (1999) Supply chain scheduling and the automotive steel supply chain. Unpublished MSc Thesis, University of Buckingham.

Hunter, A. (1990) *Quick Response in the apparel industry*'. Manchester: The Textile Institute.

Kurt Salmon Associates (1993) *Efficient Consumer Response: enhancing consumer value in the grocery industry.* Washington, DC: Food Marketing Institute.

Lee, H. L., Padmanabhan, V. and Whang, S. (1997a) The paralyzing curse of the bullwhip effect in a supply chain. *Sloan Management Review*, 38(3), 93–102.

Lee, H.L., Padmanabhan, V. and Whang, S. (1997b) Information distortion in a supply chain: the bullwhip effect. *Management Science*, 43(4), 546 ff.

Lummus, R. and Alber, K. (1997) supply chain management: balancing the supply chain with customer demand. APICS Research Paper Series, January.

Monden, Y. (1998) *Toyota Production System: An Integrated Approach to Just-in-Time*, 3rd edition. Norcross: Engineering & Management Press.

Saporito, B. (1994) Behind the tumult at Procter & Gamble. *Fortune*, 129(5), 74–82.

Sterman, J. (1989a), Modelling managerial behaviour: misperceptions of feedback in a dynamic decision making experiment. *Management Science*, 35(3).

Sterman, J. (1989b) Misperceptions of feedback in dynamic decision making. *Organisational Behaviour & Human Decision Making Processes*, 43, 301–35.

Sullivan, J and Bicheno, J (1999) Case study: application of value stream management to muda reduction in a first tier automotive component manufacturer. EurOMA International Annual Conference, Venice.

Towill, D.R. (1996) Time compression and supply chain management – a guided tour. *Supply Chain Management*, 1(1).

Towill, D.R. (1997) The seamless supply chain – the predator's strategic advantage. *International Journal of Technology Management*, 13(1), 37–56.

Umble, M. and Srikanth, M. L. (1990) *Synchronous Manufacturing: Principles for World Class Excellence.* South-Western.

Vollmann, T.E., Berry, W.L. and Whybark, D.C. (1992) *Manufacturing Planning and Control Systems*, 3rd edition. Boston: Irwin.

Wilkner, J., Towill, D. and Naim, M. (1991) Smoothing supply chain dynamics. *International Journal of Production Economics*, 22(3), 231–48.

Womack, J.P. and Jones, D.T. (1996) *Lean Thinking*. New York: Simon & Schuster.

Section 4

From Analysis to Improvement. Examples of LEAP Improvement Initiatives

From value stream mapping to the development of a lean logistics strategy

10

Case study: Tallent Engineering Ltd

Peter Gallone and David Taylor

INTRODUCTION

Tallent Engineering is one of the six component manufacturers that participated in the LEAP project. The company is located in Newton Aycliffe, Co Durham and in 1999 employed some 980 people. The company manufactures a variety of components and sub-assemblies for car manufacturers both in Britain and Europe. Typical products include sub-frame assemblies or axle assemblies for popular makes of car. Customers include Nissan UK, Ford Europe, Honda and Rover.

The plant is organized in a manner which is fairly typical for modern component manufacturers (see Figure 1):

- *Raw materials*.
  The primary raw material is steel which is sourced from steel service centres usually as slit coils averaging 2 to 6 tonnes in weight. In addition, a wide variety of components are bought in from other suppliers.
- *Blanking*. The first process is to un-coil the steel and cut it into blanks, i.e. flat sheets of steel ready for pressing. Typically these might be of a size 100 × 100 cm or 50 × 50 cm depending on the component for which they are required. The company has a blanking cell with three large presses (500 tonnes, 400 tonnes and 1000 tonnes respectively).
  The blanking unit is a 'common-user' facility in the plant in that it provides products to all of the assembly lines. In some cases ready-cut blanks are purchased from the steel service centre primarily because of restricted capacity in the blanking cell.
- *Pressing*. The second process is for the flat blanks to be pressed into shaped parts. The press shop comprises 15 presses, ranging from 75 tonnes to 500 tonnes in capacity. The press shop is essentially a common user facility, although some presses are dedicated more or less full-time to the production of specific parts.
- *Assembly*. The company has 21 assembly lines each entirely dedicated to the production of a specific sub-assembly for a specific customer. Here the pressed parts, together with any 'bought-in' components, are assembled, typically using robotic welding stations.
- *Paint*. There are two paint plants for electro-painting of finished assemblies. These are again common-user plants in that they service all the assembly lines.

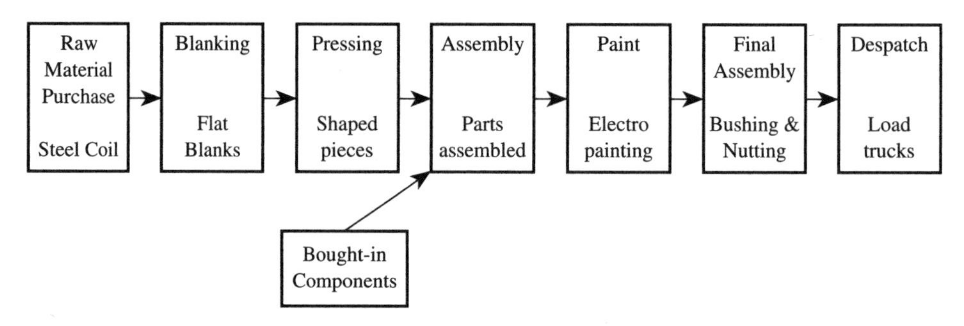

Figure 1 Outline of key processes at Tallent Engineering Ltd.

- *Final Assembly.* Subsequent to painting, a number of products require final assembly where parts are added that cannot be painted, e.g. rubber bushes, nuts and bolts. Final quality checks are also carried out at this stage.

The Lean improvement project commenced in April 1997. A team from Cardiff Business School worked in conjunction with senior and middle management to instigate lean awareness raising in the company and to carry out a rigorous situational analysis using the suite of lean mapping tools. It was agreed that the project would be focused by analysing two key product lines manufactured by Tallent Engineering – the back axle for the Nissan 906 model and the sub-frame for the Ford Mondeo.

The first task was to assemble the key staff involved in the planning and production of these two products. These included material purchasers, production controllers, shop floor staff from each of the main processes, warehouse staff, transport staff (forklift drivers), and the company Logistics Manager. These staff were given an introductory seminar on the Lean principles and approaches and on the nature of the Seven Wastes. Subsequently, they were required to complete a simple numerical ranking and scoring exercise to give a subjective assessment as to their views of the occurrence of waste in the two production systems under study. The purpose of this exercise was twofold: first, to give the researchers some preliminary indication of the potential areas of waste in the company, and second, to alert staff in the company to the nature of waste in the lean context. The aggregate results of the waste ranking exercise are shown in Table 1.

Table 1 Table of waste scores

	Average score
Transport	7.29
Unnecessary inventory	5.64
Waiting	5.57
Defects	5.5
Unnecessary motion	4.21
Over production	3.93
Inappropriate processing	2.86
Total	35

The next step involved more rigorous attempts to quantify the processes and analyse the waste in the company using the suite of Value Stream Mapping Tools.

Over a period of four days LEAP research staff, together with Tallent Engineering staff, collected the required data for the following mapping tools:

- Process Activity Map
- Value Analysis Time Profile
- Demand Amplification
- Quality Filter
- Physical Structure map
- Supply Chain Response Matrix

Figure 2 summarizes some of the key points emerging from the mapping exercise.

The results of this mapping exercise, together with some initial suggestions as to possible improvement projects, were subsequently presented to a panel of senior management from the company. This resulted in a number of immediate improvement initiatives related to specific aspects of the business. However, it was soon realized that isolated initiatives would have only limited impact. The decision was therefore taken to develop a comprehensive logistics strategy to provide a framework and direction for the initial and subsequent lean improvement initiatives. This strategy was developed by the Tallent Engineering Logistics Manager with input from the LEAP research team.

The remainder of this case study comprises a presentation of the Logistics Strategy document as it was first developed in early 1998. As would be expected, since that time the document has had various revisions as elements of the plan have been completed and as commercial and technical developments have occurred.

The approach adopted by Tallent Engineering is a good example of how the lean mapping tools can lead directly to the development of a strategy statement which then provides a framework to guide the lean improvement process over a number of years.

TALLENT ENGINEERING LOGISTICS STRATEGY

The logistics mission statement

The logistics vision is '***To develop a Lean Supply Chain based on the minimization of the "seven wastes" and the achievement of quantified targets in terms of both customer service and logistics costs***'.

The scope of logistics management

The scope of logistics management activity is illustrated in Figure 3.

The aim of logistics management at Tallent Engineering is to achieve the efficient movement of materials and information from sources of supply (inbound logistics) through the manufacturing processes (in-plant logistics) to delivery to customers (outbound logistics).

In order to achieve an efficient supply chain operation it will be necessary to consider each of the three key aspects of supply chain management, namely:

Key Points: Process Activity Map

- Operations
  - 15 short value adding operations out of 120 steps in total
  - The percentage of VA operations = 10%
- Transport
  - Internal transport is high and many steps – 5km around plant
- Storage
  - high storage time (15200 mins = 253 hours = 109.55days
  - Is overproduction of blanks and pressing necessary?
  - Has a shorter press run constraint analysis been done?
  - Can we reduce the number of stock points?
- Inspection
  - End of line rectification is part of the process
  - An inspector inspects inspection

Key Points: Supply Chain Response Matrix

Customer 1: Lead time = 29 hours
Stock time = 156 hours

Customer 2: Lead time = 118 hours
Stock time = 650 hours

- Response times are good
- Fork lifts are 'urgent' but unnecessary
- Flow difficult to follow
- Is it feasible to buy blanks?
- Is the batch size limited by the bin size?

Key Points Quality Filter Map

- High degree of late deliveries to Customer 2 – late change to schedule and demand amplification
- High degrees of late deliveries of both steel suppliers due to a move to timed deliveries or late schedule changes?
- Higher inventory of steel required due to lower service level
- Ticket-based stock control could not capture errors or true position
- Quality of finished project is good
- High in-line rectification (welding and grinding)
- Internal ppm figures are an estimate as many failures are not captured
- Quality of incoming steel – some problems
- Quality of incoming compionents is good

Key Points: Demand Amplification Map

- Dislocation between Customer 2 schedule and actual order
- Dislocation between Customer 2 demand and production
- Assembly schedule to actual assembly – close
- Can visual factory pull from customer demand?
- Is scheduling relying on personal contact and not a good system?
- Long lag between blanking and pressing
- '10 day batch' not reduced on lower customer demand
- Steel schedules subject to large delay
- Major dislocation from press to assembly
- Are Tallent buffering the blanking process because reliability of supply is variable?

Key Points: Physical Structure Map

- Surprisingly high component content
- Long tail of consumable suppliers
- Spread of spend with component suppliers – is it customer specified?
- Potential for make/buy review
- Potential for supplier direct line feed, vendor managed inventory
- Focused customer profile

Key Points Value Analysis Time Profile

- Considerable double handling of the product evident
- Delivery cost to Customer 2 – higher than manufacturing cost!
- Storage costs = 50 pence per unit
- Line One is not as cost-efficient as it first appears
- Bushing is expensive – there is a need to keep the flow going
- There is a relatively low amount of waste in the process – but what about breakdowns?
- Can steel coil stocks be standardized?
- Is the EQ line appropriate at 60% utilization?
- Unsure if the cost of stock holding is recorded adequately

Figure 2 Summary of some of the key points emerging from the Value Stream Mapping.

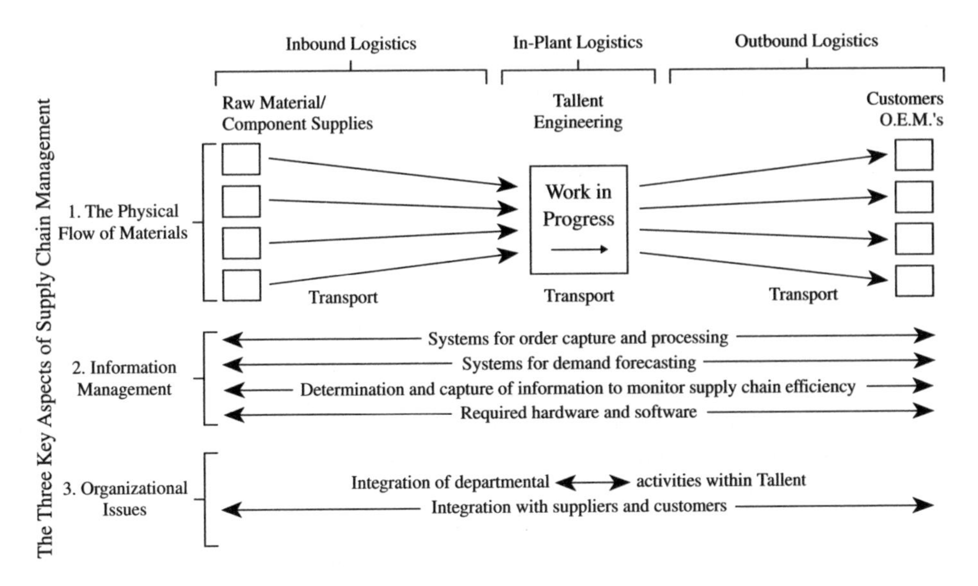

Figure 3 The scope of supply chain management.

- Managing the physical flow of materials.
- Managing the information flows which are necessary to forecast, trigger and monitor the physical flow of materials.
- Managing the organizational structure of the supply chain both within Tallent Engineering and between Tallent Engineering and its supply chain partners.

The three key objectives of logistics strategy

1. **To optimize customer service**
   The logistics system will provide products and information to customers in a way that equals or exceeds the stated requirements of each individual customer.
2. **To minimize the cost of supply chain operations**
   The logistics system will be structured and operated with a view to minimizing:

   - the fixed assets employed in supply chain activities;
   - the operating costs of the chain;
   - the inventory held within the chain.

   Cost minimization will, however, be set within the context of achieving specified customer service targets.
3. **To maximize the flexibility of the supply chain**
   The logistics system will be designed to permit Tallent Engineering to respond with maximum flexibility to:

   - customers' short-term operational fluctuations in demand;
   - longer-term strategic changes in demand either from existing customers (e.g. new models) or the wider market place (e.g. new customers/new product groups).

The overall objective of logistics management is to develop a supply chain system that out performs that of competitors and thereby creates an opportunity for Tallent Engineering to gain a competitive advantage.

The achievement of a lean supply chain

The logistics vision will be achieved through a phased approach to achieving quantified targets in relation to each of the following seven wastes:

1. **Overproduction** – the making of too much, too early or just in case. The aim is to make exactly what is required, just-in-time and with perfect quality. This is not a particular problem for Tallent as the production is continuous flow and is despatched to the customer in line with production. It can be an issue at Tallent when ordering bought-out part components as a quantity is sometimes ordered 'just-in-case' of a schedule increase. The aim needs to be to eliminate the root causes of unexpected demand changes.
2. **Waiting** – where materials (or information) are not moving or having value added and hence time is not being used effectively. The aim is to eliminate waiting. An example of this at Tallent Engineering is the waiting time between operations and storage. These could be eliminated with single-piece flow principles applied at each area.
3. **Transporting** – where materials (or information) are being transported into, out of or around the factory. Transport cannot be fully eliminated, but the aim is to minimize it through two approaches:

   – minimizing the distance between the location of operations within the supply chain both internal and external to the plant.
   – efficient use of the vehicles or equipment which will link separated locations.

   The above waste is an obvious problem to Tallent Engineering. However, as new processes are introduced the aim will now be to consider flow as a major priority in future layouts. The elimination of the degreasing process, the purchase of an additional painting facility and the opening of the new despatch facility are a significant step towards this goal. The aim must now be to ensure that what ever transportation is essential and part of the overall process is done in the most efficient way. An investigation into the Fork Lift Truck (FLT) activity including AGV is currently under way.
4. **Inappropriate processing** – using machinery and equipment that is inappropriate in terms of 'capacity' or 'capability' to perform an operation. The aim is for each operation to be performed by a machine of optimum 'capacity' (i.e. neither too large nor too small) and which is 'capable' (i.e. capable of producing defect free products). Machinery at Tallent Engineering is generally dedicated to one particular product and customer. This area of waste applies to rework.
5. **Unnecessary inventory** – which tends to increase lead time, prevent rapid identification of problems and use space as well as tying up capital. The aim is to reduce inventory in order to highlight other supply chain problems which are 'hidden' by inventory and to free up capital for more beneficial investment by the business. Tallent Engineering have made significant improvements in this area. However, we should continually strive to eliminate safety stocks carried on behalf of our suppliers. We need to educate and develop our suppliers to achieve JIT deliveries.

6. **Unnecessary motion** – involving the ergonomics of production where operators have to make difficult or unnecessary movement in order to complete a task which may affect both personal safety and output. The aim is to improve the quality of work-life for employees and at the same time improve productivity and quality. The main area for improvement at Tallent Engineering is for the loading / unloading of the painting facilities. Investigations are currently under way to load both plants from the last operation onto universal painting jigs.
7. **Defects** – in terms of *product defects* which escape to the customer, *rework defects* which are caught in-line and rectified, *scrap defects* which are caught in-line and scrapped, *service defects* in terms of delivery reliability and provision of information to the customer. The aim is to continually reduce the level of defects. We are all aware of the importance of quality in terms of product and the information and services we supply. The measure of performance for product quality is covered by the company QOS. However, we need to develop a system to measure and improve service and information defects.

Lean thinking principles

Five key principles of Lean Thinking are fundamental to the elimination of waste:

1. Specify what does and does not create **VALUE** from the customer's perspective and not from the perspective of individual firms, functions and departments.
2. Identify all the steps necessary to design, order and produce the product across the whole **VALUE STREAM** to highlight non value adding waste.
3. Make those actions that create value **FLOW** without interruption, detours, backflows, waiting or scrap.
4. Only make what is **PULLED** by the customer just-in-time.
5. Strive for **PERFECTION** by continually removing successive layers of waste as they are uncovered.

The vision of lean logistics at Tallent

Figure 4 shows the current situation in terms of the flow of material through the production process. Data collected from analysis of the Nissan 906 axle is used. At each stage of the process four measures are depicted:

- **Stock:** The amount of stock (measured in time).
- **Time:** The time it takes for a component (or batch) to travel through each part of the process.
- **Distance:** The distance materials/components travel through each part of the process.
- **People:** The number of times material/components are 'touched' as they pass through the process.

In addition, the distance between each stage of the process is shown.

Figure 5 shows how the five key principles of Lean Thinking can be used to improve the flow of materials/components through Tallent and how this will impact on stock levels, time taken, distance travelled and the number of times components are touched through the manufacturing process. The proposal is to time the planned improvements at 1, 3 and 5 years.

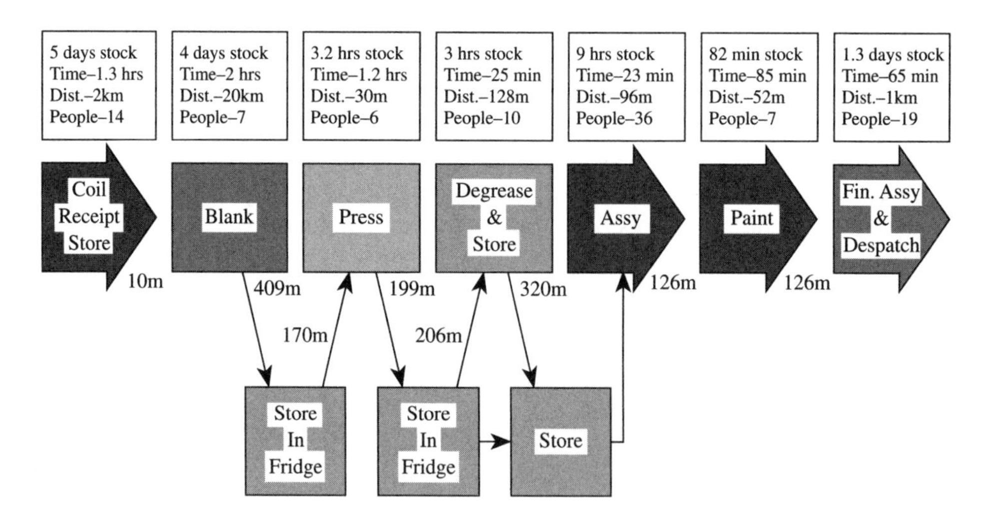

Figure 4 Current situation.

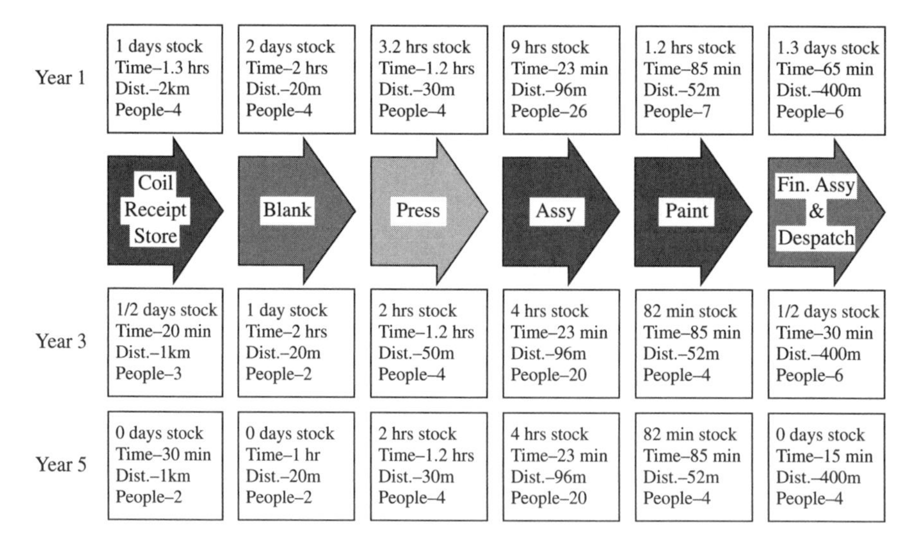

Figure 5 Lean vision.

Key elements of the strategy

Packaging policy

The general principle that will cover all packaging is that it will be supplied fit for point of use within Tallent Engineering, will be returnable, reusable, and will be funded by the supplier. There are five main areas for consideration with regard to packaging.

1. Coil deliveries
In line with our objective to achieve a lean supply chain we aim to develop partnerships with our steel suppliers which will lead to world class lead times, reduced coil stocks and storage area, and increased delivery frequencies. This will be achieved through electronic communication links between Tallent, steel processor and steel maker.

2. Blanks to press
Blanks will be presented to the press in such a way that they are suitable for loading by the press operator. Where parts are magazine fed, the magazines will be loaded to the press. It will be the responsibility of the Tool Design Department and Press Shop Management to where possible consider the use of universal magazines. All other blanks will be presented in one of the standard containers detailed below:

Container description	Weight full kg	Size / Dimensions mm
Type 1	2000	2000 × 1000 × 1000
Type 2	1500	1200 × 1000 × 800
Type 3	500	1200 × 800 × 760

Each part number will have an agreed packaging specification with a fixed stillage type and blank quantity per stillage. Pre-printed labels will be issued with palletization details. In order to allow for coil gauge variation the last stillage will have the label manually amended to show the actual quantity in the stillage.

3. Components for use on assembly lines
The containers will be supplied fit for point of use at the lineside and components will be supplied in the smallest feasible size based on the specifications below.

The container will be determined using the formula :

*Individual Component Weight * Average Hourly consumption at point of use.*
An individual container will contain a minimum quantity equal to 1 hour's usage and a maximum of 8 hours. The smallest possible container must always be the first priority. Based on these considerations one of the following standard containers will be used.

Container	Category	Size/Dimensions	Weight full kg
Type 1	Euro Box	600 × 400 × 118	15
Type 2	Euro Box	600 × 400 × 175	25
Type 3	Big Box	1200 × 800 × 760	500

Every part will have an agreed packaging specification, with a fixed container type and part quantity which will be signed off by the relevant project group.

Containers must conform to the Tallent packaging standard ref. TKAsyspack02.

Each container must be supplied with a standard odette bar-code label containing the following information:

- Tallent part number
- Description
- Supplier code
- Quantity
- Order number

4. Assembly to paint
Where possible, parts requiring painting will be loaded onto paint jigs at the end of the assembly line. These jigs will then be loaded directly on to the paint plant without additional handling. After painting parts will either:

i) be transported back to the final finishing area on the original jigs, or
ii) where after painting a part is ready for despatch to the customer, it will be loaded directly into the customer specified stillages.

5. Delivery to customer
Tallent will work with customers to offer and advise on the most cost effective stillage for transportation to customer sites. The following general principles will be applied to customer specific containers:

i) It is the responsibility of the Commercial Department for stillage design, funding, and ordering.
ii) The project team which includes logistics management will be responsible for agreeing stillage design and buy-off.
iii) After buy-off and prior to SOP, Logistics management and manufacturing management will agree a stillage maintenance procedure.
iv) It will be the responsibility of the product group to maintain stillages in line with this procedure.
v) We will ensure that all deliveries meet customer packaging requirements.

Storage policy
Storage requirements for the total site will be identified. Benefits of central or devolved storage will be applied in line with lean thinking principles. As such the overriding objective will be to ensure as much continuous flow as possible by aiming to eliminate central storage areas and having parts delivered direct from the point of manufacture to the point of use. There will be three designated unloading points and each part will have a specified unloading point based on the point of use.

The following general principles will be applied:

i) Inventory levels will be set at a part number level based on ensuring minimum quantities in line with the achievement of TX 2000 (Strategic plan) objectives.
ii) All product will be in areas protected from adverse environmental conditions to ensure that product quality is maintained.
iii) Parts will have dedicated footprints clearly identified at the lineside.
iv) Standard fixed pallet quantities by part number will be used to ensure that there is clear visibility of inventory status by part.
v) Kanban pull systems will be used to ensure that product is pulled by actual demand from the next process.

Transportation / handling policy
The adoption of the packaging and storage policies above will ensure that transportation distances and double handling are kept to a minimum along the whole process flow. This will aid the continuous effort to reduce the number of forklift trucks and waiting time.

The following general principles will apply:

i) Wherever possible, wheeled containers will be used.
ii) All suppliers will be allocated specific fixed delivery time slots and unloading points to ensure efficient vehicle turnaround and minimize waiting time.
iii) In order to operate with increased delivery frequencies the milk round collection system will be expanded in order to reduce the number of vehicles on site at any one time.
iv) Tow tugs will be used on fixed pattern cycles to move product from the unloading points to the lineside point of use.

Batch sizes policy

Batch sizes will be set with the overall objective of ensuring continuous flow and striving to eliminate waiting and storage time between operations. It is the responsibility of Logistics management to set the part classification prior to SOP.

The following general principles will apply:

i) Every part will have a batch size determined by customer demand. This will be set by product value using the criteria below:

Product class	Batch size (days of customer demand)
A1 class	1 day
A2 class	2–3 days
B class	5 days
C class	10 days
D class	20 days

ii) There will be a continuous focused effort to reduce batch sizes throughout the process by managing the parts classification categories above.
iii) Press component batches will match press blank quantities.
iv) Supplier deliveries to lineside will be made on at least a daily basis and the frequency will be progressively increased.

Supplier performance and selection policy

Supplier delivery performance will be measured and published on a monthly basis. The aim of the delivery performance measure will be to identify suppliers not achieving TX 2000 objectives.

The following categories will be measured, and % weighting applied:

Service characteristic	Weighting	Features
Delivery accuracy	43.5%	Late, Early, Over deliveries
Delivery presentation	13%	Packaging, identification, environmental
Delivery management	43.5%	Mis-labelling, Flexibility, Self sufficiency

Logistics will have an active involvement in the selection process, with a view to selecting suppliers most capable of meeting Tallent Engineering's Lean logistics strategy.

Information management policy
The logistics management function aims to optimize information flows for the whole supply chain, as illustrated in Figure 3. Through proactive discussions with our customers and suppliers we will establish systems capable of exploiting opportunities in the following areas.

Inbound:

- Suppliers
- EDI
- Bar codes

Internal:

- Bill of materials
- Internal suppliers
- Capacity planning, simulation, 'what if' packages.
- Inventory accuracy
- Visibility of information
- Shop-floor data capture

Outbound:

- Customers
- Labelling
- Demand management – responsiveness, Direct Data Links (e.g. Ford DDL)

Demand management policy
The company has a policy of zero arrears and customer service is paramount.

We aim to achieve a smooth demand level which both minimizes fluctuations in the total supply chain, and meets our objective to align build capacity with that of the customer. However, where this would result in uneconomical use of resources we will adopt a policy of producing **strategic stock** to exceed customer expectations.

CRITICAL SUCCESS FACTORS

1. That performance measures are in line with the strategy.
2. That Logistics Strategy fits with the Business and Manufacturing Strategies.
3. That the costs and benefits of the Logistics Strategy elements are proven and agreed.
4. That people in the business are aware, educated and bought-in to the Manufacturing and Logistics Strategies.
5. That external relationships are developed in line with the strategy.

SMART[1] objectives

Storage policy

Strategy Element	*Measure*	*Current position*	*Planned position*	*Critical success factors*	*Time-scale*
Continuous flow	Days supply and distance travelled	7 days	5 days	• JIT delivery from suppliers • Batch reduction • ABC/Pareto • TPM/process reliability • Devolved stores	Sept 99 Phase 1 Sept 99 Complete Continuous Sept 99
Inventory levels – set at part no.	Inventory level by component	Nil	100%	• Pareto profile of parts	Complete
Part to have default location and size	Number achieved	Nil	70%	• Space envelope required by part • Create space – lineside • Layout plan per part use	Sept 99 May 2000 Complete
Parts stored in an environmentally friendly way	Number achieved	100%	100%	• Completion of dispatch bay	June 99
Standard fixed pallet quantities by part no.	Number achieved	50%	100%	• Space envelope calculation • Supplier day • Press shop buy-in	Sept 99 New Projects complete TPI Activity 30% ongoing
Kanban	No. of items using kanban	20%	100%	• Education for mfrng management • Defined kanban system	Sept 2000 Feb 2000

Transportation/handling policy

Strategy element	*Measure*	*Current position*	*Planned position*	*Critical success factors*	*Time-scale*
Implement wheeled containers	No. of parts	20%	50%	• Supplier day involvement • PIP team use objectives as pilot	TBE No longer required.
Implement timed delivery slots and unloading points	No. of deliveries on time by supplier	30%	100%	• Simulation of existing and future plans • Supplier buy-in	Complete Complete
Implement a milk round	Transport cost	30%	60%	• Simulation of existing and future plans • Supplier buy-in	Complete Complete – no issues
Use tow tugs in plant	FLTD Reduction	6 FTE reduction = 20%	20 FTE = 50% reduction	• Minimize curves in gangway layout • Costing study	Next 6 months Complete

Batch sizes policy

Strategy element	*Measure*	*Current position*	*Planned position*	*Critical success factors*	*Time-scale*
See separate document.				• Review (manufacturing) press shop performance measurements • Develop classification of parts (Pareto) • Reduce set-ups • Education programme for press management • Supplier day – awareness, education and implementation • Introduce TPM • Introduce 5S/Problem solving	

Information management policy

Strategy element	*Measure*	*Current position*	*Planned position*	*Critical success factors*	*Time-scale*
To optimize information flows for the whole supply chain	?			• Define the IT strategy – allow the specification of a manufacturing system to occur	Complete
				• Define the critical measures at 1, 3 and 5 years	Complete
				• Stabilize demand information down the supply chain	Complete

Demand management policy

Strategy element	*Measure*	*Current position*	*Planned position*	*Critical success factors*	*Time-scale*
Minimize fluctuations in the supply chain	Adherence to plan.	85%	100% +/−5%	• Place optimum level of strategic stock at the end of production	Complete
	Customer Service	98.5% 92%	100%	• Complete demand management Probability/ variability analysis	Complete
	Supplier Performance		100%		

ACKNOWLEDGEMENT

The authors of this case would like to acknowledge the contribution given by David Brunt of LERC, Chris Butterworth of Corus and other staff at Tallent Engineering in the development of the work on which this case is based.

NOTES

1 SMART Specific Measurable Achievable Realistic and Timed.

An approach to the identification and elimination of demand amplification across the supply chain

11

David Taylor

INTRODUCTION

The Demand Amplification Effect has been well described in the literature over many years. This chapter does not primarily aim to contribute to the theory of demand amplification, which is already well rehearsed, but describes and evaluates an attempt to bridge the gap between theory and practice. A gap surely exists because it is now over 40 years since Forrester first identified and explained demand amplification, yet in spite of numerous academic papers in the interim, the effect still occurs in many supply chains. The elimination of demand amplification could remove a major cause of the uncertainty and variability which creates very fundamental problems for supply chain management. This paper describes and evaluates a 12-month project, spanning three echelons of the upstream automotive component supply chain in the UK (see Figure 1) which attempted to eliminate or at least reduce demand amplification.

During the first year of the LEAP project a comprehensive situational analysis was carried out to identify the structure of the supply chain, measure aspects of performance and identify major problems. One of the key issues arising from this review was evidence of the 'Demand Amplification Effect', which was judged by the research team to be a fundamental cause of many problems in the supply chain, including:

- Scheduling difficulties in manufacturing, particularly at bottleneck plants.
- Difficulties in managing labour requirements, which could swing from overtime to short time in response to variations in demand.
- Problems in controlling inventory levels and resulting warehouse requirements.
- Poor customer service, particularly in terms of late deliveries, divergence between ordered and delivered quantities and errors in invoicing and other paperwork. This was most apparent in relation to the performance from the steel mill to the steel service centres, but problems also existed between the service centres and the first-tier component suppliers.
- Excessive administrative effort and fire-fighting in relation to late deliveries.

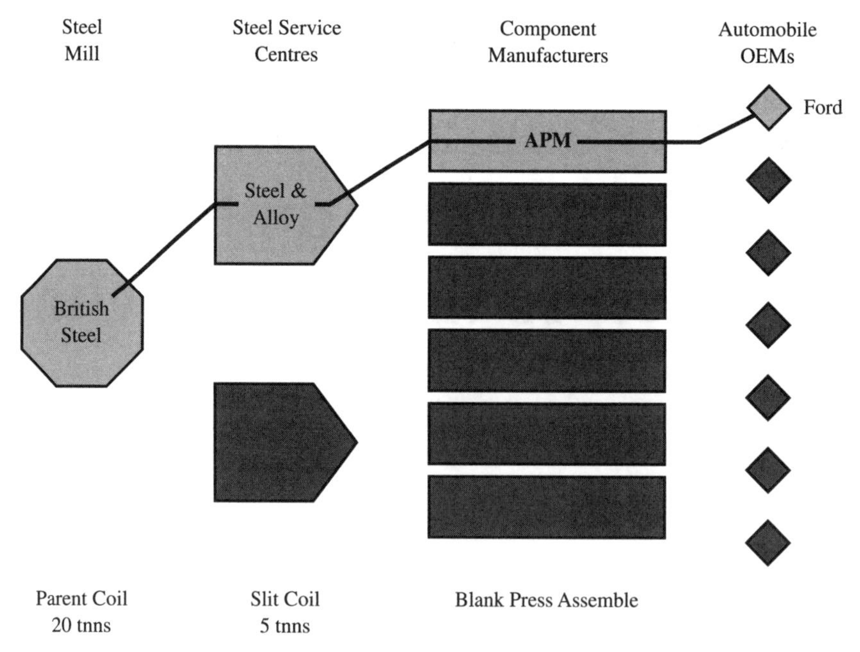

Figure 1 Tracking demand variability in the LEAP project.

It was therefore determined that a detailed project should be undertaken with the aim of trying to reduce, if not entirely eliminate, demand amplification effects.

Review of demand amplification literature

In 1958, Forrester (1) first identified the demand amplification effect whereby small changes in demand become progressively amplified or exaggerated as demand is passed from one company to another along a supply chain. Forrester primarily attributed the effect to the decision rationale of individuals responsible for demand management, which creates a tendency to over-respond to increases or decreases in demand from customers in terms of orders placed on the immediate upstream suppliers.

In 1961, Burbidge (2) cast further light on the issue by demonstrating that traditional stock control procedures based on EOQ logic would tend to amplify variations as demand passed along the chain. Burbidge coined the so-called 'Law of Industrial Dynamics' which states that 'If demand for products is transmitted along a series of inventories using stock control ordering, then the demand variation will increase with each transfer.'

Since the original work by Forrester and Burbidge there has been a succession of articles which have addressed various aspects of demand amplification. Most of these papers have concentrated on describing and/or modelling the nature of the effect or explaining its causes. For example, Sterman's work (3), based on multiple observations of play in the well-known Beer Game, categorized various causes of demand amplification. These included misperceptions of time lags and 'bounded rationality' within functional silos. A further important point made by Sterman was that players in the Beer Game tended to attribute the dynamics they experienced to external events (i.e. variable end-user demand), when in fact these dynamics were internally generated by their own actions and decisions.

The work by Towill (4, 5) and Wilkner (6) used a systems dynamics approach to develop computer-based simulations of supply chain activity and thereby test various strategies to reduce demand amplification.

In the mid-1990s Lee (7, 8) described qualitative evidence of demand amplification or as they call it 'the bullwhip effect' in a number of the retailer–distributor–manufacturer chains and then employed mathematical models to demonstrate the impact of decision strategies in creating oscillations in demand. In essence the work of Lee *et al.* is analogous to that of Forrester, but is useful in that it adds depth to the understanding of some of the causes and relates the concepts to the modern business environment.

Recent work by Wilding (9, 10), exploring the impact of 'chaos' within supply chains, demonstrates that modern computer systems designed to control supply chain activity can be inherently unstable and thereby create demand amplification effects. Various other papers describe case studies highlighting the effect in particular companies or sectors. For example, Saporito (11) examines the effects of demand amplification on Proctor and Gamble in the USA, while McGuffog (12) shows its impact at Nestlé in the UK.

A more detailed review of some of the above papers is given in Chapter 9. However, in spite of all this work it is evident that demand amplification is still rife within many supply chains. Why is this the case? Has the previous work been too theoretical or too complex to apply in practice? Is the effect uncontrollable or unmanageable, or has it just been ignored by management? This paper reports on a research project which attempted to develop and test a practical approach to eliminating demand amplification across three levels of the automotive to steel supply chain in the UK.

AN APPROACH TO ELIMINATING DEMAND AMPLIFICATION

The approach developed is not sophisticated or complex, indeed a prime objective was that it should be sufficiently straightforward to be easily applied by staff who have day-to-day responsibility for managing demand along a supply chain. A number of steps are involved:

1. Identify and quantify demand amplification.
2. Analyse the specific causes of the effect in the supply chain under study.
3. Education and awareness raising with relevant personnel.
4. Creation of a Demand Management Team from across the supply chain.
5. Development and application of detailed policies to address the effect in selected trial value streams.
6. Monitoring and evaluation of supply chain performance during the trial.
7. Roll out the policy to other value streams, modifying in light of the trial

IDENTIFICATION AND MEASUREMENT

Within the LEAP project, analysis of demand amplification was set within the context of Value Stream Management (13) which provided the methodological framework with which to approach the improvement of supply chain performance. Value Stream Management adopts a disaggregated approach, whereby individual value streams, i.e. specific products or product families, supplied to specific customers or customer groups, are studied in detail and improved in practice, as a forerunner to adopting the techniques across the wider range of products and customers. The Demand Amplification Project was therefore based initially on two high-volume parts demanded by a specific car manufacturer (OEM), produced by one of the first-tier component suppliers, with steel sourced through a specific steel service centre and originating from one steel mill (see Figure 1).

The first requirement was to develop a detailed flow chart to show all the processes through which product and information passed both between and within the three companies in the chain. This was achieved using the technique of Process Activity Mapping.

Establishing the supply chain and information chain structures

An important point to note is that the product and information flows must be tracked from stage to stage both *within* each company as well as *between* companies. This is because demand distortions can and do occur between functions within the same company, as well as when demand is transmitted between companies. Figure 2 shows that within the supply chain studied the product passed through ten processes from casting of steel slab at the steel mill to despatch of assembled components by the component supplier.

The upper section of the chart in Figure 2 shows information flows throughout the supply chain. It highlights the many different types of demand information which existed, ranging from advanced schedules giving forecasts up to six months out from the OEM, to the actual daily call-in of steel from company to company, to daily production schedules issued to individual production units. An important issue in demand amplification analysis is to identify the various aspects of demand data which exist and critically, to understand what use is made of these data. In practice it was clear that not all demand information that was passed along the supply chain was actually used by the recipients. When it comes to understanding the causes of amplification and to developing proposals to eradicate it, it is important to know just what information decision makers are actually using.

Information mapping also tracked the many processes and decision points through which the demand data passed. In practice this was very complex (the actual flow diagram spanned four pages!). Again it is important to understand clearly the administrative procedures for demand management because they may be part of the cause of demand amplification and also there may be opportunities to significantly simplify the procedures.

Quantifying demand amplification

When quantifying demand amplification the required data falls into two categories:

- Demand data – passed from company to company and from function to function within each company.
- Activity data – records of production and despatch activity at each function along the chain. This is important because the activity rates may effectively create the demand on the upstream production unit and therefore may be a root cause of the demand amplification.

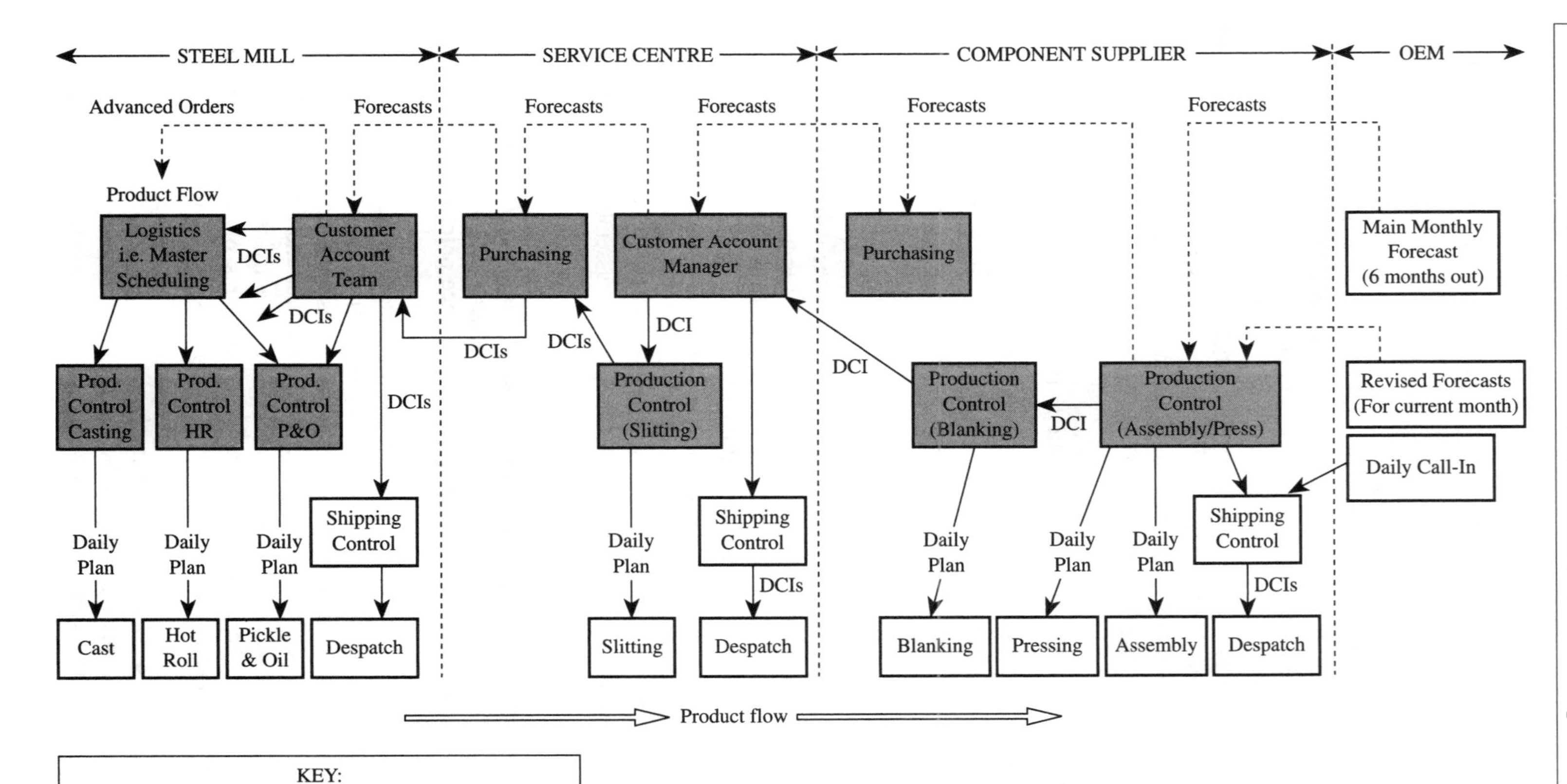

Figure 2 Simplified diagram of information and product flows.

Ideally this data should be collected historically for a 'significant period', i.e. long enough to cover a number of ordering / delivery cycles. The Demand Amplification Project commenced in June 1998 and the target was to collect the relevant data for the first 26 weeks of 1998. (see Table 1). In practice, however, it quickly became apparent that many aspects of the required data did not exist historically either because operating data was only kept for a limited period, or because it had never been recorded in the first place. A system was therefore established to collect the required data over a period of eight weeks in mid-1998 (weeks 19 to 26). However, where data was available for the whole year this was used. A further practical problem with data collection was that it required cooperation from a large number of operational staff along the chain. Staff had to monitor, amalgamate and present data regularly and consistently, often when it was difficult for them to understand the potential benefits; the inevitable result was some inconsistency and data gaps.

Features of demand amplification within the automotive supply chain

The charts in Figure 3 show the demand and production data for one of the two products that were mapped; very similar patterns occurred with the other product. All figures are given in tonnes of steel rather than number of items, in order to have a consistent measure throughout the chain.

3a Variability at the component manufacturer

It can be seen that the actual demand driving the chain, i.e. the daily call-in of parts from the OEM, was quite stable. Average demand was 13.2 tonnes per week with a relatively low standard deviation of 0.99 tonnes (Figure 3). However, Figure 3a shows that the stability begins to be lost even within the first production process at the component supplier, i.e. assembly. Instability in assembly activity is occurring even though the assembly cell has a policy of trying to operate level scheduling. It can be seen that variability in activity rates becomes further amplified between the three production processes within the component supplier, i.e. assembly, pressing and blanking.

3b Variability between the component supplier and the steel service centre

Figure 3b shows the relationship between the component supplier and the steel service centre and shows the OEM demand as the reference point which drives the chain.

The advanced schedules are quite level as these are issued by the assembly cell at the component supplier, which operates a policy of demand smoothing and level scheduling. However, the 'coil-call-in' (i.e. actual deliveries required on a daily basis) is issued by the blanking cell at the component manufacturer and this is much more erratic, as it varies in accordance with the variable blanking activity rates. The service centre must respond to the actual stated daily requirements of 'coil call-in' and therefore the *effective* demand between the two companies is very variable.

3c Variability between service centre and steel mill

The ordering and delivery pattern between the service centre and the steel mill is only apparent when viewing the data for the 26-week period because the quoted order to

Table 1 Activity data for sample product over 8-week period, showing total steel usage.

Week	*OEM Demand*	*CM Assembly Activity*	*CM Pressing Activity*	*CM Blanking Activity*	*CM Initial Schedule to SC*	*CM Last schedule to SC*	*CM Coil-call in to SC*	*SC deliveries to CM*	*SC Advanced Orders on Mill*	*SC weekly call-off to mill*	*Depatches from Mill to SC*
19	12.64	13.65	12.24	8.00	13.74	12.68	8.00	10.42	0	36	40
20	13.30	12.55	14.61	10.40	13.74	12.58	10.40	17.35	0	0	0
21	13.07	13.55	12.92	12.90	13.74	12.46	12.90	7.57	0	18	14
22	12.92	12.01	12.39	14.00	13.74	12.57	14.00	13.92	0	0	0
23	13.29	14.12	12.34	10.20	13.74	12.49	10.20	10.79	0	18	19
24	13.82	13.31	11.19	12.50	13.74	11.09	12.50	12.41	0	36	37
25	13.39	14.02	15.93	11.5	13.74	13.47	11.5	12.65	70	0	0
26	13.38	13.39	13.14	14.15	13.74	13.57	14.15	15.68	0	18	20
27											
Total	105.81	106.60	104.76		109.92	100.91	93.65	100.79	70	126	130
Average (Week 19–26)	13.23	13.33	13.10		13.74	12.61	11.71	12.60		15.75	16.25
Average (Year to date)	13.23				13.74		11.71	12.60		15.75	16.25
Standard Deviation	0.35	0.72	1.50		0.00	0.76	2.11	3.10		15.02	16.10

Notes
SC – Service Centre
CM – Component Manufacturer
Steel weight per finished part – 1.832kg.
All figures – Steel tonnes

delivery cycle time by the mill at the time was eight weeks. Figure 3c shows the service centre's orders on the mill to be irregular in both amounts ordered and order interval.

Orders are batched into 70-tonne units in order to benefit from a volume price discount offered by the mill. However, the service centre is able to call-off coils (average weight 18 to 20 tonnes) as required.

To summarize, the data collected shows significant evidence of variability in both demand and activity rates along the supply chain. This is in spite of the facts that

a. the initial demand from the OEM is relatively stable on a week-by-week basis;
b. the first production cell in the chain (i.e. the assembly cell at the component manufacturer) is endeavouring to operate a system of smoothing demand and level scheduling.

ANALYSIS OF CAUSES

Having identified and quantified demand variability, the next step towards eliminating demand amplification was to understand the specific causes of variability within the supply chain being studied. In the automotive supply chain a number of interrelated issues were identified.

Demand variability

Forrester's (1958) fundamental explanation of demand amplification suggests it is primarily a function of decision making in response to variability in incoming demand. As demand varies, decision makers at any supply unit will have a tendency to overreact to the change. Amounts produced or ordered are exaggerated in order to ensure adequate supply (or avoid over-stocking) in future periods. Although this effect was apparent in the LEAP chain it did not primarily result from external demand variability. Even though there were small fluctuations in OEM demand these were not exaggerated or transmitted down the chain because of the demand smoothing and level scheduling policy operated by the assembly cell. The demand amplification was largely *triggered* by factors other than variable external demand. However, once the variability was triggered, it was amplified as a result of the decision logic suggested by Forrester.

Other factors were found to be responsible for triggering demand variability and these included the following.

The supply variability effect

Decisions on the amount to order on upstream suppliers (internal as well as external) are partly a function of the supply record in previous periods. For example, if supply in period 1 is below that requested, the tendency is to exaggerate the order for period 2:

a. to make up the shortfall from period 1, and possibly
b. in anticipation of potential further shortfalls in period 2.

Whereas demand variability originates from the downstream partners in the supply chain, supply variability originates from the upstream units. Within the supply chain studied, three causes of supply variability were identified.

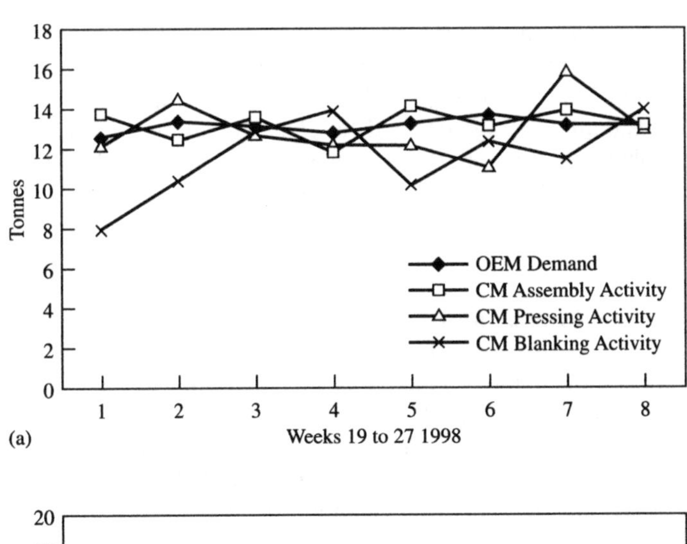

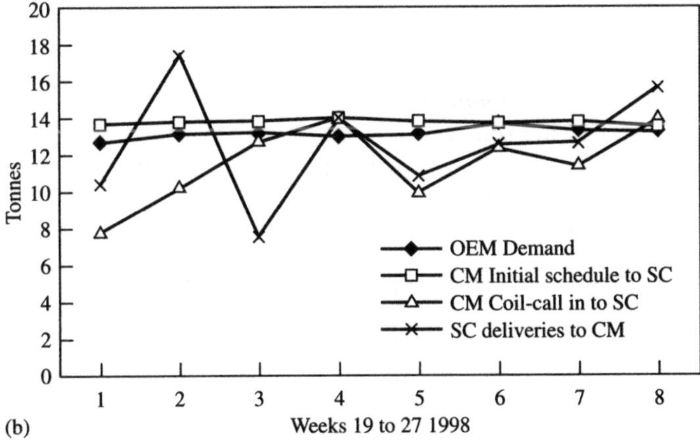

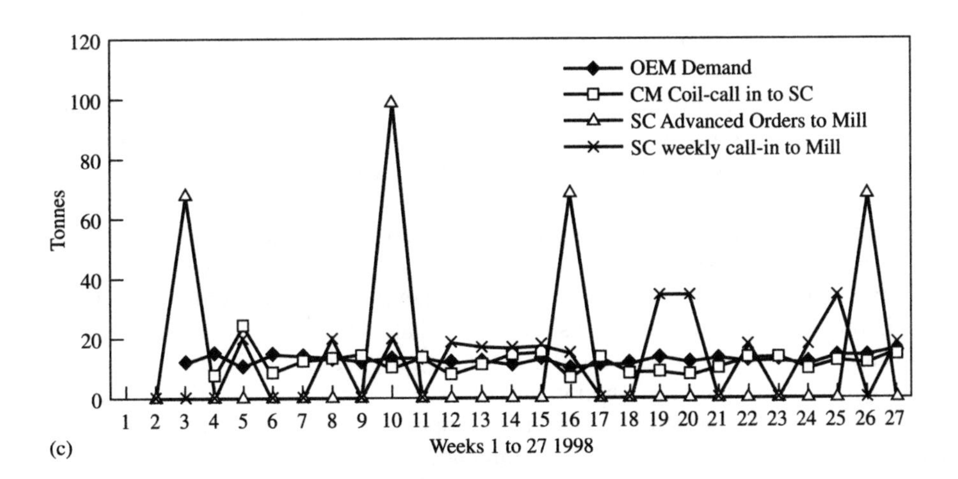

Figure 3 Charts showing aspects of demand variability, (a) Variability of component manufacturer; (b) variability between component manufacturer and service centre; (c) variability along the whole chain.

Variability in machine reliability and output

Equipment throughout the chain was prone to reliability problems. Fluctuations in output by a machine would cause a time-lagged fluctuation in the upstream demand from that machine. For example, average demand for blanks might be 1000 units per week. If in week 1 the blanking press produced only 800 units because of reliability problems, the requirement in week 2 would be to produce 1200 units and the demand for steel from the service centre increased accordingly. Variability in demand and activity thus commenced.

Variability in process reliability and subsequent product quality

Because of deficiencies in process capability, i.e. the ability to consistently produce perfect products, the volume of usable product received by any downstream units in the chain varied. Process capability was itself variable, i.e. in some weeks a machine might produce 100 per cent perfect product, in other weeks only 95 or 97 per cent. For example, if in week 1 usable output from blanking was only 950 units out of a batch of 1000, then the demand in week 2 from the press shop would be 1050 in order to cover the shortfall from week 1. Variability in demand thus commenced.

Variability in quantity of materials supplied

As the supply chain approached the source of raw materials the units of production and supply became larger and in consequence it was more difficult to adjust the quantity supplied to match variations in demand.

For example, the raw material supplied to the various units was roughly as follows:

Assembly shop – Raw material = individual pressed parts – Average weight 1 kg

Press shop – Raw material = individual steel blanks – Average weight 1 kg

Blanking shop – Raw material = slit coils of steel – Average weight 3 tnns

Steel service center – Raw material = parent steel coils – Average weight 18 tnns

The assembly shop called off quantities of individual parts (average weight 1 kg) from the press shop, e.g. 1000 per week. This demand could be fulfilled exactly provided production capacity and/or stock was available. In contrast, the amount delivered by the steel mill could not be so finely adjusted, as the minimum unit of production was one coil (average weight 18 tonnes). Furthermore, because of process variability in steel making, the size of parent coils produced by the steel mill varied from between 12 and 21 tonnes. As a consequence the size of slit coils produced by the service centre was forced to vary (it is impossible to produce a standard slit coil size when parent coil size varies, without incurring undue waste through trimming). The amount of raw material supplied to the blanking press therefore varied. And again in order to avoid wastage, the slit coil, of whatever size, would be fully blanked. This often resulted in over-production or under-production of blanks in relation to actual demand from the press shop in a given week. The natural reaction was then to adjust demand in the subsequent period to compensate for the over- or under-supply. Demand and activity variability thus commenced.

This effect was a consequence of the larger units of production further upstream in the supply chain. When product is supplied in small discrete units, e.g. individual blanks or pressings, the quantity supplied can be readily adjusted to downstream demand. As units of supply increase in size upstream, it is impossible to exactly match the amount supplied to small variations in downstream demand. Indeed here lies a fundamental challenge in taking JIT and stockless supply systems back down to the raw material end of the supply chain.

The impact of variability in equipment reliability, process reliability and quantity supplied should perhaps more appropriately be thought of as 'supply amplification effects' rather than demand amplification effects.

Functional silos and decision-making rationale

Within the three tiers of the automotive supply chain there were some 18 people directly involved in decision making that influenced production levels, inventory levels and subsequent upstream demand. Why did these people create demand amplification? Firstly, the 'functional silo' mentality was apparent whereby each person attempted to make the most rational decisions in relation to their own particular functional responsibility. For example, production managers usually aimed to maximize batch sizes to achieve economies of scale; customer account managers tended towards higher levels of stock to maintain customer service levels. Naturally managers tend to be less concerned about potential adverse effects of these policies elsewhere in the supply chain.

Secondly, at the outset of the project none of the managers concerned had any real knowledge of even the theory of demand amplification and certainly not of its actual impact within their supply chain. It was not surprising then that their decision-making rationale did not include any attempt to minimize the effect.

A third element in creating the demand amplification was what Lee refers to as the 'Rationing Game' (Lee 1997a). During the period studied, there was a situation of demand exceeding supply capability in relation to steel. Under such conditions the steel mill was forced to ration supplies between customers. In response, customers tended to over-order in the hope of being allocated sufficient supplies to meet their actual requirements; demand amplification was thus triggered.

Stock minimization policies and JIT expectations

Within all three companies in the project there was a strong pressure from senior management to minimize inventory for financial reasons. Commonly, stock levels were set in order to meet financial targets, rather than in relation to provision of carefully calculated buffers against quantified variability in demand or supply. Instead, the expectation was that such variability should be dealt with by rapid and variable response from upstream suppliers rather than by the holding of appropriate levels of safety stock. The general view was that 'in today's JIT and customer service environment, it is up to the supplier to meet our demand, no matter how variable that is'. Unfortunately such an approach fails to understand that in order for stockless / JIT systems to operate properly, it is necessary to have supply and production systems that are both capable and reliable and demand that is reasonably steady or predictable at least in the short run. The absence of such conditions, combined with the absence of safety stocks to act as a damper, triggers variability in activity and demand which is then amplified by the Forrester Effect.

Pricing policies

The effect of price discounts in relation to quantity ordered has also been identified by Lee (1997a) and was another contributor to demand amplification experienced in the supply chain under study. This effect was particularly apparent in relation to the demand pattern from the steel service centre to the steel mill. The mill offered quantity discounts on individual orders with price bands in relation to order size. This was a prime reason for the amplification in the advanced orders placed by the service centre on the steel mill. Figure 3c shows the service centre placing advanced orders usually of 70 tonnes (i.e. four parent coils) every six or seven weeks, rather than weekly orders for individual coils of 18 tonnes which would be more in line with actual demand.

In summary, therefore, when analysing the causes of demand amplification it is important to distinguish between those factors that *trigger* demand variability and those which *amplify* variability once it has been introduced.

EDUCATION AND AWARENESS OF SUPPLY CHAIN PERSONNEL

Having charted demand variability and analysed its causes the next step was to identify and involve all the personnel along the chain who influence activity rates, demand levels or stock levels. The process map showed that some 18 people across the three companies were directly involved. These people were invited to a one-day seminar, the aims of which were as follows:

- *To explain demand amplification and its detrimental effects.* This was primarily achieved by use of a simulation game developed specifically to mirror the stages and circumstances pertaining in the steel supply chain (see Chapter 17).This is similar in essence to the well-known Beer Game, but much more specific in that the context of play relates to steel mill, service centre, component manufacturer (rather than retailer/wholesaler/manufacture as in the Beer Game). Also incorporated into the game are simulations of the real problems which exist in the automotive supply chain such as quality, reliability and delivery problems.
- *To demonstrate the existence of demand amplification within the automotive supply chain.* Experience showed that explanations of the theory of demand amplification and its probable existence in a supply chain were not sufficient to instigate changes in policy. Presentation of the evidence showing the actual occurrence of the effect in the steel supply chain was absolutely critical in gaining the attention of the managers.
- *To obtain commitment from the participants to jointly take steps to eradicate the effect.* This may seem like an unnecessary statement or step. Surely companies and managers involved in a supply chain would want to eliminate demand amplification once it is shown to exist. In practice this step was not as easy as might be expected. Certain individuals were resistant to change. Some could not see the benefits to them personally or to their functional area; some were inherently reluctant to alter policies or ways of working. Others saw eradication as a personal threat because it would reduce the need for progress chasing and 'fire-fighting', which in practice comprised a major part of their working day. These concerns had to be dealt with on an individual basis.

At the corporate level the worst effects of demand amplification tend to be felt further upstream in the supply chain. It follows then that upstream units will potentially have the greatest benefits from its elimination. There is thus a requirement to demonstrate to downstream units that they will also benefit, primarily through improved service levels and possibilities of lower stock holding and the ability to smooth activity rates and resource requirements.

- *To form a small Demand Management Team.* A team comprising one representative from each of the three companies was established to jointly determine the best approach to improvement. This team decided on required policies for managing demand both within the individual companies and between the companies. The researchers put forward a number of suggestions as to possible improvement strategies, but in order to gain ownership of the solution it was important that the Demand Management Team decided which suggestions to take forward and the details of implementation.

Creating a receptive environment through the above stages was deemed critical to the success of any subsequent improvement initiatives.

DEVELOPMENT OF POLICIES TO REDUCE DEMAND AMPLIFICATION

Detailed analysis of the nature of demand amplification and more importantly its specific causes in the supply chain under study is critical in deciding what corrective measures should be taken. The Demand Management Team decided that a number of simple policies should be introduced in relation to the two sample products that had been the subject of the initial investigation. These policies were initiated over a trial period of six months and results monitored. If key performance indicators along the chain showed improvement, the policies would be extended to a wider range of products. The policies adopted addressed the various causes of variability that had been identified earlier.

Policies to manage demand variability and create demand visibility

- The weekly demand levels between the component manufacturer and the service centre were set at the level of the average demand from the OEM (i.e. 14 tonnes/week including an allowance for waste).
- Weekly demand between the service centre and the steel mill was set at one coil per week for five weeks and then nothing in the sixth week. This was to allow for the discrepancy between the usage rate of 14 tonnes and the average coil weight produced by the mill which was 18 tonnes.
- 'Standing orders' for these amounts were established between the three companies and it was agreed that these would not be altered without prior agreement of the Demand Management Team.
- Actual demand information from the OEM was to be shared directly by the component manufacturer with both the service centre and the steel mill on a week-by-week basis through the individuals in the Demand Management Team.
- Any significant changes in OEM demand away from the average of 13.2 tonnes was to be communicated immediately by the component manufacturer and the Demand Management Team would determine how the chain would react in concert in altering activity and/or stock levels.

Policies to counteract supply variability

- Production planners in each company were encouraged to set regular schedules for weekly production as close as possible to the 13.2 tonnes average OEM demand.
- Deliveries between companies were to be made on fixed days each week, allowing the possibility to regularize the timing of production and despatch activities.
- It was acknowledged that in the short to medium term, elements of supply variability would persist and this variability must necessarily be covered by safety stock until root causes of the variability could be addressed.

Stock policy

- Safety stocks were established at points along the supply chain to cover supply variability.
- A joint decision was made by the Demand Management Team as to the positioning of these stocks so as to avoid duplication at either side of corporate or functional boundaries.
- Safety stocks were initially set at a generous level to ensure no supply failures, but the aim was to progressively reduce stock levels by attacking root causes of variability.
- The only safety stock required to cover external (i.e. OEM) demand variability was a small stock of finished parts positioned post assembly, i.e. at the very end of the supply chain. All internal demand variability within the chain was eliminated by the level demand policy.

Price policy

- Price levels offered by the steel mill for bulk orders were made available to the service centre for small but regularized orders (i.e. an order pattern of one coil per week).

Decision rationale

- All decisions on demand and activity patterns were to be referred to the Demand Management Team, which would take into account the overall supply chain perspective rather than solely that of individual companies or functions.
- Performance was monitored by recording demand patterns, activity rates and inventory along the entire chain on a weekly basis.
- A monthly meeting was held by the Demand Management Team to monitor project progress.

RESULTS

Effect on demand and activity

Figure 4a shows the level demand schedules that were established between the component manufacturer and the service centre and between the service centre and steel mill. These formed the basis for planning production and despatch activity. For the first ten weeks of the trial the demand from the OEM was as expected, averaging 13.5 tonnes and varying within anticipated bounds. However, from week 10 the demand from the OEM began to fall significantly and to do so in an erratic pattern. (This was because of the recessionary conditions pertaining at the beginning of 1999 due to the Asian crisis.)

During the first ten weeks of the trial, the system operated according to plan at least in terms of a fairly regularized ordering pattern between companies. Regularization of production activity within companies proved more difficult to achieve because of the variability due to machine reliability and quality problems and scheduling problems at bottleneck plants where most products had very variable demand patterns. This was particularly evident at the blanking cell in the component manufacturer (Figure 4b). However, the demand levelling between the companies meant that this internal variability was not transmitted and amplified along the chain.

Supply from the steel mill also continued to be variable because of quality and reliability problems, nevertheless it can be seen in Figure 4c that the degree of variability was reducing over the trial. (To some extent delivery problems from the steel mill occurred because the trial was set up outside the normal order processing system and had to rely on manual intervention from account administrators. In the context of hundreds of coils moving through the steel mill each week it was easy for the two trial products to be overlooked. For example, the missed delivery in week 6 and subsequent delivery of two coils in week 7 was directly attributable to an oversight by the account administrator.)

In some respects the unexpected fall in demand which occurred during the middle of the trial period was fortuitous because it added a further, more rigorous dimension to test the proposed system. The fall in demand caused greater variability in production activity at the component manufacturer (e.g. see blanking activity record in Figure 4b), which, combined with pressure to minimize stock, resulted in reversion to a more variable demand pattern onto the steel service centre (Figure 4c). However, the upstream supply chain was able to cope reasonably well with this change, partly because there was now less stock in the pipeline due to the weekly order amounts and partly because the Demand Management Team were able to make a coordinated response to the change, by reducing activity in concert along the chain. Service centre demand to the mill was reduced by the cancellation of two coils from the mill between weeks 13 and 21. However, as the planned demand was only for one coil per week, it was relatively easy for the mill to adjust to a lower level of supply because there were no large batched orders already in the system.

Effect on inventory

It can be seen from Figure 4d that the level of inventory in the total supply chain progressively reduced over the trial period. This was in spite of the fact that demand was falling in the second half of the trial, which, under normal circumstances, might have been expected to lead to an increase in inventory levels. This positive impact on inventory was deemed to be very important in 'selling' the results of the trial to senior management.

Effect on customer service

During the trial, service levels from the service centre to component manufacturer were consistently 100 per cent on-time delivery, an improvement from 95 per cent prior to the trial. Delivery reliability from the steel mill to the service centre was less variable than prior to the trial, but was still far from perfect, as can be seen in Figure 4b. This was because the elimination of demand variability removed only one of the causes of poor delivery performance.

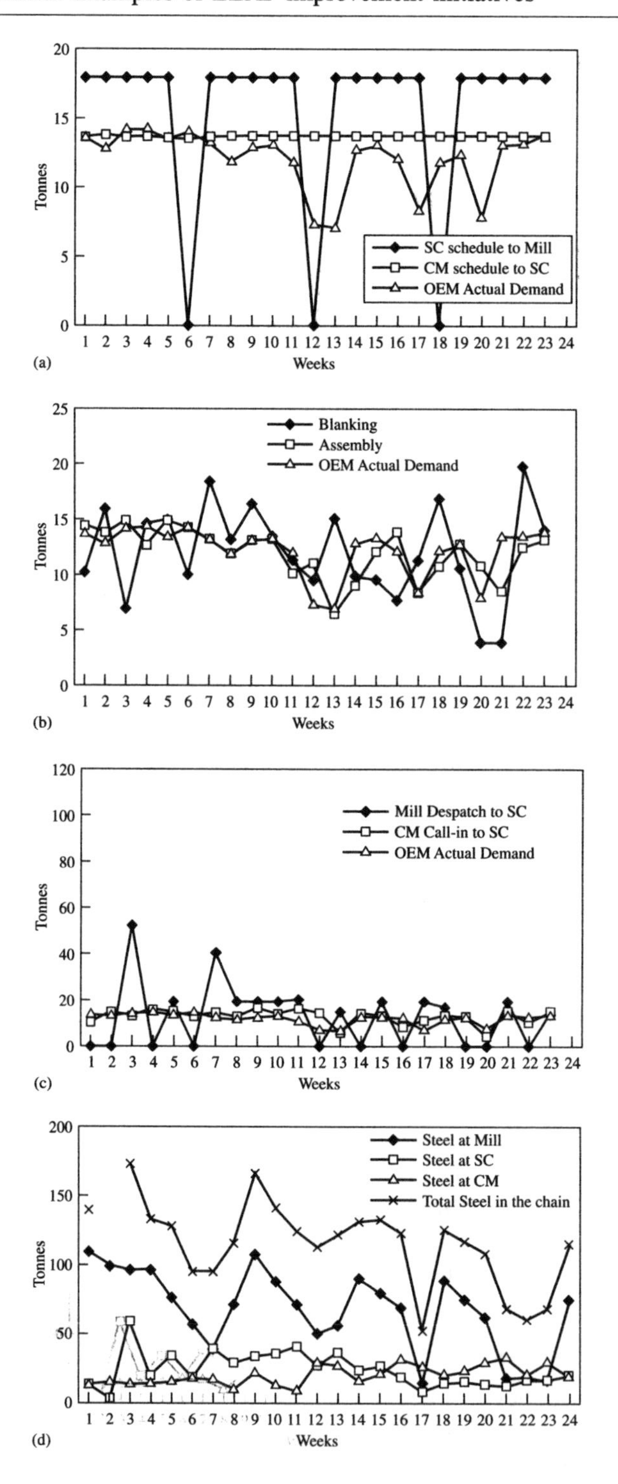

Figure 4 Performance monitors during the trial. (a) Scheduled demand between companies in the trial; (b) activity of the component manufacturer; (c) actual demand and supply during the trial; (d) inventory levels during the trial.

Evaluation of the trial by the companies

The trial was deemed a success by the participating companies. At the end of the six-month period it was agreed that the system would be continued with the two trial products. It was also agreed that computer systems at the steel mill and the service centre would be amended to permit an automated system to manage the new demand structure, without the manual intervention that was necessary during the trial. Once this is in place the new demand management approach will be extended to a wider range of products. It is anticipated that up to 70 per cent of parts supplied by the component manufacturer will be amenable to this type of system. It is when control of demand is extended to a wider range of products that the real benefit of this system will be felt in terms of regularized production scheduling, fewer machine changeovers, increased factory capacity and reduced fire-fighting, as well as greater reductions in inventory across the supply chain.

FURTHER WORK

The above discussion describes one specific attempt to overcome demand amplification in a particular supply chain. It is not suggested that this approach will necessarily have generic validity, but it is hoped that it may at least provide a starting point and framework which others can modify, adapt and apply as appropriate in other supply chain scenarios. This in fact happened within the context of the LEAP steel to automotive programme. Once initial results of the above trial began to emerge, a second demand management project was established which was carried out completely independently. It involved a different component manufacturer, a different steel service centre, different staff and different products supplied to different automobile manufacturers. The only point of commonality was that steel was sourced from the same steel supplier, i.e. British Steel Strip Products. This second project, which was planned and implemented entirely by company staff with no involvement of the university research team, is described briefly in Exhibit 1.

SUMMARY AND CONCLUSIONS

This research has demonstrated the existence of demand amplification effects in an upstream supply chain environment from OEM to raw material supplier. This complements many previous studies which have shown the effect in downstream supply chains, i.e. from OEM to retailer.

A variety of causes of demand amplification have been identified, some of which have been highlighted in previous research, but some, such as the 'supply variability effect', have perhaps added a new dimension. When considering demand amplification it is suggested that it is important to distinguish between factors that trigger demand variability, i.e. root causes, and factors that amplify the variability once it exists. The research shows that the main sources of variability in the chain studied were not the result of variable external demand, but were largely generated from variability and decision making within the supply chain.

A relatively simple and practical approach to reducing demand variability was developed and implemented. The trial had some success in reducing demand variability along the chain and at the same time in reducing inventory and improving

Exhibit 1 Demand Management: Project Two

One of the results of the demand amplification experienced in the LEAP value streams is the lack of visibility of true customer demand at the steel mill. Once this had been understood it was agreed to undertake a pilot project which would attempt to overcome this issue.

A joint project team was formed with members from the companies along the supply chain. The team agreed a set of performance measures which would be used to monitor and control the project and also to measure its success. At the same time a senior level steering group was established which met on a quarterly basis to oversee the project, review progress and agree the next steps.

The objectives of the project were straightforward:

- achieve improved customer service with less inventory;
- understand the true customer demand;
- ensure visibility of this demand along the whole chain;
- manufacture at the steel mill in line with this demand.

In this case determining true customer demand involved understanding the average call-off in the number of parts from the vehicle manufacturer, converting that into a press part requirement and then converting that into a steel requirement based on the batch cycle policy of the press shop. For example, in the case of one of the parts this resulted in a underlying demand equivalent to 50 tonnes of steel to be supplied every five days to the press shop. This information was used as the basis to calculate supply chain buffer levels and optimum manufacturing cycles.

It was recognized that the pilot project would require a relatively large amount of manual monitoring and control as it would have to run outside the normal systems for order processing and production planning. As a result it was agreed that the initial pilot would be limited to a small number of parts. Using a Pareto profile, the top five parts through one of the value streams were identified for inclusion in what became known as The Pipeline Project. These five parts accounted for almost 10 per cent of the total product through that stream.

Over a period of four months dramatic improvements in the supply chain were seen for the pilot parts. The composite measure of delivery to time along the whole supply chain increased from 70 to 100 per cent while at the same time total supply chain inventory reduced by over 30 per cent. These were impressive results and clearly showed the benefit of understanding true customer demand along the whole supply chain. However, the manual effort required to achieve these results meant that the roll-out of the pilot had to be limited. As such it was agreed that phase two would include a further 15 parts and that an automated system to manage the new demand structure would be designed to dramatically reduce the required manual intervention. Phase 2 is currently under way and the automated system is planned to be operational by the end of 2000.

Chris Butterworth, Supply Chain Manager,
British Steel Strip Products, June 1999

customer service. However, the scheme to level demand between the companies did not eliminate all causes of variability, but did help to prevent that variability amplifying and rebounding along the chain.

A number of general learning points arise from this research.

Demand levelling and flexible response

It is suggested that attempts to manage demand should have two fundamental aims. Firstly, as far as is possible, firms should try to smooth demand. The analysis here suggests that much of the variability in demand experienced by upstream partners in a supply chain is generated within the chain, rather than by real variation in end-user demand. It should therefore be possible to manage out a significant proportion of the variability that occurs along the chain.

Secondly, firms should try to develop systems capable of flexible response. In most situations end-user demand will not be stable. There is thus a need for demand management systems and supply processes that can respond quickly and flexibly when demand does vary outside expected limits. A fundamental objective of Lean production is to develop systems which provide such flexibility. Importantly, a system should be developed that permits the supply chain as a whole to flex in a coordinated manner if end-user demand changes significantly. This is in contrast to many current systems in which individual functions often respond in isolation to confused and amplified demand signals that are passed up the chain through the normal order processing procedures.

The role of inventory

The requirement for and role of inventory in the system needs to be clearly understood by mangers. There is a need to clearly differentiate between pipeline stock and safety stock and to determine appropriate levels for each. Even when demand variability is reduced by demand management policies, managers need to appreciate the requirement to keep safety stock in the chain to buffer against supply variability until root causes of that variability are addressed. Failure to appreciate the reasons why safety stock is required, when combined with attempts to operate JIT systems, is likely to increase demand variability along a chain. It would appear that senior management responsible for creating pressures to minimize inventory along the chain often fail to identify the trade- off between inventory cost (easily identifiable) and disruption cost due to variable demand (difficult to identify).

Scheduling and capacity

The biggest benefit to be gained from attempts to regularize demand will be in terms of scheduling of production facilities. When demand is highly variable scheduling of production has to be treated as a new and unique problem on a weekly or even daily basis. In the LEAP project the aim was to create a regular production schedule for as many items as possible, as this would release production capacity that was being lost due to sub-optimal plans, unplanned changeovers and disruption due to emergencies.

Personnel

The study showed that a large number of people along a supply chain made decisions that directly influenced levels of demand. An important step in reducing demand variability is to try to reduce the number of people who impact on demand. For those that remain, it is necessary to educate them towards policies that will dampen rather than exaggerate demand oscillations that do occur.

As we enter the twenty-first century there are two factors that will perhaps enable us to eliminate demand amplification from our supply chains: firstly, the capability to rapidly share demand data along the chain using modern information technology systems; and secondly, a desire to address the demand amplification problem, which stems from the increasing appreciation in industry of the benefits of improved supply chain management. Neither of these conditions existed 40 years ago, when Forrester wrote his paper, which was surely well before its time.

REFERENCES

(1) Forrester J. (1958) Industrial Dynamics: a major breakthrough for decision makers. *Harvard Business Review*, July–August.

(2) Burbidge J. L. (1961) The new approach to production. *Production Engineer*, Dec., 40(12), 765–784.

(3) Sterman, J.D. (1989) Modeling managerial behaviour; misperceptions of feedback in a dynamic decision making environment. *Management Science*, 35(3), 321–399.

(4) Towill D. and Naim M. (1993) Supply chain partnership sourcing smoothes supply chain dynamics. *Purchasing and Supply Management (PSU)*, July/August, 38–42.

(5) Towill D. (1996) Industrial Dynamics: Modeling of supply chains. *The International Journal of Physical Distribution and Logistics Management*, 25(2), 23–43.

(6) Wilkner J., Towill D. and Naim M. (1991) Smoothing supply chain dynamics. *International Journal of Production Economics*, 22, 231–48.

(7) Lee H., Padmanabahn V. and Whang S. (1997) Information distortion in a supply chain: the bullwhip effect. *Management Science*, 43(4), April, 546–558.

(8) Lee H. Padmanabahn V. and Whang S. (1997) The bull whip effect in supply chains. *Sloan Management Review*, 38(3), 93–102.

(9) Wilding, R. (1998) Chaos theory: implications for supply chain management. *The International Journal of Logistics Management*, 9(1), 43–57.

(10) Wilding, R. (1998) The supply chain complexity triangle – uncertainty generation in the supply chain. *The International Journal of Physical Distribution and Logistics Management* 23(6), 599–616.

(11) Saporito B. (1994) Behind the tumult at Proctor and Gamble. *Fortune*, 129(5), 74–82.

(12) McGuffog, T. (1997) 'Effective management of the UK value chain. *Proceedings of the Logistics Research 2nd Annual Conference,* Institute of Logistics, UK.

(13) Hines, P., Rich. N., Bicheno. J., Brunt. D., Taylor, D., Butterworth, C. and Sullivan, J. (1998) Value Stream Management. *International Journal of Logistics Management*, 9(1), 25–42.

ACKNOWLEDGEMENTS

This chapter is based on papers that were first published in *The International Journal of Logistics Management*, 10(2) and *The International Journal of Physical Distribution and Logistics Management* Vol No 10(2), 55–70, 1999.

Kaizen and kaikaku 12

John Bicheno

INTRODUCTION

The mapping activities described in earlier chapters reveal two types of improvement opportunity – short and medium term. Short-term opportunities, the 'low hanging fruit', may be dealt with by kaikaku or 'blitz' events described in this chapter. These events have been popularized in recent years, particularly by some consulting organizations, but their short-term success has led several companies to run literally scores of such events per year while often ignoring what Juran calls 'holding the gains'. The LEAP project has therefore attempted to make balanced use of such events. Moreover, the attractiveness of blitz events should not overshadow other valid approaches to continuous improvement. How the different approaches to continuous improvement, including blitz events, fit together in the context of Lean Operations is the subject of this chapter.

AN IMPROVEMENT CLASSIFICATION

Continuous Improvement has become a major theme, perhaps the major theme, in manufacturing over the past 15 years [1,2]. Although Kaizen has become well established since Imai's classic work, more recently it has become clear that there are two elements to improvement, namely continuous improvement and breakthrough improvement. Thus Juran [see 3] refers to 'breakthrough' activities, using 'project by project' improvement, to attack 'chronic' underlying quality problems as being different from more obvious problems. Davenport, in the context of business process re-engineering, has referred to 'the sequence of continuous alteration' between continuous improvement and more radical breakthroughs by reengineering. And Womack and Jones [4] discuss 'kaikaku' resulting in large, infrequent gains as being different from kaizen or continuous improvement resulting in frequent but small gains.

A traditional industrial engineering idea is that breakthrough or major event improvement activities are not continuous at all, but take place infrequently in response to a major change such as a new product introduction or in response to a 'crisis'. This may be termed 'passive' or 'reactive' improvement. With increasing competitive pressure in manufacturing has come a need to be 'proactive' about improvement. This may also be termed 'enforced' by which is meant not leaving events to chance but forcing them to happen. Examples are the requirement for automotive suppliers to reduce prices by 5 per cent annually or the 3M policy that dictates that 30 per cent of revenues will come from new products every year. In product design traditional 'event pacing' can be thought of as passive whereas 'time pacing' [5] is definitely proactive.

There are therefore four approaches to improvement, as shown in Figure 1. There is, or should be, a place for all four types in every organization. Adopting lean manufacturing does not mean ignoring other forms of improvement to concentrate on kaizen and kaikaku. Passive approaches are a useful supplement and should continue. However, if all improvement is of the passive, reactive type the company may well slip behind.

Before describing each type in more detail, examples of each of the four categories will be given. Classic passive or reactive incremental improvement approaches are the suggestion scheme and the quality circle. It is possible that such schemes will result in breakthrough gains, but this would be the exception. Reactive breakthrough improvement includes classic industrial engineering or work study, especially where such methodologies are as a response to externally imposed change. Examples include factory layout, and new technology introduction. The classic example proactive incremental or enforced improvement is the 'water and rocks' inventory analogy so often associated with lean manufacturing. Thus Bill Sandras discusses deliberate inventory withdrawal in moving from 'loose kanban' to 'tight kanban' [6]. Finally, the classic proactive breakthrough approach is the kaikaku or 'kaizen blitz' event, where productivity gains of perhaps 30 per cent in a week seem to be the norm [7].

PASSIVE INCREMENTAL

The classic type of passive incremental improvement is the suggestion scheme, with or without rewards, and with or without team emphasis. The classic team-based passive incremental improvement method is the Quality Circle. Contrary to popular conception, the reward-based suggestion scheme is alive and well at many Japanese companies. At Toyota's US plant, for instance, awards are based on points and range from $10 to $10,000. Toyota has the attitude that all suggestions are valuable, so the company is prepared to make a loss on more mundane suggestions to develop the culture of improvement. But at the other end of the Pareto the company reckons that

Reactive/Passive Incremental	Reactive/Passive Breakthrough
Productive 'Enforced' Incremental	Proactive 'Enforced' Breakthrough

Figure 1 Types of improvement Activity.

the top 2.5 per cent of suggestions pay for the entire reward programme, even though a good number of suggestions at the bottom end are loss-making taking into account the implementation time. Thomas Edison is reported to have said that the way to have great inventions is to have many inventions. Toyota and others insist that all suggestions are acknowledged within 24 hours and evaluated within a week. Non-acknowledgement and non-recognition have probably been the major reason for suggestion schemes producing poor results and being abandoned.

Likewise, team-based Quality Circles are an integral part of the Toyota Production System. At Toyota, QC presentations to senior management occur almost every day. At Japanese companies QCs often meet in their own time. Management involvement and support are crucial elements. Edward Lawler [8] has described a 'cycle of failure' for many Western QCs. The following sequence is typical. In the early days the first circles make a big impact as pent-up ideas are released and management listens. Then the scheme is extended, usually too rapidly, to other areas. Management cannot cope with attending all these events, and is in any case often less interested. In the initial phases, the concerns of first-line supervisors, who often see QCs as a threat to their authority, are not sufficiently taken care of in the rush to expand. Some supervisors may actively sabotage the scheme; others simply do not support it. Then, as time goes on, with less support from management and supervision, ideas begin to run out. The scheme fades. And it is then said, 'QCs are a Japanese idea which do not work in the Western culture.' (By the way, it was Deming who introduced quality circles to Japan, albeit that Ishikawa refined the methods.)

From the foregoing we learn a few important lessons: (1) not all improvements will pay, but creating the culture of improvement is more important; (2) give it time, and expand slowly; (3) recognition is important – management cannot often be expected to give personal support, so establish a facilitator who can; (4) do not underestimate potential opposition; (5) react, and fast, to suggestions; and (6) give them the tools and techniques, and probably the time.

Several of the LEAP companies are in the early stages of self-directed work teams. British Steel (Corus) is a noteworthy example. Creating an awareness of waste amongst such teams has enabled them to focus more clearly on improvement initiatives. Likewise, as has been noted elsewhere in this book, mapping activities carried out by the teams themselves not only help teams 'put on their muda spectacles' (as Dan Jones would say) but help prioritize incremental improvement projects.

PASSIVE (OR REACTIVE) BREAKTHROUGH

Many traditional industrial engineering and work study projects are of the passive breakthrough type, particularly when left to the initiative of the IE or work study department. Of course, IEs also work on enforced breakthrough activities initiated by management or by crisis, but passive breakthrough activities, led by IEs, have probably been the greatest source of productivity improvement in the twentieth century. Many of Taiichi Ohno's activities could be classified as passive breakthrough (apparently he was a great experimenter in the dead of night) [9].

Most of the LEAP companies have industrial engineering functions of some type. Unfortunately, some of these functions were not adequately schooled in the principles of Lean. In at least one case a major new layout was undertaken in parallel with the LEAP project, but with very little communication between the two groups. The result was a layout with many features of mass manufacturing, which the operating team attempted to put right as follow-up incremental activities. The possibilities were limited, however, due to the fact that major robot cells were already in place. There is an important lesson to learn here. That is, that groups responsible for major breakthrough initiatives have huge long-term impact on the productivity of the company. It is important therefore that such people have priority in learning Lean principles and work side-by-side with Lean implementation groups. Involvement is key. Industrial engineering groups must be active participants in mapping activities and in Lean visioning activities.

ENFORCED BREAKTHROUGH: KAIKAKU

Kaikaku means 'instant revolution' and aims at spectacular and very rapid productivity improvement in a focused area. As Imai has pointed out, it is often easier to achieve a 20 per cent improvement than a 5 per cent improvement! Kaikaku exercises have gained the serious attention of managers in recent years, if for no other reason than the scale of the gains achieved. One of the most stimulating and frequently discussed chapters on Womack and Jones' already classic book *Lean Thinking* tells the story of a kaikaku exercise at Porsche Cars. The 'Ohno disciples' who led the project made a dramatic entry into the troubled Porsche works, announcing, 'Where is the factory? This is obviously the warehouse!' (It was of course the factory), and the very next day, without further analysis, proceeded to walk around and simply cut off the inventory storage shelving to chest height to allow visibility! Shock horror at a company famous for its meticulous engineering. The payoff came six months later when the first Porsche ever produced without rectification rolled off the assembly line.

Another illustration of kaikaku in action, alas missed by many, was Sid Joynson's excellent BBC TV series *Sid's Heroes* [10]. Sid put forward the idea that a 30 per cent productivity improvement in two days was frequently possible and proceeded to demonstrate just that in a variety of manufacturing and service companies. Sid Joynson does not use the word kaikaku, preferring to call it simply 'kaizen', but such achievements are undoubtedly in the breakthrough category. In the US the Association for Manufacturing Excellence (AME), a non-profit society, puts forward that a 40 per cent gain in productivity is possible during their one-week 'Kaizen Blitz' activities. The AME Kaizen Blitz format emphasizes training during week-long exercises, and aims to build up an in-company critical mass of people to enable future improvement activities to be done in-house. Several consulting companies also promote blitz events.

Cardiff's LEAP programme has adopted a two- or three-day format for their kaikaku exercises. This is only possible because mapping of the value stream has been thoroughly done beforehand using a range of mapping tools which are far more comprehensive than those generally used in a five-day blitz. The findings have also been presented to management prior to the event. Mapping often identifies 'low-hanging fruit' and these are the areas that should be tackled first. The identification of potential areas for kaikaku projects comes directly out of mapping as described in earlier chapters.

Perhaps the secret of such kaikaku events has to be a 'just do it' blanket authority, combined with high expectations, and the involvement of middle or senior managers in the kaikaku team. The level of confidence of management to allow, support, and even participate in such events is strongly influenced by the quality of mapping and subsequent presentations. In most kaikaku exercises permission is not sought for each individual change during the two days, but management must agree beforehand to support any procedural or layout change as broadly identified by the mapping exercises, provided it does not cost more than, say, £100. The emphasis in a kaikaku is on radical, fast, low-cost improvement. The emphasis is not on perfection, but on improvement. This is important: the philosophy is that all improvement is good, that we will not get it right the first time but there will be improvement and we can return again and again.

The participants in a LEAP kaikaku comprise two or three experienced external people (LEAP staff), three company managers (preferably from different areas) and at least three operators from the area concerned. Operator involvement is essential for hands-on knowledge and also for buy-in. It is usually necessary to build some buffer stock in preparation for the exercise, although the aim is to keep working during the exercise albeit at a reduced pace. The first morning is spent on data collection, although the team benefits from prior analysis during company mapping. If necessary, team members receive a very brief introduction to waste analysis and kaikaku aims on the day before the exercise proper. Three teams are formed; one maps inventory and material handling, one maps the process and quality or the scheduling procedures, and one looks at layout including measuring and mapping all distances and movements. Checklists have been developed for each team. The first afternoon is spent on brainstorming and analysis. The emphasis is on improvements that can be made immediately, although a note is made of improvements that can only be made later. The team identifies at least one improvement activity for every team member to be responsible for. An attempt is usually made to actually implement some changes, however small, during the first day. This sends out an important message: this is for real, this is now. Overnight, maintenance people may be called upon to change the locations of power points, airlines, etc. Then, during the second day, the changes are actually implemented. Then they are checked and adjusted. Finally, if possible, the new standards are established.

Every kaikaku event is based around the elimination of waste or muda. The starting point, therefore, is that all participants must have a good knowledge of the 'seven wastes', and as Dan Jones would say, be wearing their 'muda spectacles'. Kaikaku also emphasizes 'Gemba' or the workplace [11]. Know the workplace. Discuss at the workplace. Analyse at the workplace. Do not be tempted back into the office.

Very often, a kaikaku will be based around a 5S housekeeping approach (or more strictly a 3S since the last two stages are longer term). An English version of 5S is 'CANDO' [12]. C is for cleanup and throw out, including 'red tagging' items that are earmarked for subsequent disposal. This must be done without compromise. Following cleanup, A is arranging – 'a place for everything'. Here waste must be minimized: stillages in designated thought-out locations, buffer stocks considered, walkways delineated or reviewed, shadow boards established, clear signals (kanban?) established. All movable machines should be considered for relocation – moving them close together and into cells. Then N for neatness – 'everything in its place'. Procedures must be fixed. A five-minute end or beginning of shift cleanup routine established – note, not just any daily cleanup, but a specific routine must be worked out for each workcentre. This 3S programme should lay the foundation for the remaining two Ss: D for discipline and O for ongoing improvement.

In addition to 5S, some kaikaku events emphasize 'one-piece flow' as an objective. This type of kaikaku may not be appropriate unless the 5S foundation is in place. If scheduling and flow are to be the priorities, the starting point is the 'takt' time, that is, the average customer demand during a shift divided by the planned actual working time during the shift. The takt time establishes the drumbeat around which all work operations should be balanced. To achieve this, layout may have to be changed and a pull system (kanban?) established. All this in 2–5 days!

Of course, to get it right the first time is unrealistic. So, the kaikaku team must be prepared to run the redesigned system, test it, and make adjustments. As has been said, some changes simply cannot be done in the time frame, in which case clear responsibilities (who and when) must be established.

The stages of a kaikaku event: outline

In this section typical stages of a kaikaku event is given. Detail is not appropriate here, and may be found in the references [7, 13].

1. Identify the area
2. Identify the focus of the event
3. Identify the event pattern and suitable times
4. Gain management support and commitment
5. Select the team
6. Establish takt time
7. Preparation: materials, inventory
8. Awareness and training, clarifying objectives
9. Establish block diagram
10. Establish sub-teams
11. Mapping and data collection
12. Initial analysis
13. Initial changes and testing
14. Further changes and measurement
15. Standardization
16. Follow-up

A brief overview of a typical kaikaku event

The following composite case study is based on several actual kaikaku events undertaken during the LEAP project.

Following a mapping exercise at an automotive part supplier it was decided to stage a 1+2+2-day series of kaikaku events. Initial mapping had revealed a large proportion of movement wastes and double handling and apparently large inventories.

The findings of the original mapping exercise were presented to a group of senior and middle managers. They decided that a 1+2+2-type event would be most suitable for their situation which included a relatively low awareness level of JIT / Lean Manufacturing / Waste concepts, fairly complex product routings, and a number of machines that were fixed to the floor. The target area for the exercise was identified as being one where there appeared not only to be good opportunity, but where the product range was limited to a family of products with similar routings.

The team to carry out the event was identified. It comprised the external facilitator, the kaizen facilitator from the company, the cell supervisor, an industrial engineer from another site in the same group, an accountant from the factory, and three cell operators. All had to commit to being completely free of other duties for the duration of the events, although the industrial engineer and accountant were permitted to attend half-hour morning meetings.

Day 1

The first day, or half-day to be precise, included an introduction to lean manufacturing concepts. This was done by playing a JIT / Lean game. The seven wastes were introduced and talked through, and the desirable aims established: improve flow by reducing batch size (one-piece flow would not be possible), reducing waste, and encouraging flow. An outline of the forthcoming event was also given. A degree of team building was also the aim at this stage.

Days 2 and 3

The first two-day event began the following week. After an initial briefing the team split into three sub-teams, each tasked to complete their study by the end of the first morning. The three cell operators were each involved in a different initial analysis team. The layout team made a cut-out model of the cell and adjacent areas during the first morning, and measured total distances moved. The inventory and material handling team quickly got into questioning the amount of parts stored in the cell (several weeks supply in some cases), and the material movement quantities and methods. The process team chose to get into the ergonomics of the workstations and assembly methods that were the apparent cause of some quality problems. Data collection included measuring inventory, estimating lead times, timing of selected activities (some in greater detail than others), measurement of flow length, and taking photographs.

Immediately after lunch the sub-teams each made a brief presentation to fellow team members. Brainstorming and prioritizing took place during the first afternoon. The team spent a good part of the first afternoon at the cell, discussing alternatives *in situ*. Some immediate changes were decided upon and a number of airlines and electrical connection points were moved overnight. During the second day, the changes that were actually implemented included: relocation of a small press, the swapping around of a bicycle shed and a storage area, the relocation of a material store in the cell enabling a smoother flow to take place, changes to the assembly process itself, the repair of a door which was causing a draught, a drastic reduction of batch quantities (changeover was already very short), and, as a result, new material handling procedures. Interestingly, some of the women cell operators were initially opposed to any change on the basis of safety, but became enthusiastic supporters when changes proposed by their fellow operators were put in place. The afternoon of the third day was spent prioritizing actions, and assigning responsibilities. The list ran to several pages.

Days 4 and 5

During the intervening period, alterations to some work desks were made by maintenance, racks were acquired (from another part of the plant), preparations to move a machine were made, and a variety of materials (e.g. paint), signs, and tools were acquired. Briefings were also given to senior management. The operators who had participated in the initial days were encouraged to discuss what had happened and what was planned with their workmates.

On days 4 and 5 the agreed-upon changes were made. The major change was to layout, including the relocation of a semi-automated assembly machine. New material handling arrangements were begun. Then the line was run and observed. Some further changes were necessary. A number of new ideas emerged, some of which were immediately attempted. During day 5, after further adjustment, the team attempted to write new standards. These could not be completed. In any case all the operators were to be involved, not just the three operators on the kaikaku team. Estimates of the savings in space, inventory, throughput time and material handling, were made, and 'before and after' photographs assembled.

In the experience of the LEAP research team, the reaction to kaikaku exercises has often been one of astonishment. 'I've worked here for ten years and seen more change over the past two days than over the whole past ten years' is not an untypical comment. Kaikaku exercises have also been found to be excellent for team building and cross-functional communication. They are fun. So if you are concerned with building teams, good advice seems to be: don't take them away on an outward-bound exercise, turn them loose on a kaikaku!

Womack and Jones have pointed out that kaikaku activities can continue to be effective if repeated at intervals of a few months, but often by emphasizing a different aspect. For instance, housekeeping may be the focus of the first kaikaku event held on an area, but the next time the focus may shift to flow, safety, or quality.

KAIZEN (ENFORCED INCREMENTAL IMPROVEMENT)

The classic 'enforced improver' is Toyota. Toyota has managed to keep continuous improvement alive for over 20 years, and in times of push to redouble their efforts to take yet more waste out the system. According to Womack and Jones, Toyota's Ohno was 'the most fearsome opponent of waste that the world has ever seen'. The Ohno legacy lives on at Toyota. Waste elimination is not a matter of chance that relies upon operator initiative, but is driven. There are a number of ways in which this is done:

- *Response analysis.* At Toyota operators can signal, by switch or cord, when they encounter a problem. At some workstations, there are a range of switches covering quality, maintenance, and materials shortage. When an operator activates the switch, the overhead Andon Board lights up, highlighting the workstation and type of problem. People literally come running in response. But the sting is in the tail: a clock also starts running which is only stopped when the problem is resolved. These recorded times accumulate in a computer system. They are not used to apportion blame, but for analysis. Thus at the end of an appropriate period, say a fortnight, a Pareto analysis is done which reveals the most pressing problems and workstations.
- *Line stop.* A Toyota classic, closely related to response analysis, allows operators on the line to pull a cord if a problem is encountered. Again, the Andon Board lights up. Again, the stoppage is time recorded. But the motivation to solve the problem is intense because stopping the line stops the whole plant. This means application of the 5 Whys root cause technique.

- *Inventory withdrawal.* Many will be familiar with the classic JIT 'water and rocks' analogy, whereby dropping the water level (inventory) exposes the rocks (problems). This is done with a vengeance at Toyota. Whenever there is stability, deliberate experimentation takes place by withdrawing inventory to see what will happen. Less well known is that this is a 'win-win' strategy: either nothing happens in which case the system runs tighter, or a 'rock' is encountered, which according to Toyota philosophy is a good thing. It is not any rock, but the most urgent rock. Deliberate destabilization creates what Robert Hall has referred to as a 'production laboratory' [14].
- *Waste checklists.* Toyota makes extensive use of waste checklists in production and non-production areas alike. A waste checklist is a set of questions, distributed to all employees in a particular area, asking them simple questions: 'Do you bend to pick up a tool?', 'Do you walk more than two yards to fetch material?', and so on. Where there is a positive response, there is waste. The result is that individuals and teams never run out of ideas for areas requiring improvement.
- *The 'stage 1, stage 2' cycle.* At Toyota there is a culture that drives improvement. This culture or belief stems from the widely held attitude that each completed improvement project necessarily opens up opportunity for yet another improvement activity. For want of a better phrase, the writer has termed this 'stage 1, stage 2' [15] after a list of JIT 'stage 1' activities that lead to 'stage 2' opportunities which in turn lead to stage 1 opportunities, and so on. The list of possible chains is very large, but an example will suffice. Thus set-up reduction (stage 1) may lead to reduced buffers (stage 2), which may lead to improved layout (stage 1), leading to improved visibility (stage 2), leading to improved quality (stage 1), leading to improved scheduling (stage 2), and so on and on.

In the LEAP project, some of these ideas have just begun to be implemented. Their long-term success, however, depends upon ongoing management commitment. Top management involvement is not called for, but the presence of a full-time kaizen facilitator, whose responsibility is continuous improvement, is almost certainly a necessity.

CONCLUSION

The emergence of kaikaku or radical, enforced breakthrough is one of the most exciting opportunities for productivity enhancement to emerge in recent years. Every company should have a regular programme of kaikaku events. Yet, kaikaku is but one of four possibilities, the others being 'true' kaizen and traditional reactive, incremental and reactive, breakthrough. The four types of continuous improvement activity are not alternatives; they can and should be made to work together. Although this paper has set out a few pointers, much research is needed to understand the best strategies for each of the cells.

REFERENCES

[1] Japan Management Association (1985) *Kanban: Just in Time at Toyota.* Productivity Press, Chapter 6.
[2] Imai, Masaaki (1986) *Kaizen: The Key to Japan's Competitive Success.* New York: McGraw Hill.
[3] Bicheno, John (1997) *The Quality 60: A Guide for Service and Manufacturing.* Buckingham: Picsie Books.

[4] Womack, James, and Jones, Dan (1996) *Lean Thinking*. London; Simon and Schuster.
[5] Eisenhardt, Kathleen and Brown, Shona (1998) Time pacing: competing in markets that won't stand still'. *Harvard Business Review*, March–April, 59–69.
[6] Sandras, William (1989) *Just in Time: Making it Happen*. Oliver Wight.
[7] Laraia, Anthony C., Moody, Patricia and Hall, Robert (1999) *The Kaizen Blitz: Accelerating Breakthroughs in Productivity and Performance*. New York: John Wiley and Sons.
[8] Lawler, Edward E. III (1996) *From the Ground Up: Six Principles for Building the New Logic Organisation*. San Francisco; Jossey-Bass, Chapter 6.
[9] Reingold, Edwin M. (1999) *Toyota: People, Ideas and the Challenge of the New*. London: Penguin Books.
[10] Joynson, Sid and Forrester, Andrew (1996) *Sid's Heroes: 30% Improvement in Productivity in 2 Days*. London; BBC.
[11] Imai, Masaaki (1997) *Gemba Kaizen*. New York: McGraw Hill.
[12] Bicheno, John (2000) *The Lean Toolbox,* (2nd edn). Buckingham: Picsie Books.
[13] Bicheno, John (1999) *STORMFLOW: A Kaikaku Kit*. Buckingham: Picsie Books.
[14] Hall, Robert (1987) *Attaining Manufacturing Excellence*. Falls Church, VA: Dow Jones – Irwin / APICS.
[15] Bicheno, John (1994) *Cause and Effect JIT*, 2nd edn., Buckingham: Picsie Books.

From value stream mapping to shop floor improvement: A case study of kaikaku

13

Chris Butterworth

INTRODUCTION

This case study describes a mapping and improvement project undertaken within one of the LEAP first-tier component manufacturers. The company manufactures pressed steel automotive components and supplies these in some cases as bare metal pressings or mainly as completed sub-assemblies which are painted and delivered ready for final assembly to the vehicle. The main physical flow consists of blanking, pressing, welding, painting, final assembly and packing.

VSM TOOLS USED

Value Stream Mapping was undertaken to highlight opportunities for the application of lean thinking within one of the main value streams (1) within the company. A representative product was chosen which generated a high percentage of total revenue, was supplied to a strategic customer, and was typical of the generic value streams on this particular site.

The Process Activity Map (PAM) followed the selected part from receipt of raw material from the steel service centre to despatch of the finished product to the OEM customer (car manufacturer). The physical flow was walked with the mappers following exactly the route taken by the material as it was gradually transformed through the various stages of production. The information flow was also mapped. Data was recorded in terms of time in minutes, the distance travelled in metres, the number of times the product was handled and the level of inventory at each stage. As the data is being recorded each step is categorized as either an operation, transport, inspection, delay or storage. Each step is also categorized as either 'value adding' or 'necessary non-value adding' or 'non-value adding'. Value adding steps are defined as those, that physically change the product (or service) in some way, which the customer would be willing to pay for. In this case, for example, they would include welding and assembly of the part. A necessary non-value adding step includes activities which are necessary in the current process but which do not add value. For example, this could include the transportation of product between operations located in different buildings. It is necessary because the operations have been located in this way but it should be questioned and removed if possible. A non-value adding step would, for example, be storage or a rework activity.

From the data collected for the Process Activity Map, it is possible to build up a Value Analysis Time Profile (VAPT) (see Chapters 3, 4 and 8). This map plots time along the horizontal axis and cost along the vertical axis. Using a simple spreadsheet configuration each step on the process activity map is allocated a cost. Two cost lines are created, one of which captures all the costs incurred while the second includes only those costs related to the steps that add value. The cost of each step is the cumulative costs of materials and includes a proportion of overhead allocation such as the cost of the equipment used and the human resources employer. Transportation costs are captured for forklift trucks and cranes based on the transportation times recorded. Cost is also allocated for storage and delay based on an annual charge of say 25 per cent of the value of the goods, which is applied pro rata to the actual storage time mapped. This figure should cover all related storage cost such as heating, lighting and space and will depend on the particular product being mapped.

Not all costs will be captured in this manner; for example, general overheads relating to design or finance will not be included. However, where it is felt to be appropriate it is possible to allocate an estimate of these overhead costs to the physical costs captured on the VATP and compare the total figure to the selling price.

SUMMARY OF MAPPING RESULTS

Analysis of the Process Activity Map and the Value Analysis Time Profile highlighted a number of opportunities for waste reduction. The total elapsed time from receipt of raw material to despatch of the finished part was some 408 hours, while the value adding time was a small fraction at less than 1 per cent. This is discussed in more detail below.

THE PROCESS ACTIVITY MAP

Table 1 summarizes the PAM. It shows the number of steps or discrete actions from receipt of raw material to despatch of the finished part and classifies them by time (duration) and distance. The number of 'touches' refers to the number of times the product was handled by someone. This is a useful indicator of how well a particular product is 'flowing' through the operations. As in this case where the number of operations is 16 and the number of touches is 129 there is a clear indication of multiple handling and poor process flow.

Table 1 Process Activity Map Summary Table

	Flow (steps)	*Distance (m)*	*Duration (ms)*	*Touches*
Operation VA	16	9	84	29
Transport	33	1856	93	67
Inspection	6	3	23	6
Delay	7	0	534	27
Storage	6	0	23746	0
TOTAL	68	1868	24480	129

Another way to present the PAM results is to graph the elapsed time for each classification as a percentage of the total elapsed time. This very clearly illustrates the size of the improvement opportunity and provides a high-impact visual representation of the results (Figure 1).

As Figure 1 shows, almost 97 per cent of the elapsed time was spent in storage. The next largest amount of elapsed time was 534 minutes or 2.2 per cent in delay. These two were differentiated on the basis of 'storage' being time in a planned storage location and 'delay' occurring when a part was waiting for the next operation but was not in a planned storage area, e.g. left waiting on the factory floor. Detailed analysis of the PAM shows that an average of four days' storage time was in goods receipt before the first operation was started and that lengthy delays occurred between each operation. This was largely due to poor coordination of activities between different operations and different production planners. Each tended to plan and operate in isolation, creating a push system through the whole process resulting in long delays and a batch and queue operation. It also contributed to significant multiple handling and excessive transportation. The percentage of value adding time was 0.2 per cent of the total elapsed time. Clearly they were massive opportunities to improve, but one of the issues always faced when looking at data in this way is the 'so what?' factor. What does it really mean that 97 per cent of time is spent in storage? In order to address this issue a Value Analysis Time Profile was constructed. The Value Analysis Time Profile is shown in Figure 2.

THE VALUE ANALYSIS TIME PROFILE

By allocating a cost against each physical step in the value stream and differentiating between those that add value and those that just add cost it is possible to illustrate potential improvement opportunities and the cost benefits. The starting point for the map in this case is the purchase price of the raw material for a single main component. All other costs are pro rata for a single component. Waste reduction opportunities from the VATP should be viewed in three dimensions:

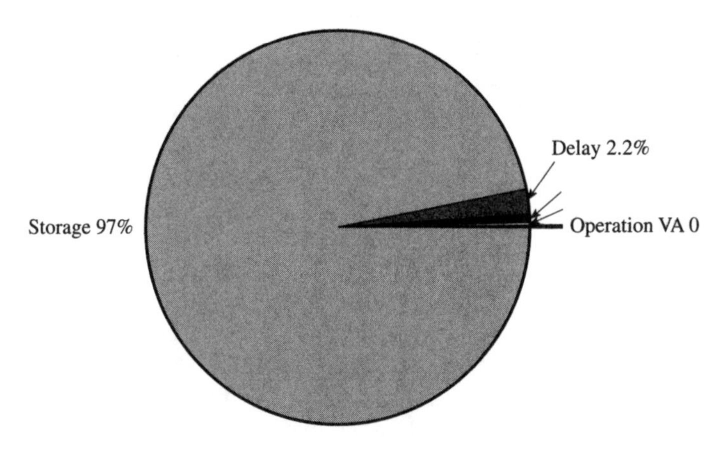

Figure 1 Process activity summary. % of elapsed time.

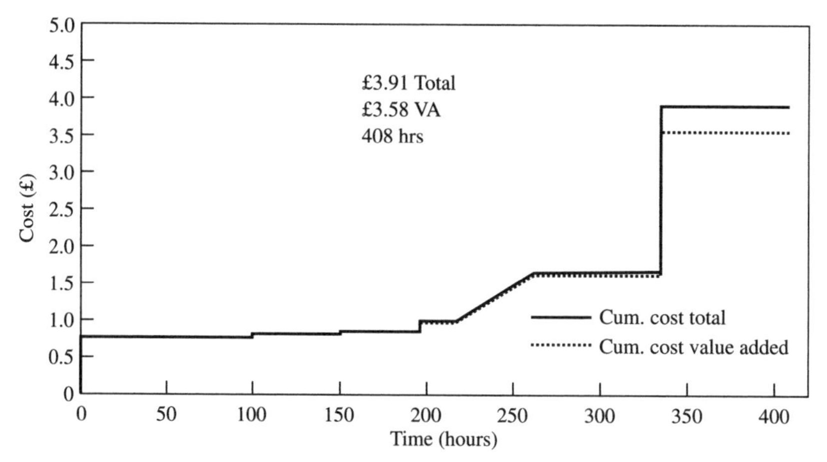

Figure 2 Value Analysis Time Profile prior to improvement activity.

a) Cost can be reduced along the horizontal axis by reducing the total time taken from receipt to despatch.
b) Cost can be reduced by decreasing the non-value adding activities.
c) Total cost can be reduced by activities that reduce the vertical axis, e.g. material cost.

This map shows that a potential saving of 33 pence (over 8 per cent) per part existed in 'Dimension b' alone if all non-value-adding activities could be eliminated. While this was unlikely, even a 50 per cent reduction would yield a saving of over 4 per cent of total cost which would equate to over £38,000 per annum on the total volume produced. As the map shows only one relatively low-value part, it could be argued that the VATP does not visually illustrate the true cost of holding inventory and this is one reason why it should be viewed alongside the PAM summary. Also, in this particular case the painting was subcontracted and the map shows the charge made and the elapsed time from despatch to the painters to return of the part to the plant.

Figure 2 also shows a large amount of value being added in the assembly stage. As is often the case, this is largely the result of expensive bought-out components being added to produce the final assembly. These have been added at purchase price. With this proportion of cost they clearly warrant close control and investigation for cooperation with suppliers to explore opportunities for cost reduction. While this was noted, it was beyond the scope of this particular exercise.

OTHER MAPS

The other tools used in this case were the demand amplification map (DAM) and the physical structure map. The DAM helped to illustrate the disconnect in terms of batch sizes between the press shop and the assembly line and between the press shop and raw material. DAMs are discussed in more detail in Chapters 9 and 11 in this book. The physical structure map confirmed that the company had already undertaken considerable efforts to rationalize its customer and supplier bases (Figure 3).

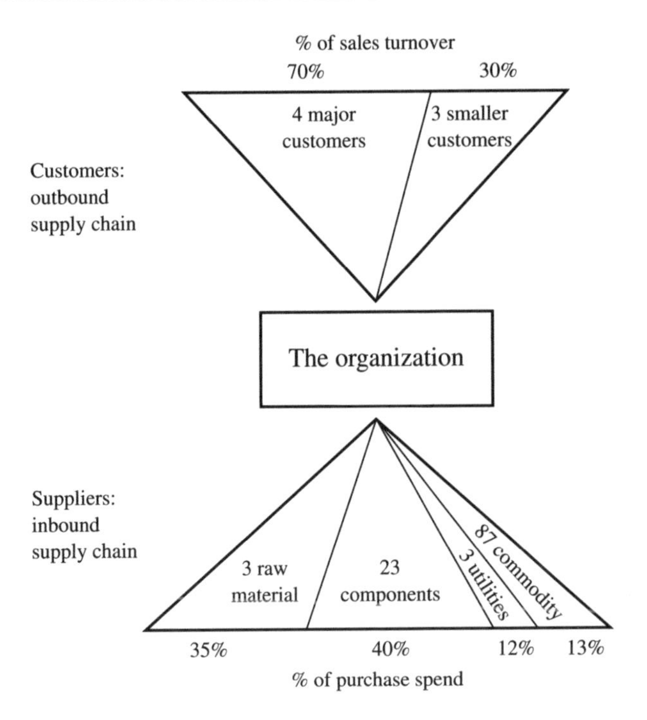

Figure 3 Physical structure map.

SUBSEQUENT IMPROVEMENT INITIATIVES

Following a presentation of the mapping results to the directors it was agreed that a kaikaku exercise (see Chapters 1, 5 and 12) would be undertaken to address some of the main issues identified. Kaikaku means 'instant revolution' and aims at spectacular and very rapid productivity improvement in a focused area. The LEAP programme adopted a two-day format for kaikaku exercises, but this was only possible because mapping of the value stream had been completed beforehand. Operator involvement in kaikuku events is essential for hands-on knowledge and also to ensure 'buy-in' from staff.

The Kaikaku Activity

The data from the value stream mapping showed relatively high levels of inventory and a total throughput time of just over 408 hours from receipt of raw material to despatch of the finished assembly. The maps also showed poor linkages between the processes in that product was pushed from one process to the next, creating excessive waiting time and transportation.

It was agreed that a multifunctional team would be put together for an intense two-day kaikaku activity focused on reducing inventory and throughput time. The teams consisted of a team leader, two operators, a materials handler, a process improvement engineer and a materials manager. The group was split into three small sub-teams looking at detailed data collection in:

- inventory and material handling;
- the manufacturing process;
- layout.

The teams spent the first morning collecting more detailed data and confirming for themselves the results of the value stream mapping. In the afternoon the various findings were pulled together and the opportunities for action brainstormed. These opportunities were then discussed in more detail by the team and an action plan agreed.

Potential actions where sorted into two groups: activities that could be carried out immediately by the team and activities that needed additional support. Fifteen significant improvements where identified that could be implemented immediately. Indeed the team was so keen that some improvements where actually implemented that same afternoon.

Summary of immediate improvement activities

A 'to do' list was agreed for activities that would take place that afternoon and the next day. Responsibility was given to someone for each activity and action plans agreed. A sample of the actions is given below.

- Sort, clear out and organize workspaces.
- Rearrange despatch racking.
- Relocate manual welder.
- Reorganize assembly area layout.
- Move remote assembly benches to central assembly area.
- Create new sub-assembly storage point next to assembly area.
- Erect protective weather screening at new storage point.
- Move forming machines to reduce transportation.
- Move sample board so it is more visible.
- Move test vice nearer to operator.
- Weld up a new rack for bought-out finished parts, e.g. nuts and bolts.
- Train forklift drivers in new movement procedures.
- Change packing instructions to standardize quantities through all processes.

A further action list was drawn up with responsibilities and timings for those actions that couldn't be done immediately. The team leader took the responsibility to ensure that these actions were undertaken. These are listed below:

List of improvement actions for later

- Introduce Kanban process initially between despatch and final assembly to create a visible link to actual customer consumption.
- Plan activities in the press shop to reduce batch sizes to further reduce inter-process inventory.
- Follow up discussions with paint subcontractor on standardization and quality control of packaging.
- Examine possibility of suppliers providing some of the bought-out parts pre-assembled.
- Installation of workbench containers for small bought-out components to cut down operator travelling time and unnecessary motion.
- Install permanent weather screening at new storage point.

What happened

That same afternoon a '3S' activity (Sort, Simplify, Sweep) (4) was carried out in the welding and assembly areas which filled a medium-sized rubbish skip and created much-needed space. At the same time packaging instructions where changed in each process to standardize pallet quantities based on the customer's daily requirements. Previously no fixed quantities where specified until final despatch, making inventory control and visibility difficult and increasing transportation.

The next day a controlled location was created next to the assembly line for the storage of painted parts waiting final assembly. The team had found that these were stored in four or five different locations and that a major element of downtime on the assembly area was waiting for a forklift truck (FLT) to deliver the required parts.

The team calculated the maximum inventory required for each part and marked out a specific area, which would take the maximum number of pallets. Part numbers were stencilled on the floor, and on the rear wall at the eye level of the forklift truck driver. The space was restricted but the team agreed that this would be useful to control overproduction. Material to the maximum agreed levels was moved to the new area and an assessment made of the remainder. To the team's astonishment they found that one process could be stopped for two days in order to bring inventory down to the levels identified.

The other benefit of this action was that the assembly line could collect their own material using a hand truck rather than waiting for a forklift truck. It also meant that forklift truck movement was significantly reduced, as there was now only one location for the painted sub-assemblies rather than the previous five locations.

In the assembly area itself the workbenches were reorganized and two additional assembly benches fitted into the same space. These were previously in a separate building and resulted in excess transportation. Work in the area was made easier with 'bought-in materials' (e.g. bolts, screws, etc.) clearly labelled and located nearer to the operators.

The layout of the area was reorganized for 'flow', allowing material in at one door and straight out through another door into the despatch area without unnecessary delay and transportation.

Within the welding area, a test vice which had to be used on regular frequency was relocated next to the operator rather three metres away. This meant that the tests could be conducted within the cycle time of the robot welder, which was not previously possible.

One part variant required two additional forming operations and these machines where located in such a way as to restrict flow and actually interfere with final assembly operations. The team decided to move the machine into the space created in the welding area as the process was actually required prior to welding. The last action on the first afternoon was to move these machines with a combination of muscle, sweat and pallet trucks. But it was worth it!

Before they went home for the night the team briefly reviewed what had been achieved in a single day. The most telling comment came from one of the operators. 'We have been able to put more changes in place today than we have managed on the last two years. I feel great!'

The results

At the end of the second day an attempt was made to summarize the main benefits of the work that had been undertaken:

- Downtime in the assembly area was cut by 60 per cent, as waiting for material to be delivered by the forklift truck was eliminated.
- Space saved was over 40 sq. metres, meaning that flow could be improved and new work incorporated into the existing space.
- Transportation was cut by over 50 per cent releasing over five hours of FLT time per week. This had an additional benefit of reducing waiting time in other areas of the factory.
- Housekeeping was dramatically improved with clear locations for all parts, leading to increased operator efficiency and greater pride in the workplace.
- Inventory was now clear for all to see and could be visually managed. Better links were established between each process and an opportunity identified to cut inventory by up to 30 per cent.

New value analysis time profile

After the kaikaku exercise had been completed a revised VAPT was constructed taking into account the changes that had been made. This is shown in Figure 4.

As can be seen, the changes had a significant impact on the costs of the component. The saving is a reduction in cost of 23 pence per part or some 6 per cent. Over the annual volume of the part this equates to an annual saving of over £50,000. This figure excludes additional savings which were made on other parts as a direct result of the kaikaku activity as many of the improvements were applied directly to similar products.

The total elapsed time from receipt of raw material to despatch of the finished part was reduced by over 35 per cent, largely through the reduction of storage time. The level of material in the new storage point next to the assembly was used as a natural

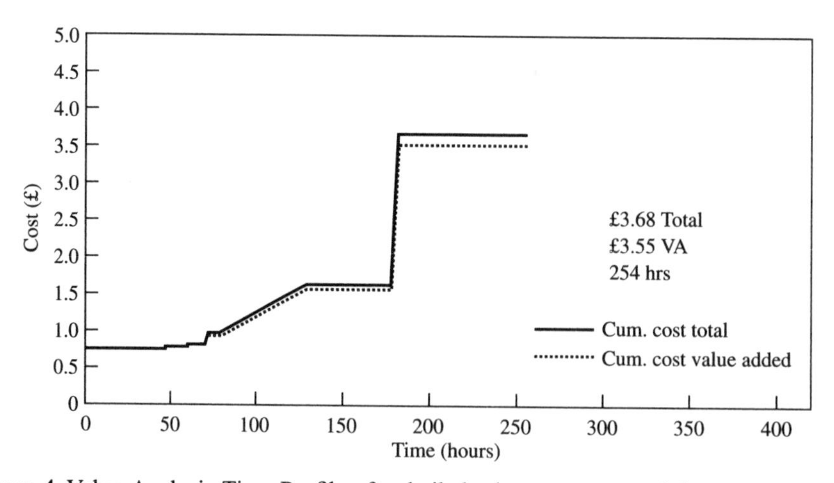

Figure 4 Value Analysis Time Profile after kaikaku improvement activity.

control point for the amount of material released further back in the process. While a kanban system was not introduced as part of the work it was recorded as a future activity the team would investigate.

CONCLUSION

One of the common criticisms of improvement activities focused on a particular area is that it can be difficult to see just what impact they have on the 'bottom line'. This is valid and no improvement programme should consist solely of isolated activities. However, the VATP provides a tool which allows managers not only to cost the benefit of changes but also to have a view of what potential benefits exist before any work is undertaken. This can be a very useful aid in determining where to focus scarce resources.

The PAM on its own showed that potential for improvement existed but the VATP can be very useful in convincing management of the cost benefit of undertaking the work. The use of the kaikaku approach to realize the benefits means that the process changes are owned by the operators and line managers and mean that they are much more sustainable that if they had been seen as imposed by 'outsiders' or senior management.

ACKNOWLEDGEMENTS

Thanks and acknowledgement are given to John Bicheno who was actively involved in the work described in this case study.

REFERENCES

[1] Womack J. and Jones D. (1996) *Lean Thinking*. New York: Simon & Schuster.
[2] Hines P. and Rich N. (1997) The seven Value Stream Mapping tools. *International Journal of Operations and Production Management*, 17.
[3] Brunt, D., Sullivan, J. and Hines, P. (1998) Costing the value stream. Logistics Research Network Annual Conference, September.
[4] Bicheno J. (1998) *The Lean Toolbox*. Buckingham: Piscie Books.
[5] Bicheno J. (1999) Kaizen & Kaikaku. *Logistics Focus*, April.

14 Developing an exemplar line at LDV

Malcolm Taylor and Nick Rich

INTRODUCTION

Since the 1980s it has been Japan, and Nagoya[1] in particular, that has defined 'world class' levels of performance in car assembly and parts manufacturing. This 'Mecca' for high performance and high quality manufacturing has drawn aeroplanes full of British managers seeking to find the inspiration for their own factories. Many of these managers have brought back with them simple practices and pinned their hopes to these solutions but very few British companies have understood the 'logic' that creates the foundation for these powerful industrial businesses. This logic is not exclusively Japanese but neither is it obvious or necessarily common sense. Solutions such as kanban are exactly that – they are solutions to a problem or need. What the legions of British managers have imported is someone else's solution to a problem that probably was not understood by the visiting manager in the first place. What has been missed is the less visible dimension of 'world class' manufacturing or, put simply, the 'logic' from which car companies, like Toyota, draw their power.

Washwood Heath is, geographically speaking, a long way from Nagoya but it is the home of the van company LDV. Just like Toyota, LDV sponsors a football team known locally as 'the Villa' and just like Toyota, LDV is a company with 'a mission'. This mission covers all elements of the operations at Washwood Heath but this chapter will concentrate on LDV Press Operations. The chapter will detail how the teams in the factory have improved their part of the business, sustained this improvement and worked closely with suppliers to get more British-made steel into more British-made vans.

LDV AND PRESS OPERATIONS

LDV is an assembler of light commercial vehicles in the United Kingdom. The business is located at a large complex in Washwood Heath in the Heartlands area of Birmingham. The company produces in the region of 15,000 vans per year and holds a major slice of the British market for light commercial vehicles. The products made at the site have included models such as the 'Morris Minor', 'J Series' van, 'Sherpa' van and more recently the 'Convoy' and 'Pilot' ranges of vehicles. To support the vehicle assembly and final operations, LDV has one of the largest press shops in the UK. These presses produce parts for the 'body in white' operations and also manufacture products and sub-assemblies for a number of other passenger vehicle assemblers. With almost one hundred presses within the factory, ranging in tonnage and technological complexity from the simple 'old presses' to fully automated lines, the press shop represents a large-scale business in itself. Just like any other large business, the press shop has had to adapt to the current, highly competitive, marketplace, even though the output of the press operations is used internally. The press operations have not taken for granted the need to support a highly performing assembly track with a highly performing press operations system. Benchmarking and awareness of practices at other press operations are therefore key to maintaining the external 'market' focus of the LDV press facility. In this respect, the press operations is a 'business within a business', fully capable of generating and delivering its own strategies to support the assembly of LDV vans to levels of performance that would normally be associated with car production.

A HISTORY OF THE SITE

LDV is a long-established company whose previous incarnations include Freight Rover and until its receivership Leyland Daf Vans. Out of this receivership rose LDV in the form of a management 'buy-out'. Many employees invested their own money in the new company. A period of rebuilding followed the 'buy-out', restructuring the workforce to suit a reduced volume base, and from this point the company has consolidated and grown.

More recently, in 1998, LDV began a reciprocal relationship with Daewoo Company of South Korea and the press operation began to actively benchmark itself on a truly global scale and against companies with newer facilities.

THE CHALLENGE

Within the press operations the challenge was to introduce a cell-based manufacturing system that would deliver certain operating criteria and would ultimately be self-managing.

THE CAPTIVE LINE: BAY 7

The first team-based cell to be developed, one of many that would follow, took place in a captive line and was located in Bay 7 of the works. This line of four presses was the first production line to be given the challenge and demonstrate that 'world class' levels of performance could be achieved at the site. The location of the cell was quite fortunate and the line of four presses stand next to an automated line within the shop. This setting allowed the 'new LDV' to be shown visibly to all employees within the factory and reinforced the view that, while automation was a necessary means of improving the business, it would be accompanied by self-managing teams. The key difference between these two cells is that the robots would 'move standard metal' while the teams would engage in problem solving and variety of production.

As a pilot stage of the implementation process, the captive line represented a 'gamble' that held some risks and uncertainty. What looked good and logical on paper is not always the way the plans translate into practice. The pilot stage would therefore determine whether the large press shop with decades of history could make the 'step up' in performance to take the business into the twenty-first century and sustain the improvement process.

The directors of the company selected Malcolm Taylor, a member of the senior management team, to implement the new way and to work with the team as coach and facilitator of the change process. Malcolm, with 15 years' worth of experience at the Washwood Heath site, knew well the problems of the 'old way' having worked within the system himself. He also knew the tools and techniques used by other leading press shops in the world after experience with a consultancy, and fact-finding visits had taken him to see how other pressing facilities had implemented and sustained change.

With a wide remit, a financial budget, and autonomy to shape the production system from a blank piece of paper, an extensive plan was developed to guide the 'roll-out' process. The plan was developed over a three-month period and included building in the flexibility needed to cope with the unforeseen issues that would inevitably arise as the programme was implemented.

The translation of the plan into a physical improvement process at the captive line began in June 1998. At this time, the presses were lifted from their press beds and completely refurbished by the press manufacturer.

This press refurbishment programme included an opportunity. This opportunity was used to request the upgrade of certain asset sub-systems and also the introduction of mechanisms that would make operations and maintenance easier to conduct. With the presses withdrawn from production, the business took an unprecedented decision – to use this time to recruit internally and train the new teams in the logic of 'world class' manufacturing both in the classroom and also through practical instruction in the press facility.

A team member profile was developed and this profile was communicated to the workforce along with the expectation of the line and volunteers were invited to apply for positions on the new line. The number of applicants per vacancy was very high. Each applicant was interviewed and a series of Belbin tests were undertaken with all candidates, to try to understand how compatible their personalities were with team working. The information gained from these exercises allowed the company to select the first of the new teams in the press shop.

The completion of the initial team integration exercises, together with the recommissioning of the presses, allowed the development of the new production system to begin. The production system was not an 'off the shelf' or simple copy of what had been seen elsewhere. It was a system that was based on a concrete logic that did not prescribe solutions but served to direct the implementation of the cell by the team. At the most basic level, the logic that underpins world-class levels of manufacturing performance and the LDV production system is simple and straightforward. Stated simply, the logic is

1. Improve *Safety* in the factory as the most rudimentary discipline in the factory.
2. Improve *Morale* in the factory in terms of the attendance of the shop floor and also the inputs of the team members to improve the production processes.
3. Next comes *Quality* in terms of the quality performance of the process area (inputs to outputs). This key step in the logic is related directly to improved productivity and for each percentage increase in quality performance there will be an incremental rise in productivity. By consequence, as quality and productivity rise,

the unit costs and costs of site operations will reduce. Costs are not the key driver in the logic system yet, though.

4. If quality is good then *Delivery* is the next stage of logic. This suggests that as quality rises and costs begin to fall, then manufacturing precise and quality assured products in smaller and smaller batches will improve the business. As a result, stock turns can increase and buffer inventories can be reduced.
5. Finally, there is *Cost* reduction. With a quality process and a high level of delivery reliability, the final stage is to look for ways of streamlining the operation and removing costs.

It seems slightly ironic that for most manufacturers in the UK, it is cost reduction that remains the focus and has always been a primary goal of the traditional 'metal moving' systems. The five simple steps in the LDV logic were used as 'gateways' whereby the most fundamental systems of safety and morale could be established and proven before the team could progress to higher levels of performance. It is perhaps this logic path that has been missed or ignored by the British managers that have visited Japan. These managers have witnessed the current stage of the evolution process (at the highest levels of productivity and quality) but all too often these managers have forgotten to track the history of events that has created these systems. Indeed, while cost reduction is the output of the logic (and an output at most stages of the evolution) it is the highest order goal that can only be pursued when the production system is reliable and robust. To remove costs (such as inventory) from an unstable system will ultimately ransom customer service and lead times. In addition, the logic of a mixed 'score card' prevents the bizarre decisions and behaviour associated with myopic cost reduction. The QCD logic promoted in Japan and adopted by most of the world-class press facilities may well be misleading and better stated as QDC.

The other benefits of the step-logic approach involved the inability to compromise the previous stage of the programme. To illustrate this point, cost reductions would not compromise the safety in the factory. This was not necessarily the case of the traditional systems of manufacturing in the United Kingdom whereby to increase output corners were often cut (such as removing press guards). The cutting of corners leads to two inevitable consequences – the first is the payment of penalties imposed by Health and Safety agencies and the second is the cost to the individual in terms of personal injury.

The production trial phase, scheduled as part of the master project plan, provided an opportunity to practise the new procedures and 'debug' the detailed issues from within the system. At this stage the teams engaged in tool change overs and collapsing the theoretical set-up times defined by the champion. Every shift of every day during this stage would set up and remove many different tools (70 in total) that would be used for full production. The constant repetition of the changeover process and the production of a small number of components increased the familiarity of the team with the tooling. Armed with an education of problem-solving skills and the newly instilled belief that the team was infinitely more powerful than the individual, the process was streamlined and new standard operating procedures were documented and revised within the team and between the shifts.

It was at this stage, prior to production, that the issue of the 'team leader' was suggested. In the past, levels of supervisory management were used but in the new system the role of the team leader was not defined. This was a deliberate measure taken by the senior management champion in the belief that the teams should take part in discussing and creating this role themselves rather than imposing a structure on the team. Many British companies continue to wrestle with this issue today but in the spirit of the 'new working way', LDV elected to involve the team. The inevitable questions raised by the team was the need for a leader, the role of the leader (working or supporting the line) and the remuneration of the leader for the increased level of responsibility. It was initially agreed to work without this position until the team were happy a 'natural leader' had emerged whom the team would be happy to elect democratically.

HOUSEKEEPING: BANISHING ABNORMALITY AND CREATING THE COMMON WAY

An activity that particularly benefited from team-based problem-solving skills and the passion to improve motivation and safety involved the development of visual standards for the environment of the presses. Traditionally, in the press industry, the concept of housekeeping was regarded as low priority or a response to a customer visit. The issue of housekeeping is, however, closely linked to the two foundation processes of safety and morale. A cluttered and dirty environment is both dangerous and depressing for workers. Using a standard approach, and resources provided by the champion, the team undertook many small projects to create a disciplined system of housekeeping and layout. Working from the presses backwards, the teams allocated spaces (that were clearly marked out and painted) for the presentation of tools next required for production. Then tracing the tooling route backwards, the main store was completely redesigned and located at the head of the line; in addition, bins and working tools were brought into the area and once again designated positions were clearly marked out for them. Absence of a part of the production system from its designated space therefore meant that the article was being used or that a procedure had not been followed. The latter action would prompt the team to find out why standards were not being adhered to. Indeed, no element of Bay 7 escaped scrutiny and change. In parallel, and to prevent slippage, audit sheets were created and a large 6ft by 4ft housekeeping board was erected. This board contained the standards and also more pictures of what the team decided was 'acceptable'. In addition, each press was equipped with a flip chart to record improvement ideas and problems whenever they occurred.

DEVELOPING THE MEASUREMENT SYSTEM: QCDSM

These ideas and the team's performance were reviewed each day. Also, the champion held 'one-to-one' meetings with the team members on a bi-weekly basis to ensure that all concerns were exchanged in an open and receptive environment. To ensure that the teams were supported during the shift, the champion would also 'walk the lines' during the shift in order to assess the production performance of the team and also to reinforce the 'new way of working'. This activity was not regarded as an intrusion but simply a means of keeping the communication channels between the team and the management open in the short term. As the system began to evolve and production levels grew, the teams developed a formal reporting system whereby KPIs would be displayed in the area and updated by the autonomous team. This information, based on the processes of quality, cost, delivery, safety and morale, formed the basis of the management reporting system whereby trends could be tracked (on a monthly rather than daily basis).

The reporting system also represented another stage in the development of 'autonomy' by the team. Traditionally, close supervision and constant interference characterized the way in which the teams were managed. Under the new system, management attention would be 'pulled' to the team through the interpretation of the performance indicators. With the measurement system came responsibility and self-management whereby actions taken by the team could be seen to have a demonstrable impact on the performance indicators. The ability to relate action to improvement was therefore an important activity and a process that was based on fact and not opinion.

NORMALITY AND ABNORMALITY: PRESS OPERATIONS

The next stage in the development of the Bay 7 team was to improve the process of production. With a proper and disciplined environment, problem-solving techniques and a formal reporting structure, the teams were equipped to influence the quality, cost and delivery processes. During this part of the team development, two major initiatives were undertaken. The first initiative concerned the development of a production 'system' – of which the presses were just one element. The second initiative involved the improvement of the production targets themselves.

Previously, the planning of production, for the presses, had been a remote activity that involved both production and material scheduling. The new system, which had no constraints for the team, implied that given a demonstrated level of responsibility, more and more decisions could be deployed to the team. The system within which the presses operated involved an inbound buffer of steel blanks and an outbound buffer of pressed panels. The scheduling of production, to meet forecast demand from the customer, was used to push metal through these elements in time for consumption in the old system. However, the production of large batches for all of the tools was not a possibility for the team. Instead low-volume and high-variety production was needed. The team, in parallel to all those specialist employees at the site, began to develop their own production system based on low finished goods inventory and working to replenishment of these buffers. This form of 'pull' system implied that customer demand would trigger the selection of press dies based on consumption and triggers in the inventory system. However, a high variety of pressed products (some requiring all four presses and some requiring only two) meant that the system would need to be robust and not cause chaos when demand was sent back to the captive line. With a number of meetings between the champion, teams, and material scheduling, it was decided that an effective system could be built and the team could draw their schedules from the central planning department in the short term. In the longer term, the team would undertake the responsibility themselves and be guided by the champion who would act as mentor to the team.

The performance improvement initiative was an integral part of the production control system. With this initiative the team sought ways, based on collected data, to improve the quality and delivery performance of the line without inflating the costs of operation. The team took many calculations, including how long it took to fill a stillage, the average time taken for the customer to empty the same stillage, the average time taken to change the press, and the number of stillages required to make a standard size batch of product. These simple measures, and many more, formed the basis of the production system and replaced the old way whereby metal just arrived for pressing.

During this stage, the Overall Equipment Effectiveness (OEE) measure was introduced (Nakajima 1988). This measure multiplied the percentage of time that the press was available and had work to do, with the performance of the press in terms of actual stroke rate achieved versus the theoretical speed that could be achieved and finally the right first time rate of production. This measure denotes the 'value adding' rate of the press and also highlights the weakest link of quality, performance or availability for improvement.

In order to influence this OEE figure, the teams would need to employ their technical understanding of the asset, its functionality and also to track these losses as well as the other quality, cost and delivery indicators. That is not to say that the problem-solving toolkit was redundant but in order to continue to improve further the toolkit would need to be supported by a greater understanding of the technology employed (presses and materials handling equipment). The OEE figure therefore took the team to a new stage of performance whereby their progress could be benchmarked against the generally accepted 85 per cent threshold of 'world class' performance. It also meant that the captive team could compare themselves with the fully automated line pitting human against robot. This provided some attraction for the team as the automated line had, since its introduction, dominated the factory 'tour' conducted for visitors. The comparison of OEE levels was therefore used, by the team, as a means of sustaining improvement if only to 'level up' to the figures achieved by the fully automated line.

A part of the initial training programme, designed by the senior management champion, was activities that related to the functionality of the technology at the captive line. These early lessons provided the foundation upon which the general engineering awareness and 'senses' of the team members could be developed to allow the team to take on the routine maintenance tasks in the factory. As it happens, the improvement of these visual and other senses proved invaluable during the ramp-up of production when one of the flip charts recorded an unusual vibration at the crown of the press. This problem, small and almost undetectable for an inexperienced or untrained operator, was followed up with the problem-solving team and the maintenance department. The source of the problem was found to be a 'wearing loose' of a major sub-asset. If the problem had been left, the press would have failed and production would have been lost for many days. The final element of the production system was therefore the development of routine and front-line asset care activities. These activities included asset cleaning, asset lubrication and inspection routines for the team and the retention of major and planned activities by the central maintenance department.

TAKING IT TO THE NEXT STAGE

With a reliable and effective manufacturing system that was stabilized and controlled, the next stage of improvement for the captive line lay outside of the Washwood Heath factory. To 'step up', the teams began to map and interrogate the material flow from suppliers to the head of the line. This investigation was used to understand the level of waste associated with poor material control and, just as the teams had learned that fluctuating demand caused a problem to their own work patterns, they sought to understand their own impact with suppliers. In order to remove waste and exploit the new system of working the teams visited and began to understand the problems associated with supply.

Steel and Alloy Processing Limited of West Bromwich, a long-term supplier to the Press Operations, was approached to solicit advice and to gain the involvement of the service centre in the next phase of improvements. Traditionally, Steel and Alloy Processing had made many improvement proposals to the Press Operations but with the traditional fortunes of the customer business there were limited opportunities to work jointly and for mutual benefit. Under the new system of controlled manufacturing on the captive line, the ingredients were right for supplier integration.

After negotiations and reciprocal visits by teams at both factories, it was decided to implement a fax-ban system whereby the traditional schedules of the materials planning department would be supplemented with a 'call off' for materials that would be processed within the next three shifts. In this manner, the system would also benefit the service centre and would allow products to be slit just prior to despatch. At LDV it meant that the team now controlled and took responsibility for the provision of steel, not only for their shift but the others that would follow it. This final element of the production system brought the end of the 'old way'. Metal did not simply arrive – it was requested by an 'empowered' team in the press shop who planned to consume in response to the finished parts withdrawn by the customer.

REFLECTIONS

From a standing start in 1988 when the company had struggled to survive, the teams of managers and operators in the press shop have managed to achieve outstanding results. While the programme was a pilot stage it was not an experiment. The project was planned and executed by the teams and their champion. The success of the area, which has itself created some arrogance in the team, has been a learning opportunity for the business and also provided the internal model through which the process was 'rolled' horizontally to other areas of the press shop. The proof of the success in Bay 7 has been achieved, not through short-term uplifts in performance and simply 'turning up the lights' but through a continuous and self-sustaining process of improvement. Each improvement has been woven into the production system for the team and written into standard documents to prevent slippage. These movements forward and their documentation have acted as a 'ratchet' to prevent slippage and to allow the team to build from each level of improvement made. In parallel, each improvement has also created the opportunity to deploy even greater levels of responsibility to the cell and the different shift teams.

Below is a summary of the teams improvement in key elements of performance in Bay 7:

Indicator	*% Improvement*
Set-up time	Reduced by 75%
Throughput	Increased by 40%
Utilization	Increased from 40% to 75%
Quality	Consistently less than 50 PPM

From a wider perspective, the company is a 'brown-field' site and a particularly large-scale business. The LDV experience and results prove that it is possible to make lasting changes within an existing factory and to make the 'turnaround'. Companies with a history, particularly those with a history of strained labour relations, tend to have an organizational inertia when it comes to change. This organizational inertia is resilient and deep-rooted even though most employees know that 'there must be a better way'. Decades of customary practice, poor internal factory relationships and 'seeing it all before' creates a potent blend of reasons why a business cannot change. In the modern environment, especially for manufacturers in the UK and the Midlands in particular, survival depends upon making these changes quickly. Businesses therefore face a clear choice: to survive they must improve productivity and manufacturing performance or lose out to the competition abroad. Being 'big' in the Midlands is no means of defending global market share unless 'big' refers to the improvements and numbers of improvements that the company is sustaining. This chapter, just one highlighted study from the company, may mean that, with continued improvements, Nagoya is not too far away after all.

ACKNOWLEDGEMENT

The authors would like to acknowledge those managers, employees and union at LDV Press Operations and Steel and Alloy Processing Limited whose combined improvement effort provides the content of this chapter.

NOTES

1 Nagoya is the location for Toyota Motor Corporation and its major suppliers.

REFERENCE

Nakajima, S. (1988) *Introduction To TPM.* Portland, Oregon: Productivity Press.

The lean press shop – a consideration of batch sizing, layout, and set-up procedures

15

Ann-Kristin Hahn and Jens Nießmann[1]

INTRODUCTION

Preliminary LEAP research identified the large-batch-oriented press shop scheduling procedures as a key decision point in the value stream, where the demand patterns of the vehicle manufacturers are significantly altered and poor batch-sizing approaches lead to unnecessary and wasteful inventory. Batch production, of course, is inevitable where multiple products need to be processed on one facility, while changeover times are significant and cannot be avoided. This is the situation typically found in press shops. This case study shows how the detrimental effects of batch production could be reduced by introducing lean operations at a first-tier component supplier as part of the LEAP project. The work has been focused on finding a proper batch-sizing technique, improving the press shop layout, and defining standard operations for the changeover procedures.

Pressing operations are done in a special manufacturing environment, which is characterized by the following features:

- The layout can be classified as job shop (or process) layout.
- Little time is needed to complete a press operation for a single piece (typically a few seconds).
- Set-ups that have to be carried out in order to change production from one part to another are quite time-consuming (typically 30 to 90 minutes). They require a multiple of the actual operation time on one item.
- When doing a changeover, the die, a rather big and heavy tool, has to be changed. Special equipment is needed to lift and transport the die.

PROBLEMS OF LARGE BATCHES

The quality and efficiency of the scheduling system has a direct and crucial impact on the inventory levels in a production system. Large batches inevitably cause excess inventory unless the entire batch quantity is immediately 'consumed' by the external or internal customer. Although having zero inventories is a non-achievable goal, it points in the right direction. Usually, carrying inventory is associated only with bound-up capital and the stocking costs. But inventory causes a wide range of far more detrimental effects. Large batches increase WIP inventory, extend production lead times and work against regular flow. Workstations become blocked for a long time, which diminishes the flexibility to reschedule and react to demand changes. Defect or scrap detection costs rise with larger batches. If transfer batches are large, it takes longer until a defect will be detected at the next operation. Furthermore, if an error is detected, very often the whole batch needs rework (or to be thrown away). With smaller batches costs of rework and scrap decrease.

Inventory always is a symbol of problems (1). Even worse, inventory is not only a symbol or outcome of problems, but is a source of waste itself and has detrimental effects on the whole production system. Monden (2) is often cited, calling 'inventory the root of all evil'. With good reason, Womack and Jones (3) describe unnecessary inventory as one of the seven 'deadly' wastes obstructing lean manufacturing.

Since scheduling affects workforce and facility utilization as well as lead times, inventory levels and even quality issues, it is a key factor for manufacturing productivity and competitiveness. Effective scheduling can reduce inventory, improve on-time delivery, shorten lead times and improve the utilization of critical resources. Moreover, scheduling not only affects the performance of individual companies but has spreading effects throughout their supply and distribution chains.

IMPROVING PRESS SHOP SCHEDULING

Erratic schedules were identified during the mapping exercises as being a major source of instability. Given that demand instability and press reliability are already being addressed, a remaining problem in participating companies is simply that schedulers are unsure of how to construct a good schedule, or even what a 'good schedule' means other than ability to meet customer demands. We will assume that a good schedule means meeting delivery requirements with minimum inventory. Inventory has to be seen as both cause and effect of productivity problems at the same time. Therefore, inventory reduction is a primary concern in lean manufacturing.

Unlike most of the existing models do, batch-sizing decisions should not be ruled by trade-offs between set-up costs and costs of carrying inventory. The reason why such 'cost-minimizing' models should be avoided is that the involved types of costs are hard to quantify in practice. Even worse, costs of carrying inventory are likely to be underestimated, which leads to bigger batches resulting in increased inventory, longer lead times, increased rework and scrap. So far, no research has been undertaken to develop a reliable procedure for evaluating the real costs caused by inventory. If at all, statements about these costs are given in vague percentage ranges, which are expected to start well above 50 per cent (4). Nevertheless, the usage of EOQ-type (Economic Order Quantity) models is still widespread and was found to rule the MRP settings at the studied LEAP companies. As a representative for this class of models the basic EOQ/EBQ (Economic Order Quantity/Economic Batch Quantity) model will be briefly discussed here.

Using EOQ/EBQ models is the classic approach to determine order quantities and set order intervals (replenishment cycle) for a single product. Almost countless modifications have been developed in order to adapt the model to various real-world conditions. The approach has two major advantages. One is its simplicity. The other is the flatness of the function in the EOQ 'zone', which is the area around the curve's minimum, where the total costs are rather insensitive to deviation from the EOQ.

Criticisms deal with the model's weakness concerning its implied assumptions and more recently address the question about its underlying rationale in the light of JIT and lean manufacturing. Some of the assumptions, applied to keep the model simple, strongly limit its applicability in practice (e.g. a constant and known demand or the constant cost per unit without quantity discounts). Most of the EOQ model's modifications have been developed to overcome these limitations.

Other questions surround the assumptions concerning the involved types of costs. In the first place, the costs of carrying inventory should be scrutinized. Points of criticism are the assumed linearity, the validity of the used interest rate and the fact that the costs of inventory are purely related to the items' value. Storage space is assumed to be unlimited and cost free. Therefore, the more bulky and inexpensive the considered items are, the more risky it is to use the EOQ model [JOSHI]. As a consequence, it is very likely that users of the EOQ/EBQ formula take inappropriate, underestimated costs of inventory into account. Increasing the holding costs *ceteris paribus* has three effects: First, it increases the total costs of any order or batch quantity. Second, it will shift the optimum (minimal costs) of the total cost curve to the left. In other words, with the revised inventory costs the EOQ will decrease. Moreover, this effect is amplified, when set-up time reduction results in smaller set-up costs. Thus, one should be aware that whatever the outcome of an EOQ/EBQ calculation is, the real optimum can be expected to be a significantly smaller number. The third effect of steepening the carrying costs curve is that the EOQ 'zone' shrinks. As a consequence the model's advantage of insensitivity diminishes (Figure 1).

The JIT and lean thinking philosophy has raised the question whether an optimal batch size could or should be calculated at all. The ultimate target is to achieve one-piece flow, batch sizes of one. Thus, batches should always be as small as possible. Set-up costs should be shifted into fixed costs and should not be part of any batch calculation at all (5). Put another way, 'the amortization of setup time is irrelevant when attempting to achieve a competitive advantage in customer responsiveness and inventory position' (6).

Inventory should be reduced not only because it ties up the company's capital, but because of its detrimental effects on quality and lead times. These effects are hardly measurable in terms of money and thus cannot be integrated into any kind of EOQ model.

However, the vast number of modifications that have been elaborated in numerous publications suggests that the EOQ approach is widely accepted and applied despite all criticisms. The history of the model and the enormous attention that is paid to it show two things: First, people rather prefer working with a simple model than using more sophisticated methods. Although methods from operations research, such as linear, non-linear or integer programming, or even AI approaches might – at least in theory – deliver better results, users prefer simpler models and tend to deny any savings predicted by more complicated optimization models (7). Second, designers and users of any kind of decision-supporting models or systems should deeply examine and understand the models' prerequisites and limits. All assumptions should be scrutinized because models can never reflect all aspects of real-world conditions.

Besides the problem of quantifying costs, the existing batch-sizing models share mostly one or more of the following drawbacks. The most severe is the failure of models such as the EOQ/EBQ, POQ, etc. to account for any interaction that exists among the individual items to be scheduled for production (i.e., they neglect that

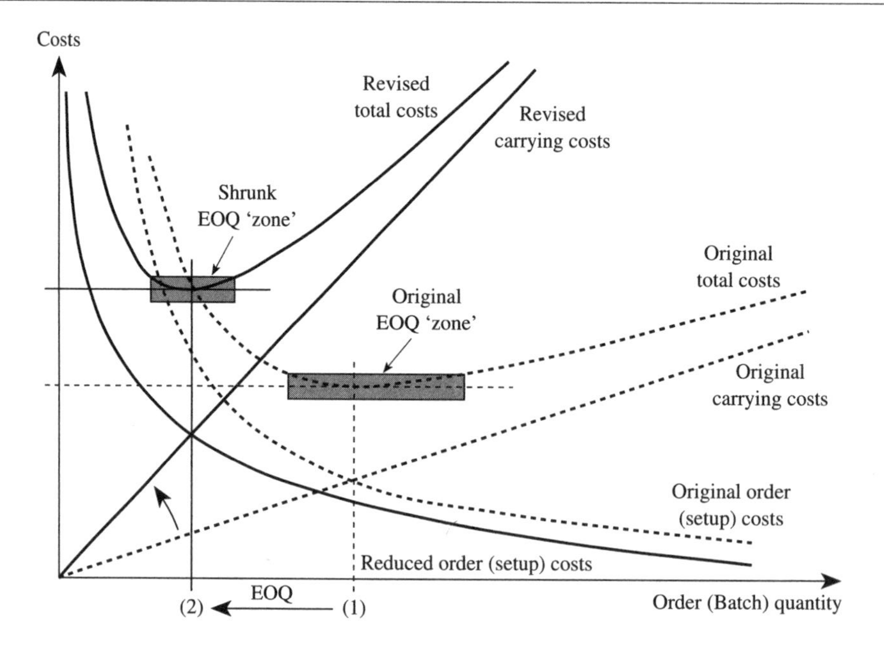

Figure 1 EOQ model including 'revised costs'.

products compete for limited capacity). This can result in unfeasible schedules. Models like the Hahn, Bragg and Shin decision model (8) or the Common Cycle (CC) model (9, 10) do account for constrained capacity but cannot calculate individual solutions for each type of product. Consequently, the opportunity of inventory reduction by running the high-volume items (runners and repeaters) more often than the low volume (strangers) is missed.

Bestwick and Lockyer (11) propose a heuristic approach called *coverage analysis* to solve the constrained order cycle problem. Items are classified according to their annual value of demand before individual replenishment numbers for each class are calculated. While holding the total number of replenishments constant, low-usage items are ordered less frequently and high-usage items more frequently. The applied algorithm has been based on the same old EOQ idea of finding minimal costs by trading order costs against costs of carrying inventory. However, two simplifying assumptions have been made, which allow the model to find a cost-optimal solution. Firstly, it is assumed that total replenishments are limited to a number that is below the figure that would result from summing up the 'optimal' numbers suggested by EOQ calculations for each type of item. Secondly, to solve the underlying equations, the simplifying assumption is made that the same percentage i (expressing the costs of inventory in the EOQ approach) and the same order or set-up costs do apply for all items and that these variables could therefore be expressed through a constant of proportionality.

Although no evidence or further reference is provided, Bestwick and Lockyer claim a typical saving of 25 per cent on investment in stock by using coverage analysis. The model might be an useful tool for inventory reduction, provided that set-up costs are uniform and costs of carrying inventory are constant for all products.

However, the assumption of uniform order costs could be suitable for purchasing decisions, but does fail to consider product-individual times and costs for setting up production facilities.

By scanning the vast body of literature on batch-sizing techniques, coverage analysis has been the only model found trying to minimize inventory by determining product-individual replenishment intervals in consideration of a constrained number of replenishments. Another important feature is that neither the costs of carrying inventory nor order respectively replenishments costs need to be known as long as they are constant for all items.

To sum up, no model was found to be capable of finding an optimized batch-sizing solution in terms of inventory reduction while considering individual set-up times and taking into account that capacity is a set-up frequency constraining factor.

In line with 'lean thinking', the aim of proper scheduling is also to make the schedule as regular as possible. The starting point has been to do a Pareto analysis of parts moving through a press shop. Typically, the resulting distributions have extremely long tails. The scheduling problem is therefore to run the repeaters as frequently and regularly as possible and the 'strangers' less frequently. It should at least be possible to improve the schedule performance significantly, if not to optimize.

Consider Figure 2: This is representative of a more 'advanced' press shop schedule in as far as regularity has been achieved. The left-hand diagram shows complete regularity, which is the objective of some schedulers. The right-hand diagram maintains the same number of changeovers but reduces inventory and improves regularity, and usually delivers better performance of the big runners at the expense of some minor components. The example shows that batch-sizing and sequencing decisions affect both inventory levels and operations regularity. Special attention has to be paid to the situation of constrained capacity where only a limited amount of time is available for set-up actions. Under such circumstances batch-sizing decisions have to put primary emphasis either on inventory reduction or on regularity.

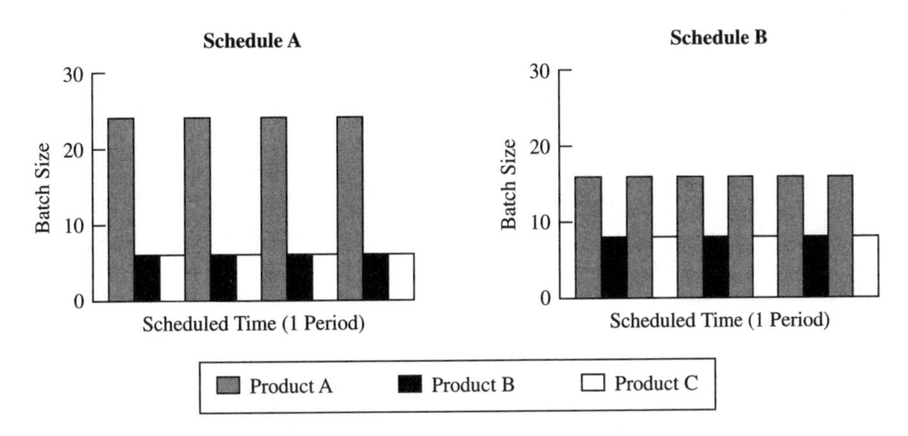

Figure 2 Lean scheduling with the same number of set-ups.

BATCH SIZING AND SCHEDULING: CASE STUDY

As at many other companies, the press shop scheduling at the case firm is based on manual spreadsheet calculations (12). The production quantities are deducted from the FAS issued by the assembly production controllers. The FAS includes the demand of press parts going into assembly as well as 'pure' press parts. All the press shop scheduling is done by one of the shift leaders, who repeats this task once every two weeks. Batch sizing and sequencing are simultaneously combined into a more or less 'trial and error' approach. The basic idea is to have batch sizes equivalent to six days' demand. This '6-day-policy' is 'home-grown' and understood to equal one week's demand plus one day safety stock. The scheduler tries to apply this policy for all the parts, regardless of their demand volume or the costs per unit. Due to the time-consuming set-ups and the constrained capacity it is not possible to run every part once a week. Therefore, the sixth day in the batch sizes is really more of an inevitable necessity rather than a 'safety' consideration.

When creating the press production plan, the scheduler follows the list of parts in his or her spreadsheets from top to bottom and consecutively schedules the start times for each job by marking the run times (including set-up times) on the time scale. The list does not include any form of ranking, because it is composed arbitrarily without any priority rule. On the spreadsheet, each weekday is represented by 24 columns dividing the day into its hours. As a result the schedule looks like a stairway descending from top-left to bottom-right where each step marks one press run. When all parts have been scheduled once, the remaining time is usually not sufficient to schedule a second press run for each part (theoretically, 6 out of the 10 working days in the scheduling period have been planned). Consequently, a decision has to be made regarding which parts to run twice and which only once during the two-week horizon. The selection of the 'one run' parts is mostly based on expected raw material availability and on the current press part inventory. Finishing the schedule is then a simultaneous activity of rescheduling the 'one run' parts and spreading the second press runs of the other parts over the second week. This fortnightly scheduling rhythm, in contrast to the scheduling approaches found at other press shops, does not use rolling schedules. Instead in the broad sense a fixed period requirements model (FPR) is applied.

In practice it became obvious that the planning for the second week expresses a vague intention rather than a reliable production schedule since daily rescheduling occurs towards the end of the two-week period. There are two main reasons forcing the rescheduling. The first is simply the lack of raw materials (i.e. blanks) due to late delivery. The second reason is missing stillages. Most of the press parts have to be stocked in designated stillages or containers. If not enough stillages have been returned from the vehicle manufacturers, the production has to stop. The more often this happens, the more useless the generated schedules become, as production turns into a matter of what can be done and what has to be done most urgently to recover promised delivery times.

The length of the scheduling period, two weeks or 10 working days, has been set arbitrarily and is supposed to bring visibility to the press shop production and reduce the scheduling efforts – but in practice just the opposite has been achieved.

The applied scheduling approach generally fails to consider or achieve:

Regularity: The replenishment interval of each part varies by chance, depending whether two or only one press run is scheduled.

Demand visibility: Due to the absence of regularity and the arbitrary sequencing and rescheduling in the second week of the schedule, the exact times of raw material requirements are unpredictable.

Reliability: The scheduling of the second week is more or less a wasted effort since rescheduling is almost taken for granted.

Inventory reduction: The batch sizing is done by a fixed batch size policy, totally regardless of demand volume and product costs.

The inefficient performance of the scheduling system could be seen as both cause and effect of the production problems. As will be described in the following section, the stillage problem and the short or late deliveries from the steel suppliers are a recognized phenomenon at many first tier companies. At the case company these issues have very detrimental impacts on production scheduling and execution. Other LEAP research (see Chapters 9 and 10 and reference (13)) has shown how irregular and unstable demand patterns affect and aggravate the delivery performance of upstream suppliers. Therefore, the question had to be raised whether the poor press shop scheduling could be blamed on external factors or was it in itself contributing to a vicious circle – impeding regular and on-time steel delivery and stillage return.

A PROPOSED NEW APPROACH TO PRESS SHOP SCHEDULING

Even operating at the status quo (i.e. without any set-up time reduction or capacity extension) a more 'advanced' press shop scheduling could offer potential savings through inventory reduction: analysis of scheduling procedures showed that the batch-sizing technique – perhaps better referred to as batch size policy – was certainly not optimal and therefore caused wasteful inventory. What was found could be called 'the waste of equal treatment', an ignorance of the fact that items have different demand volumes and different values. As already stated, items with high yearly demand (£-usage) should be run more frequently at the expense of some minor-demand parts.

Appropriate batch sizing should take account of the fact that parts differ in demand volumes and in unit costs. By doing so, it should reduce, if possible minimize, the overall inventory level, thereby achieving cost savings and taking the press shop one step further towards lean manufacturing.

The expected cost savings will be twofold. On the one hand, capital will be freed up at once. On the other hand, the costs of carrying inventory will decrease, in terms of storage space costs and material handling costs. For press parts it is reasonable to say that these factors also correlate with the value of the stocked parts, since the value is mostly a function of the raw material cost, which in turn indicates the weight and size of the parts. Hence, the main objective of appropriate batch sizing is the minimization of the total value of stocked parts. The objective *is not* to minimize the total of set-up costs and costs of carrying inventory. This would call for an EOQ-type trade-off of these costs, requiring that the costs coefficients are exactly determined, which is not the case in practice.

The 'Six-days batch sizing' is illustrated in Table 1. The data is taken from the B-Line, which comprises five large presses between 350 tonnes and 500 tonnes (see

Table 1 Inventory levels according to 6-day batch sizing policy

Internal Part-No.	*Demand Value* [£]*	*Set-up Time [minutes]*	*No. of Set-ups**	*Total Set-up Time [minutes]**	*Batch Size [days covered]*	*Value of average Inventory [£]*
wo4511/1	105600	120	6,67	800	6,0	7920,00
wo4501-1/2	103200	60	6,67	400	6,0	7740,00
wo4515-16	43200	120	6,67	800	6,0	3240,00
wo3585/1	42000	60	6,67	400	6,0	3150,00
wo4509-10	37920	60	6,67	400	6,0	2844,00
wo3612	33440	60	6,67	400	6,0	2508,00
wo3643/1	30600	60	6,67	400	6,0	2295,00
wo3634/1	28560	60	6,67	400	6,0	2142,00
wo3613	28160	60	6,67	400	6,0	2112,00
wo3566-7	12360	60	6,67	400	6,0	927,00
wo3568-69	10680	60	6,67	400	6,0	801,00
				5200		Total: £35,679,00

* per 40-day period

Figure 3). As the production process for most of the parts pressed on the B-Line is a flow through the presses B1 to B4, these are scheduled as a single machine, whereas the B5 press is scheduled separately. This case study ignores the fifth press.

Table 1 also shows the resulting inventory level according to the company's 6-day policy (Demand values ('£-Usage'): = Cost per Unit × Demand per Period). To optimize the batch sizes a new algorithm has been developed and applied, the results being shown in Table 2.

The approach obviates the necessity of knowing the usual cost coefficients for set-up costs and costs of carrying inventory, which are very difficult to determined in practice. The batch sizes are calculated for 40-day periods (i.e. 8 weeks) since this has

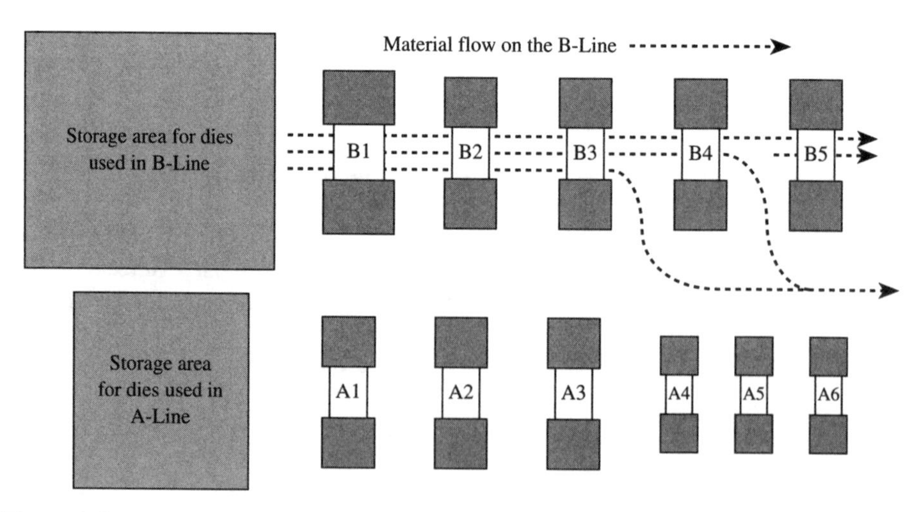

Figure 3 Large press bay layout.

Table 2 Reduced inventory levels according to new batch-sizing algorithm

		Theoretical Optimum:		*No. of Set-ups rounded to integers:*		*'Full week' replenishment cycles:*	
Internal Part No.	*Demand Value [£]*	*No. of Set-ups*	*Average Value of Inventory*	*No. of Set-ups*	*Average Value of Inventory*	*No. of Set-ups*	*Average Value of Inventory*
wo4511/1	105600,00	8,4	6319,57	8	6600,00	8	6600,00
wo4501-1/2	103200,00	11,7	4417,54	12	4300,00	8	6450,00
wo4515-16	43200,00	5,3	4042,01	5	4320,00	4	5400,00
wo3585/1	42000,00	7,5	2818,16	8	2625,00	8	2625,00
wo4509-10	37920,00	7,1	2677,78	7	2708,57	8	2370,00
wo3612	33440,00	6,6	2514,63	7	2388,57	8	2090,00
wo3643/1	30600,00	6,4	2405,48	6	2550,00	8	1912,50
wo3634/1	28560,00	6,1	2323,91	6	2380,00	8	1785,00
wo3613	28160,00	6,1	2307,58	6	2346,67	8	1760,00
wo3566-7	12360,00	4,0	1528,80	4	1545,00	4	1545,00
wo3568-69	10680,00	3,8	1421,11	4	1335,00	4	1335,00
Totals:			£32,776,56		£33,098,81		£33,872,50
Total Set-up Times:		5200		5160		5280	
Potential Savings:			8,13%		7,23%		5,06%

been found to be the time-window in which the vehicle manufacturers forecast their demand with fairly high reliability and stability. The 40-day batch-sizing horizon also coincides with the length of stocking time thought to be critical for unpainted press parts, after which these could become rusty. Hence, no parts should be produced in batches covering more than two, or at most three months of demand.

Table 2 indicates possible savings of about 8 per cent in inventory. However, a LEAP study comprising four independent press shops, all at first tier companies, revealed that on average inventory reductions of 10 to 20 per cent can be achieved by using the newly developed batch-sizing algorithm (14). In practice, the potential savings may be even higher, as possible impacts on the safety stock were not considered in the above study.

STANDARD OPERATIONS FOR CHANGEOVERS

By defining (and applying) clear standard procedures the time needed for one changeover can be reduced significantly. The most important point is to convert as much of the total work as possible into external (i.e. off-line) activities. These external activities (for example, all those activities that are necessary to prepare the changeover procedure) can be separated from the 'core' (i.e. on-line) changeover activities, thereby decreasing the actual machine downtime. In the case study, the changeover that was observed during the first visit to the press shop took approximately 25 minutes for the changeover of just one press. The applied changeover procedure involved all operators working at the respective press line and the forklift driver. Lacking a standard procedure, some of the workers stood idle for a considerable amount of time during the changeover.

The typical changeover process is as follows. The new dies are taken out of the storage area by an overhead crane, then transported to the aisle between the two press lines. Transport to and from the respective press is done by forklift. Small conveyors linking the presses have to be removed when carrying out the changeover in order to make the presses accessible for the forklift. The entire procedure is sketched in Figure 4.

PROPOSED IMPROVED CHANGEOVER PROCESS

Using a changeover field located in front of the storage area to prepare the changeover and to convert formerly internal set-up activities into external activities can achieve a decrease in set-up times. Another benefit of the changeover field is that it contributes to visibility by showing that a changeover is to be done soon when dies are moved to the field. Additionally, the intermediate stop at the changeover field ensures that the dies are always presented in the right order and in the right direction. The changeover field should ideally provide enough space for all dies to be taken to the presses as well as for those to be removed from the presses.

Figures 5 and 6 show how the changeover procedure should be organized. This procedure can be applied independently from the number of dies that have to be changed at the same time. It only is only necessary to ensure that the changing field is

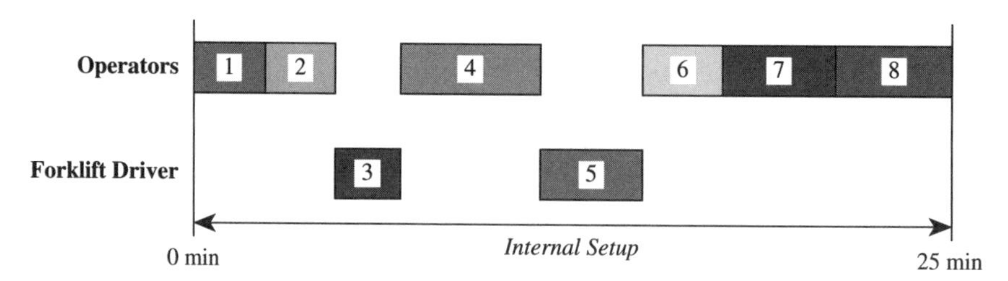

Figure 4 Current changeover procedure.

1. Remover convenor
2. Loosen die in press
3. Take die out of Press
4. Fix chains of crane at new die in store and transfer die out of store
5. Bring new die by forklift to press
6. Fix new die in press
7. Put conveyor back into place
8. Bring old die by crane to store

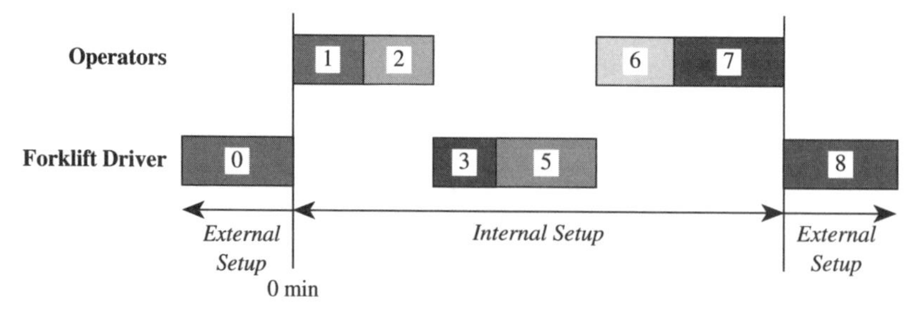

Figure 5 Improved changeover procedure.

Numbers have the same meaning as in Figure 4, additionally 0. Arrange all dies to be inserted next on the changing field.

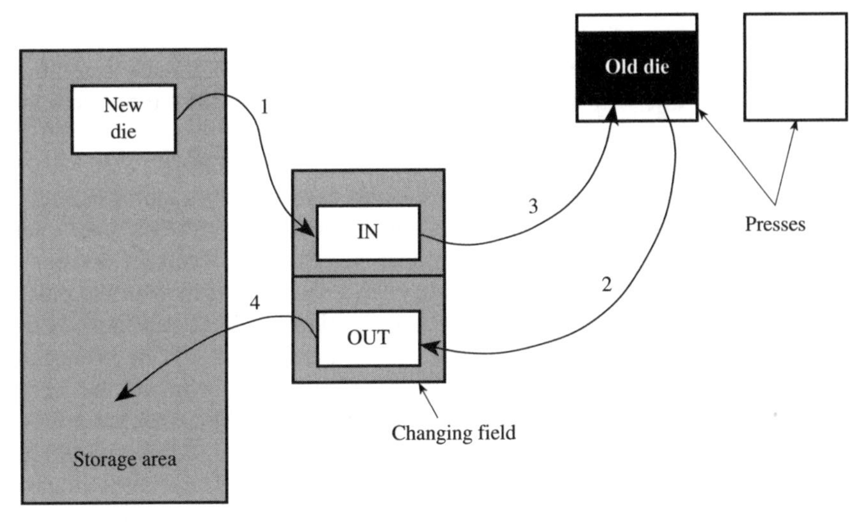

Figure 6 Improved changeover procedure.

big enough to take all dies that are part of the changing procedure. In the 'worst case' at the company, when dies 4515/16 are changed against dies 4509/10 or 3556/57, a field of the size 7.60m × 3.60m is required.

The changeover procedure could be further refined in terms of transport distances during the internal set-up if each die had been arranged next to the press which was going to use it. However, due to space shortage at the aisle between the two press lines, this was not possible.

LAYOUT CONSIDERATIONS

The layout of the press shop area deals on the one hand with the arrangement of the presses themselves and on the other hand with the layout of the die storage area. The latter can be easily achieved in the short term and can have a big impact on set-up reduction. According to the press shop group leader, the following issues have to be considered when rethinking the layout of die storage area:

- All dies must be easily accessible because the operators have to fix the chains of the overhead crane to them to make them ready for transport.
- For safety reasons not more than two dies can be stacked. As the height of one die is approximately 90 centimetres, it would be too dangerous for the operator to climb onto a tower of three dies to fix the chains of the crane. Additionally, high towers destroy visibility.
- Not all of the area can be used for storage purposes. First, the toolmaker and his workstation located in the left corner of the hall cannot be moved at the present time. Second, the crane cannot cover the whole area: a distance of 1.50 metres to the walls has to be taken into consideration.
- There can be four or five dies belonging to one set of dies used to produce an item. Dies that are used together should be stored together.

- The arrangement has to be fixed, so that each die has a proper place where it belongs. Furthermore, the arrangement must be displayed visibly to eliminate searching.
- An aisle of 1.50 metres width has to be left free to fulfil safety requirements.

In order to facilitate the new layout there are a number of requirements:

1. Data for all dies that are used in the large bay press shop (A-line *and* B-line since the storage area keeps all dies) has to be collected. This data should include heights, widths and lengths in order to prepare the decision about the arrangement in the die storage area. The physical data have to be combined with the data that resulted from applying the new scheduling algorithm. The scheduling algorithm can be used to show the frequency with which products should be produced. In order to reduce the total time required to access the dies, they will be arranged according to their respective frequency of usage, i.e. those dies that are used most frequently should be located in the front of the storage area. When collecting the physical data for the dies, it was found that more dies were kept in the storage area than were actually used. In aiming for a really lean store, dies that were no longer used should have been removed to make the whole area more tidy and orderly but this was considered to be impossible since there was no other storage location available.
2. The next step is to work out the arrangement of the dies in the area. This is done by placing scaled models of all dies found in the area on a scaled field representing the area that can be used for storage purposes. A plan of this was drawn, which is shown in Figure 7 with a list of the dies in Table 3.

Table 3 Numbers representing dies in Figure 7

1	3612,13 Op.30	**21**	3634 Op.20	**41**	3589	**61**	3643
2	3612,13 Op.20	**22**	3669 with one smaller die on top	**42**	3589	**62**	3643
3	3612,13 Op.40	**23**	3557,58 Op.20	**43**	3589	**63**	3585 Op.40
4	3612,13 Op.10	**24**	3557,58 Op.30	**44**	3589	**64**	3585 Op.10
5	3612,13 Op.50	**25**	3557,58,59,60 Op.10	**45**	3589	**65**	4501,02 Op.30
6	3670 Op.40	**26**	3687	**46**	3568 Op.30	**66**	4501,02 Op.20
7	3670 Op.30	**27**	3498	**47**	3568	**67**	4501,02 Op.10
8	3670 Op.20	**28**	3605	**48**	4515,16 Op.30	**68**	4501,02 OP.40
9	3670 Op.10	**29**	4071,72	**49**	4515,16 Op.20	**69**	4509,10 Op.20
10	4511 Op.50	**30**	3681	**50**	4515,16 Op.40	**70**	4509,10 Op.40
11	4511 Op.60	**31**	4233,34	**51**	3643	**71**	3566,67 Op.40
12	3634 Op.40	**32**	3562,62 Op.10	**52**	3643	**72**	3566,67 Op.20
13	3634 Op.10	**33**	3571,71 Op.10	**53**	3585 Op.30	**73**	3566,67 Op.50
14	4503 Op.20	**34**	3559,60 Op.20	**54**	3585 Op.20	**74**	4512 set of 4 dies
15	4503 Op.10	**35**	3559,60 Op.30	**55**	3568,69 Op.10	**75**	3684
16	4511 Op.40	**36**	4515,16 Op.50	**56**	3568,69	**76**	4509,10 Op.30
17	4511 Op.20	**37**	4515,16 Op.10	**57**	3568,69	**77**	4509,10 Op.10
18	4511 Op.30	**38**	3679	**58**	3568,69		
19	4511 Op.10	**39**	3498	**59**	3566,67 Op.30		
20	3634 Op.30	**40**	3667,68 set of 4 dies	**60**	3566,67 Op.10		

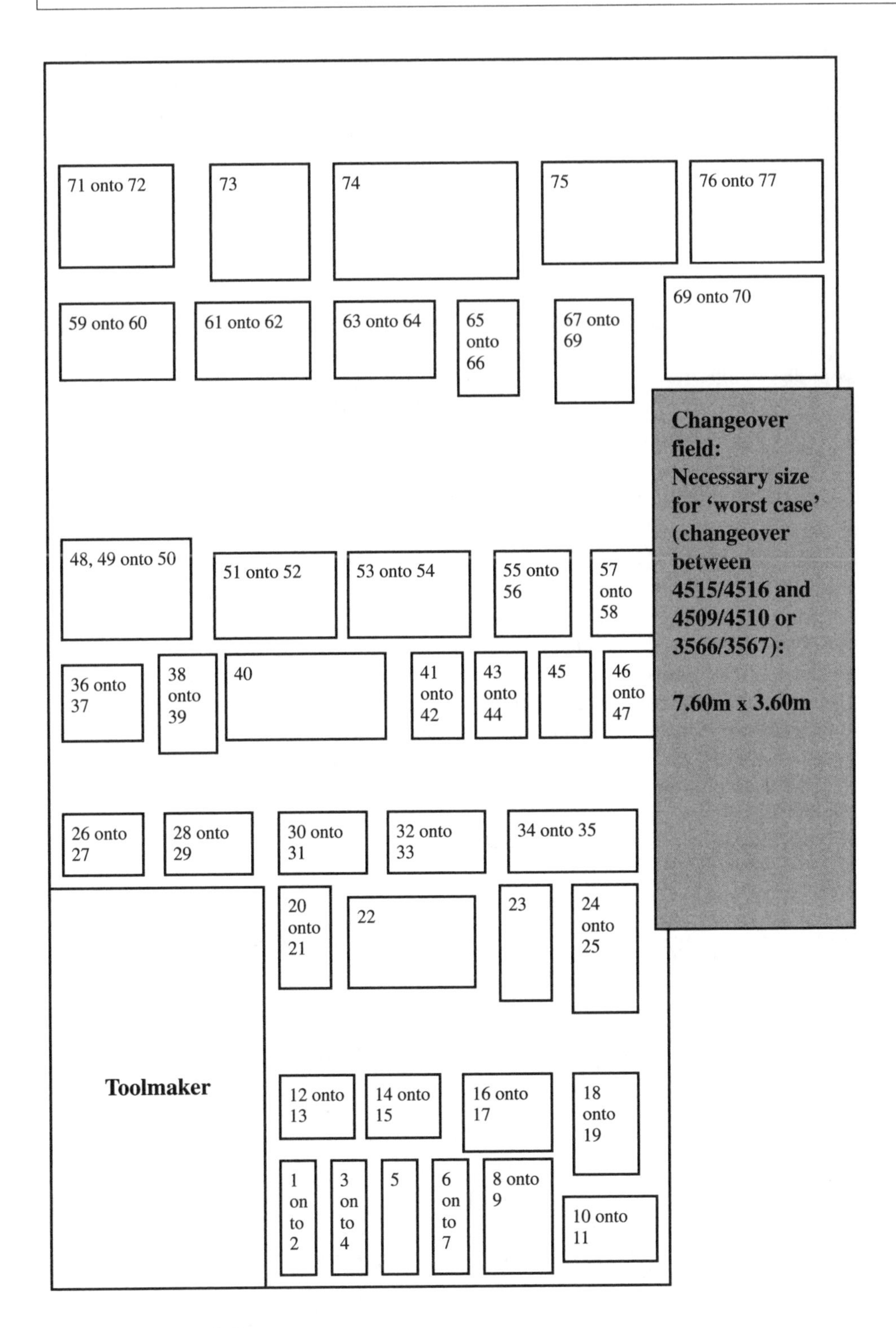

Figure 7 Improved die arrangement.

3. Having worked out the arrangement the area has to be prepared for the move, which includes activities such as cleaning the area. After cleaning, the floor should be repainted, ideally with colour-coded fields indicating which dies are to be kept in which place. Then the dies can be moved to their new location.
4. Finally, the new arrangement has to be displayed on a board so that the planned location of the dies is easily visible. Visibility is an important factor when trying to reduce set-up times.

After completing the refurbishment of the area, the changeover should be practised several times.

Redesigning the layout for off-line changeovers is a relatively easy, short-term improvement. A more fundamental change in relation to achieving the Lean vision would be to alter the layout of the presses themselves. Several parts flow through the B-Line, starting at press 1, going over presses 2 and 3 to press 4. More desirable than a straight-line layout would be arranging the presses of the B-Line into a U-shaped cell (15) shown in Figure 8.

This layout would open up the following advantages:

- **One-piece flow:** The production can flow piece by piece according to the pull of internal and external customers. The need for material handling is reduced because operators walking from press to press transport parts. No automated material handling devices are used any more.
- **Set-ups:** Set-ups are facilitated and can be conducted within a shorter time period. The presses are now much easier to access for the forklift; it is no longer necessary to remove the small conveyors that currently link the presses.
- **Operators:** Not every press has to be staffed with a worker. By synchronizing each process and conducting one-piece flow production fewer people could achieve the same production volume. Operators know the entire process and handle every press.
- The U-shaped layout improves **visibility**.

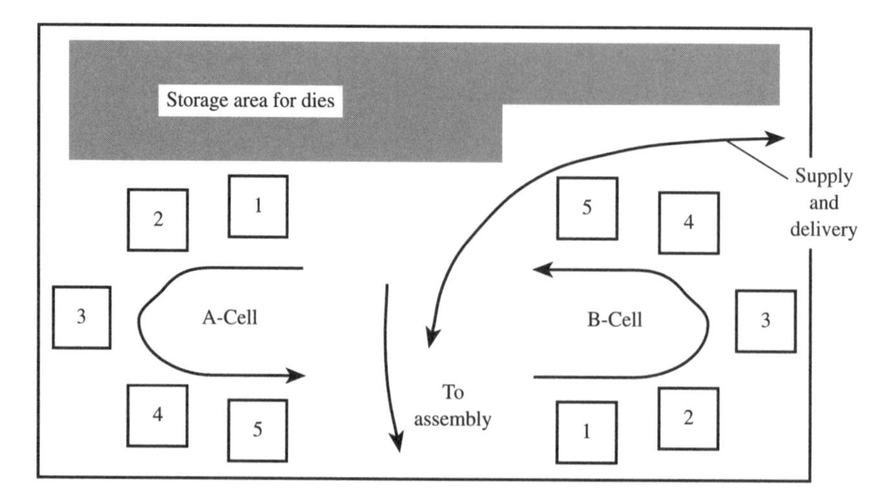

Figure 8 Long-term goal for the press shop.

- Operators work together more closely and can help each other, if any problem occurs. **Communication** is fostered.

POTENTIAL SAVINGS

If set-up times could be decreased by introducing a standard procedure for set-ups, the inventory could be reduced, as shorter set-up times allow more frequent set-ups while still dedicating the same amount of time for production. The relation between changeover time and the average value of inventory is shown in Figure 9.

SUMMARY

In summary, research within the LEAP programme identified large-batch-oriented press shop scheduling procedures as a key decision point in the value stream and an area where demand can be distorted internally within a firm. While batch processing is necessary, given the current assets used by the first-tier component suppliers, the research presented in this paper shows that a number of measures can be taken to limit the size of batches and manage changeover times.

In the research presented in this chapter, the aim was to reduce the inventory levels of press parts in order to free up the capital tied up in stocked parts and to reduce annual costs for carrying inventory. Although many companies still use a fixed interest rate of 25 per cent to calculate carrying costs, considering all carrying costs, the rate is surely at least 50 per cent, perhaps as high as 100 per cent. It is further estimated that typically more than 50 per cent of expenditures in manufacturing are inventory costs (13, 16, 17). In addition, the removal of inventory should be reduced because of its detrimental effects on quality and lead times.

In order to reduce inventory levels it was necessary to implement a combination of lean tools and techniques – standard operations to reduce changeover times and layout changes to improve flow. In addition, a new scheduling algorithm has been developed that helps companies optimize batch sizes and hence reduce inventory levels.

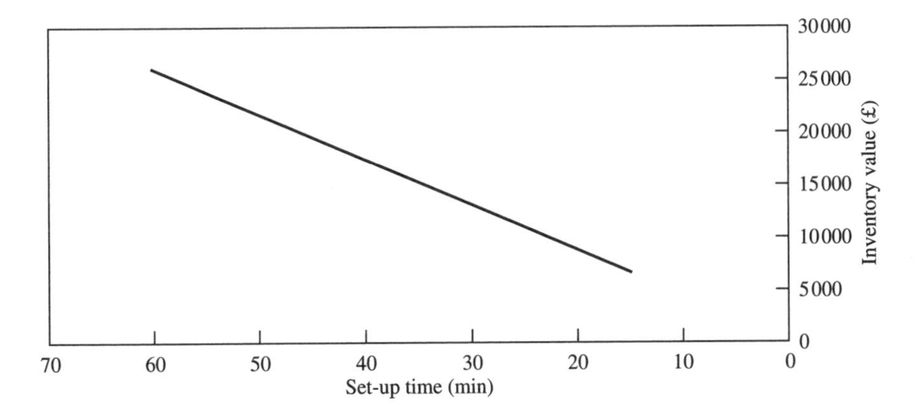

Figure 9 Inventory value in relation to set-up times.

NOTES

1 The authors were seconded to the LEAP project during 1999 and this chapter is based on work presented in their MSc theses.

REFERENCES

(1) Hall, R. W. (1983) *Zero inventories*. Homewood, Illinois: Dow Jones-Irwin.
(2) Monden, Y. (1998) *Toyota Production System*, 3rd edition, Engineering and Management Press.
(3) Womack, J. P. and Jones, D. T. (1996) *Lean Thinking: Banish waste and create wealth in your company.* New York: Simon & Schuster.
(4) Schonberger, R. J. and Knod, E. M. (1997) *Operations Management: Customer-focused principles*, 6th edn. Chicago: Irwin.
(5) Hall, R. (1983) *Zero Inventories*, Homewood, Illinois: Dow Jones-Irwin.
(6) Mahoney, R. M. (1997) *High-Mix, Low-Volume Manufacturing.* New Jersey: Prentice-Hall, Inc., p. 118.
(7) Schonberger, R. J. and Knod, E. M. (1997) *Operations Management: Customer-focused principles*, 6th edn. Chicago: Irwin.
(8) Hahn, C.K., Bragg, D. J., and Shin, D. (1988) Impact of the setup variable on capacity and inventory decisions. *Academy of Management Review*, 13(1), 91–103.
(9) Sugimori, Y., Kusunoki, K., Cho, F., and Uchikawa, S. (1977) Toyota Production System and Kanban System – materialization of just-in-time and respect-for-human system. *International Journal of Production Research*, 151, 553–64.
(10) Pinto, P. and Mabert, V. (1986) A joint lot-sizing rule for a fixed labor cost situation. *Decision Sciences* 17, 139–50.
(11) Bestwick, P. F. and Lockyer, K. (1982) *Quantitative Production Management*. London: Pitman Books Ltd.
(12) Nieβmann, J. (1999) Press shop scheduling – a new approach to repetitive batch sizing with constrained capacity. MSc Dissertation, University of Buckingham.
(13) Holweg, M. (1998) Supply chain scheduling and the automotive steel supply chain. MSc Dissertation, University of Buckingham.
(14) Nieβmann, J. (1999) Press shop scheduling – a new approach to repetitive batch sizing with constrained capacity. MSc Dissertation, University of Buckingham.
(15) Hahn, A.-K. (1999) Layout design for lean manufacturing. MSc Dissertation, University of Buckingham.
(16) Bicheno, J. (2000) *The Lean Toolbox*, 2nd edn. Buckingham: Picsie Books.
(17) Schonberger, R. J. and Knod, E. M. (1997) *Operations Management: Customer-focused principles*, 6th edn. Chicago: Irwin.

Section 5

Achieving Organizational Change

Parallel Incremental Transformation Strategy: an approach to the development of lean supply chains[1]

16

David Taylor

INTRODUCTION

The vision of the lean supply chain is increasingly clear; the techniques and tools for analysing supply chain performance are increasingly well defined; but it is also clear that the transformation of supply chains from current status to lean is a complex and protracted task. This paper describes the Parallel Incremental Transformation Strategy (PITS), an approach to triggering and sustaining the supply chain improvement process. The application and development of the PITS methodology is described in two scenarios: the upstream automotive component supply chain in the UK – the LEAP project, and secondly, a company operating a global supply chain in the footwear sector.

The objective of the LEAP programme was to develop a lean supply system from the major raw material supplier (British Steel Strip Products), through two steel service centres, to six first-tier component suppliers which provide sub-assemblies and components to the major car manufacturers in the UK. Improvements in supply systems are targeted both within the individual companies and between supply chain partners.

The footwear company manufactures materials used in the construction of shoes, including inner soles, linings and heel/toe stiffeners. Customers are shoe manufacturers. Markets are worldwide, but with a concentration in Europe and Asia. Manufacturing takes place primarily in the UK. Distribution is through distributors in foreign markets and direct to manufacturers in the UK. A key element of the company's strategy is to improve the competitive position through enhanced supply chain performance.

PARALLEL INCREMENTAL TRANSFORMATION STRATEGY (PITS)

The traditional approach to business improvement is sequential: data collection, data analysis, strategy formulation, followed by operational improvement (Wass and Wells, 1994). However, the complex nature of supply chains may render this sequential approach impotent. The PITS approach recognizes four dimensions within supply chains that impact upon and complicate the planning and improvement task:

- Time lags: the extensive nature of supply chains means that data collection analysis and strategy formulation are inevitably time consuming (Hines *et al.*, 1998a). How can the commitment and enthusiasm of the supply chain personnel involved in the improvement process be generated, harnessed and sustained over a prolonged period?
- Functional silos: are frequently cited as barriers to supply chain improvement (Dimancescu, 1992). How should initiatives to create horizontal structures be developed?
- Hierarchical structure: it is well recognized that involvement of staff from director level to shop floor is necessary to achieve lean chains (Dimancescu *et al.* 1997). How can this be achieved?
- Limited appreciation of the scope of supply chain management and the benefits of improved supply chain performance amongst many senior and operational managers (Morehouse, 1993).

The Parallel Incremental Transformation Strategy is a methodology which has been developed to address these issues. It aims to create a supply chain improvement programme in such a way as to educate, involve and enthuse personnel from all levels and parts of the supply chain in numerous, self-generating and self-sustaining incremental improvement initiatives.

The vision of the lean supply chain is increasingly clear; the techniques and tools for analysing supply chain performance are increasingly well defined; but it is also apparent that the transformation of supply chains from current status to lean is a significant task. The PITS approach addresses the transformation process through six initiatives that, once started, proceed in parallel towards the goal of a lean supply:

Initiative One: Education and awareness

Initiative Two: Waste analysis

Initiative Three: Creating an appropriate organizational structure

Initiative Four: Value stream mapping

Initiative Five: Incremental improvement

Initiative Six: Evolutionary development of supply chain strategy

INITIATIVE ONE: EDUCATION AND AWARENESS

The requirement

The initial step in the PITS process is to develop an education programme, the aim of which is to explain the scope of supply chain management and the potential commercial benefits of improved supply chain performance. However, at the outset a number of realities should be acknowledged.

- The probable impetus for supply chain improvement is likely to have come from one area of the company – possibly a middle or senior manager involved in one of the supply chain functions. The rest of the company may be indifferent, ignorant or even antagonistic to the initiative.

- Although the terminology of supply chain management, lean production and lean supply is increasingly in use within companies, it is clear that many managers have only a limited understanding of these concepts and the implications of their application.
- Supply chain professionals must be aware that they are not 'preaching to the converted'. Lean supply concepts are well understood by those directly involved in their development and application. There is, however, a danger of forgetting that the majority of managers are not steeped in supply chain management or lean thinking. Furthermore, it is quite likely that CEOs and board members may view lean supply concepts with some scepticism, perhaps as just the latest panacea being offered by academics or consultants (Morehouse, 1993).

For these reasons it is important that an education and awareness programme is conducted, which commences at the outset and continues throughout the supply chain improvement initiative (Hines *et al.*, 1998a).

Initial sessions should involve senior and middle management. In the footwear company this commenced with a one-day seminar attended by the Managing Director, other board members and the managers responsible for the functional areas of the supply chain such as purchasing, manufacturing, distribution and customer service. In the LEAP project educational seminars were delivered to similar groups in each of the individual companies and also to an over-arching 'Supply Chain Steering Group' comprised of the Managing Director or Operations Director from each of the participating companies.

The scope of activity

The first objective of the educational programme is to give a clear, if basic, understanding of two key issues:

- the scope of supply chain management and its objectives;
- the principles of Lean Thinking and Value Stream Management.

The second objective is that participants should begin the process of applying these concepts within the organization so that educational activities are not just viewed as academic exercises.

Supply chain management

The scope of supply chain management is explained through discussion of the Supply Chain Management Model shown in Figure 1. This highlights the three key issues which must be addressed in managing supply chains, namely, the physical flow of materials, the flows of information and the organization/control of the chain. Participants are then asked to construct a schematic diagram (or diagrams) of their own supply chain based on this model. In practice this may take some time, particularly in complex chains. For example, in the footwear company three teams were formed which each spent a whole afternoon gathering information on the exact structure of the firm's supply chains in Europe, Asia and the Americas. Few companies appear to have these diagrams already prepared, which in itself is an indication of the lack of focus on supply chain management.

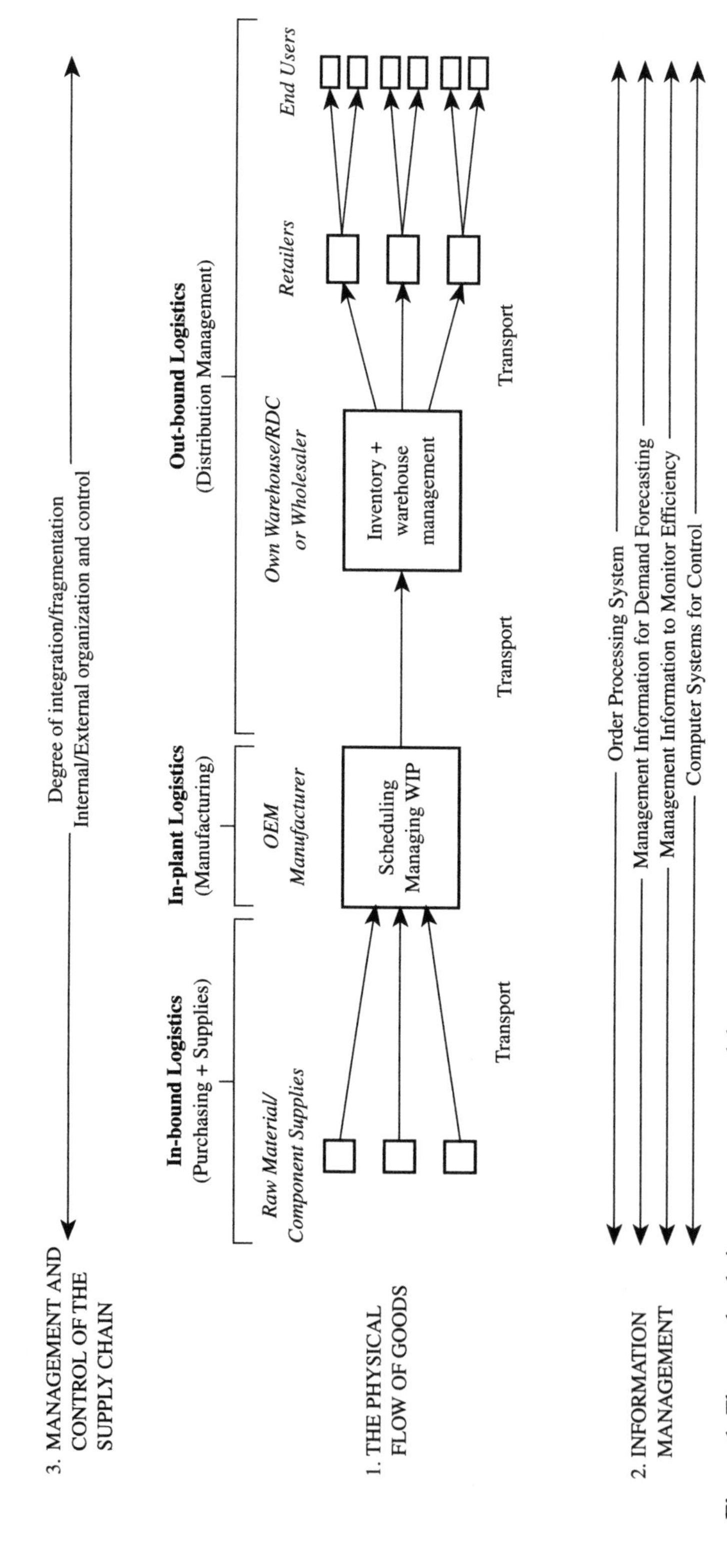

Figure 1 The supply chain management model.

The objectives of supply chain management are also presented in a simple format (Figure 2) and participants are asked to consider the effect of supply chain activity on overall corporate performance. A particularly useful exercise at this point is to ask managers individually to define and rank their perceptions of the five most important elements of customer service. It is usual to find many different criteria and ranking amongst the group of managers, which highlights the need for the company to develop a clearer understanding of the customer's service requirements and to identify common goal posts for supply chain activity.

1 Optimize Customer Service
- Delivery performance
- Time to serve
- Quality

2 Minimize Costs
- Operating costs;
- Fixed assets employed
- Inventory

3 Maximize Flexibility/Agility
- Time to respond
- Time to market

Figure 2 The objectives of supply chain management.

Lean thinking and Value Stream Management

The basic ideas of Lean thinking are introduced to managers by brief explanations of the following issues (see Figure 3):

- The Five Lean Principles
- The Seven Wastes
- The Value Stream Mapping Tools
- Value Stream Management

The first three of these issues are amplified in subsequent steps in the PITS methodology. The concept of Value Stream Management (VSM) requires understanding and action at the outset, as it is fundamental to the lean approach.

> The value stream is the specific set of actions required to bring a product (or service) through the three critical tasks of any business: the problem solving task i.e. product design and launch; the information management task i.e. order taking through to delivery scheduling; the physical transformation task i.e. the process from raw material acquisition to product delivery to the customer. (Womack and Jones, 1996, p. 19)

VSM, in contrast to more traditional approaches to supply chain management, is a *disaggregated* approach. It requires identification and analysis of the operation of the value streams for specific products or product families to specific customers or customer groups. The approach is as follows:

Five principles which are fundamental to the elimination of waste:

- SPECIFY what does and does not create VALUE from the customer's perspective and not from the perspective of individual firms, functions and departments.
- IDENTIFY all the steps necessary to design, order and produce the product across the whole VALUE STEAM to high light non-value-adding waste.
- MAKE those actions that create VALUE FLOW without interruption, detours, backflows, waiting or scrap.
- ONLY MAKE WHAT IS PULLED by the customer just-in-time.
- STRIVE FOR PERFECTION by continually removing successive layers of waste as they are uncovered.

(a)

Overproduction – the making of too much, too early or just in case.

Waiting – where materials or information are waiting to produce to the next process. They are not moving or having value added.

Transporting – where materials (or information) are being transported into, out of or around the factory. Transport cannot be fully eliminated, but the aim is to minimize it.

Inappropriate processing – using machinery or equipment which is inappropriate in terms of 'capacity' or 'capability' to perform an operation.

Unnecessary inventory – which ties up capital and space and prevents identification of problems.

Defects – defined in terms of product defects, rework defects, scrap defects or service defects.

Unnecessary motion – the ergonomics of the work-place.

(b)

Tool	Origin
• Process Activity map	*(Industrial Engineering)*
• Logistics Pipeline map	*(Time Compression)*
• Demand Amplification map	*(Systems Dynamics)*
• Product Variety Funnel	*(Operations Management)*
• Quality Filter map	*(New Tool)*
• Decision Point Analysis	*(New Tool)*
• Physical Structure map	*(New Tool)*
• Value Analysis Time Profile	*(New Tool)*
• Overall Supply Chain Effectiveness	*(New Tool)*

Create a quantified multi-dimensional assessment of supply chain performance

(c)

Figure 3 Initial lean education. (a) The five principles of lean thinking; (b) the seven wastes; (c) tools for value stream mapping.

1. Identify key product flows to key customers using Pareto techniques and select one or two 'A-class' value streams for detailed mapping and analysis.
2. Devise improvements to these key value streams, which will not only enhance the service level to the targeted key customers, but also provide a model for subsequent improvement of other value streams.

A second important activity for managers during, or subsequent to, the initial educational seminars is therefore to identify the company's key value streams and to prioritize one or two for detailed study and analysis.

In the LEAP project this was a fairly simple task, as the first-tier suppliers each produced a relatively small number of sub-assemblies for one or more of the UK car manufacturers. Value streams were easily identified and tracked back to the steel producer. In the footwear example, the seemingly simple request by the researcher to identify key product flows to key customers triggered a major data collection and Pareto analysis exercise of markets, customers and products which had hitherto not been carried out by the company.

After the initial education programme, a fundamental principle of the PITS approach is that as much as possible of subsequent lean education should be carried out by managers within the company. Managers are directed as to what to teach and where to acquire the knowledge, but they, rather than outsiders, actually deliver the teaching to other members of the company. In this way the knowledge is securely captured within the company and its presentation is firmly rooted in the company's operating situation.

The outcomes of Initiative One

Considerable time has been devoted to describing the education Initiative because it is extremely important in laying the foundations for the subsequent lean supply chain initiatives. Effectively it prepares and fertilizes the ground in which the 'lean seeds' will be planted and will hopefully flourish. As a result of Stage One activities:

- senior and middle management should have developed a clear view of the scope of the supply chain management and the principles of the lean approach;
- the company's supply chain structure has been identified along with key value streams which are targeted for detailed investigation;
- managers will have been equipped and encouraged to start educating their own staff in order to spread the awareness of lean supply concepts across the company.

INITIATIVE TWO: WASTE ANALYSIS

The aim of this activity is to help managers to identify waste in individual value streams and hence, find an appropriate route for waste removal or at least reduction. The focus on waste (or 'Muda' in Japanese) was pioneered by Toyota's chief engineer Taiichi Ohno (Shingo, 1989). He demonstrated that a systematic attack on waste would also provide a systemic attack on poor performance within manufacturing and supply systems. Ohno identified seven fundamental forms of waste which are summarized in Figure 3b. Although Ohno's work was originally developed within a manufacturing context, the wastes are equally relevant across other activities in the supply chain.

Initially, waste analysis is carried out with the managers responsible for supply chain functions. Participants are asked to rank the 'Seven Wastes' and weight them in terms of their relative importance by allocating a notional 35 points. A structured interview of approximately 30–45 minutes duration is then carried out with each participant, in which detailed questions are asked as to the rationale behind the scoring, the areas of the supply chain in which waste occurs and the nature of the perceived wastes.

The results of the numerical waste analysis are presented by tabulation of the aggregate scores (Table 1). The comments from the waste interviews are also tabulated and categorized in relation to the Seven Wastes as shown in Table 2. It is also useful to tabulate comments in relation to the areas of the supply chain in which the waste occurs, for example inbound flows, manufacturing, warehousing, order processing, as this gives a focus for improvement for particular operational groups within the chain (Table 3).

Although these results are based on subjective opinion they nevertheless are of significant value. Possibly for the first time, they present a systematic and structured collection of managers' views of the areas in the company where there are problems and scope for improvement.

The outcomes of the initial waste analysis exercise are as follows :

- Wastes which are clearly identified are targeted for immediate elimination: 'picking the low hanging fruit!' (Hines *et al.*, 1998b).
- Some areas of waste are targeted for more detailed analysis and investigation through the value stream mapping process.
- The exercise is the first fitting of 'Muda glasses' for the participants. As people become more familiar with waste analysis they become more skilled at seeing waste within the workplace and taking steps to eliminate it.
- Managers who participate in the first waste analysis exercise are encouraged to replicate the process with staff in their own departments. The ideas and application of waste identification and removal can thus be cascaded throughout the company. For example, in the footwear company the Manufacturing Director conducted waste analysis with factory supervisors and middle managers, who in turn did the same with shop floor staff. In this way all manufacturing staff were alerted to waste. Subsequently, shop floor teams were established and charged with the task of waste removal as a focus for continuous improvement.

Table 1 Automotive component supplier waste analysis – aggregate scores

Waste Type	*Average score*
Waiting	7.2
Defects	6.3
Unnecessary inventory	5.1
Inappropriate processing	4.7
Transportation	4.6
Overproduction	3.8
Unnecessary motion	3.3

(Maximum possible score per issue = 10)

Table 2 Summary and classification of comments regarding the seven wastes. An example from the LEAP project

Defects	*Unnecessary inventory*	*Overproduction*	*Inappropriate processing*	*Waiting*	*Unnecessary motion*	*Transport*
• High internal defect rates • Many defects stem from poor tooling • Lack of attention to detail by operatives • External ppm = 5000 Internal ppm = not known • Paint shop defects = 0.1% • Defect levels recorded on Nobo's but information lost after 1 day • J.C. work – overproduction – defects because not spotted early • Monthly scrap and rework report is available with causes • Estimate could save £300,000 by eliminating major defects Causes: Tool design at new product development stage • Lot of quality problems with tools (mainly heavy presses). Don't seem to find root causes • Defects in assembly caused by lack of operator awareness or care.	• Inventory on site = 4 days. If take out obsolete stock and trials for new products = c 2½ days • Who owns the inventory? • Large inventory of parts from Japan – long lead times • Lot of inventory before and after paint • Inventory levels = RMats/Blanks – 2 days In assembly – 3–4 days Paint/despatch – 4 days • Requirement to keep presses going – unnecessary stock after presses and unnecessary raw materials.	• Causes • Finishing off last 100 blanks in a box • Overproduction mainly in BU3 • Due to lack of flat schedules from cust. • BU5 feels BU3 & 2 may over-order blanks, e.g. 1 week's supplies – so tie up blankers • Occurs in order to keep workforce busy	• Rover – manual keying of 120 purchase orders per week • Scope to improve Assembly and Final Assembly operations • Stems from poor design of tools and processes – needs more value-engineering at front end • Lot of problems in defects/transport/stock stem from poor engineering of processes at the outset • Inappropriate processing mainly relates to having to do rectifications/rework.	• Waiting for FLTs • Waiting for loading on paint plant • Delays in assembly + earlier – culminate in paint and dispatch • For pallets/stillages either from internal or external sources • BU2 – waiting for tool room maintenance tool room – no system for prioritising work • Waiting for setting of presses because only 4 setters	• Block stacking of dies – unnecessary motion • Cranage now available to handle all dies but not yet introduced.	• Too many FLTs in small area – safety issue • Small pallets and hand trucks would eliminate FLTs • Poor layout – lack of space for tool storage – lot of transport • Aim to eliminate FLTs by 2000 • Are 17 FLTs • Lot of waiting for FLT in BU3 – only 1 FLT

Table 3 Footware company: summary of key problems across the supply chain

Inbound flows	*Manufacturing*	*Warehouse*	*Transport*	*Inventory*	*Order processing*	*Distribution channels*	*Product users*
• Purchasing works as an operational function • How efficiently we don't know • Doesn't appear to be any attempt to manage inbound flows and suppliers more strategically	• Factory produces a high degree of defects – which have significant knock-on effects in terms of inventory and late deliveries • Schedules don't appear to be closely related to real demand • Factory is possibly working to its own agenda • Who is in control?	• Operates as an inefficient buffer area • Receives poor service from factory and shipping • Gives poor service • Who is in control?	• Inefficiencies and delays due to paperwork • Company doesn't have control of most export transport • How much emergency transport is used?	• Lot of stock, but wrong stock • Stock is covering many rocks • Central stock is probably a necessary interim measure but may not be best in long run • Huge stocks out in the distribution channels • Who is responsible for inventory overall?	• Processes orders, handles queries, but with very little visibility along the chain • Acts at an operational fire fighting level • No one is trying to make this into a strategic role	• Sell products • Hold lots of stock (probably wrong stock) • Scream for supply • But don't transmit real demand info to HQ • Currently their demand represents a mixture of stock replenishment and real demand	• Very little contact with these real customers • No information on issues which are of real concern to these customers and their likely future requirements • Little direct contact from HQ

INITIATIVE THREE: ESTABLISH AN ORGANIZATIONAL STRUCTURE FOR SUPPLY CHAIN IMPROVEMENT

The development of a lean supply chain requires understanding and commitment from three groups within the organization and/or supply chain: senior management; lean supply champions; management and operational staff involved in supply chain processes. Each of these groups must be established and tasked with different activities as summarized in Figure 4.

Senior management: the lean steering group

Supply chain improvement initiatives are inevitably far-reaching, cross-functional and challenging to pre-existing operational practices, business policies and functional hierarchies. It is critical that senior management (i.e. board level) clearly understand the strategic competitive benefits of the development of a lean supply chain, so that they can give the support necessary to make the changes and be prepared to resolve the trade-offs and conflicts which will inevitably occur. In the footwear company, all main board directors participated in this strategic steering group. Their initial task was to educate each other in 'Lean Thinking', so that they would be in a position to make informed decisions on the supply chain improvement proposals which would be presented to them. Five seminars were organized to explain and discuss the Five Lean Principles, with each session being led by a different director. In the LEAP project, a cross-company steering group was established with senior management representation from each participating company. This group ran a variety of educational events ranging from expert speakers, to self-teaching, to use of a supply chain simulation game (see Chapter 17).

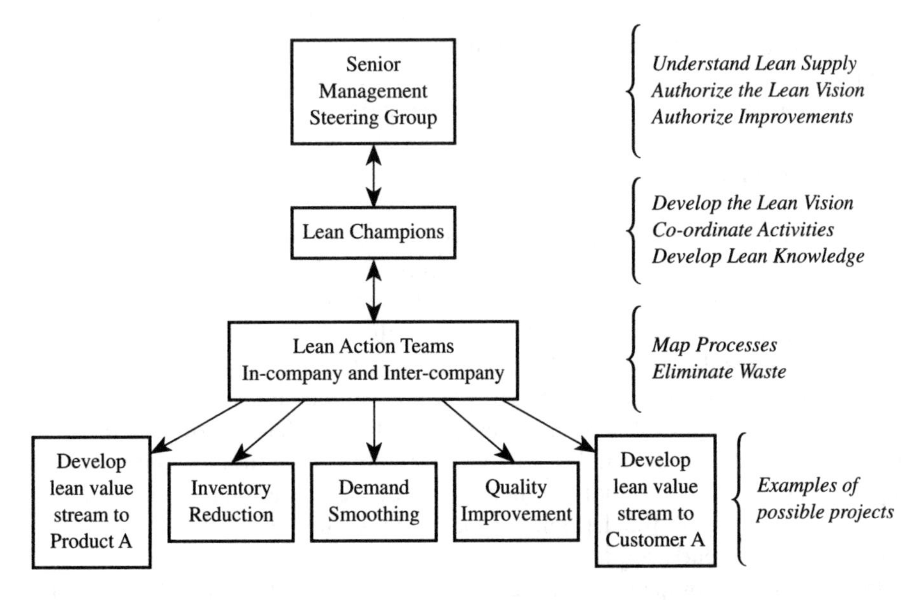

Figure 4 The three-tier structure for lean transformation.

Lean supply chain champions

In each company there is a requirement to establish a small nub of people who are committed to the development of a lean supply chain. Their task is to become the in-house experts. They are researchers and propagators of lean knowledge and are responsible for the development of the overall supply chain improvement strategy and its orchestration. The background of these people varies from company to company, but when selecting them, it is important to assess their willingness to learn and adopt new ideas.

Operational teams

Operational staff, responsible for the various elements of the supply chain, are the third group which will be involved in the improvement process. It is important to involve operational staff as much as possible in the various stages from waste analysis, through mapping, to specific improvement projects. These are the people on the ground who will see the waste and recognize many opportunities for improvement. The key factor here is to try to capture their commitment and enthusiasm in order to channel their initiative to help towards the improvement process. In practice many different operational groups will be established to deal with issues. (For examples, see Figure 4.)

INITIATIVE FOUR: VALUE STREAM MAPPING

The value stream mapping approach uses a selection of mapping tools to create a multidimensional assessment of supply chain performance (Figure 3c).

The nature and output of these tools are well described in Chapter 3 of this book, 'The seven Value Stream Mapping tools'. Assessing the most effective methods of applying the tools and benefiting from their output was an important objective of the LEAP research project.

Collection of data using these tools is necessarily a time-consuming process. The required base data may not exist in an appropriate form and may involve significant data manipulation or, in some cases, the establishment of new data collection systems.

The approach adopted therefore is to trigger different operational groups to collect the data from their particular projects. The Lean Champions first acquire knowledge of the mapping techniques and then apply the techniques with the relevant operational teams. The selected key value stream is the focus of the initial mapping. The results of each individual map should be analysed by the participants in the mapping process and, wherever possible, immediate improvement activities centred on waste removal should be instigated. Rather than waiting until data from all maps has been assembled, it is preferable to initiate some immediate improvement activities so that the momentum of the operational team is maintained and the relevance of the mapping process to waste removal and supply chain improvement is clearly apparent.

In the footwear company, the selected value stream was the highest-selling product to the largest domestic customer. The starting point was to construct a Process Activity Map to show every detail of the process from the customer placing an order, through the sales order processing system, production scheduling, product manufacture and delivery to the customer. The mapping was carried out by members of the relevant lean operational teams. In this way the operational groups were directly involved in the data collection, analysis and subsequent improvement activity and were also equipped with the mapping technique, which could subsequently be applied to other value streams. Other maps were constructed using teams drawn together on the basis of people who had access to the relevant data and, more importantly, were in relevant positions to act upon the outcome of the analysis.

The unifying point in the mapping are the Lean Champions. They are the initial mentors and project coordinators. They also bring together the results of the mapping activities, which can quickly develop their own momentum within the various areas. Indeed it is apparent that once these mapping tools are given to operational groups and the potential benefits clearly explained, many staff are keen to apply them, as it gives a methodology and focus for grass-roots improvement activity which may previously have been absent.

INITIATIVE FIVE: INCREMENTAL IMPROVEMENT ACTIVITIES

Spontaneous improvements

The process of data collection and the results of the maps invariably highlight many wastes, inconsistencies and opportunities for small-scale or local improvement in supply chain activities. Staff who had ‘Muda glasses’ fitted in Stage Two now have objective data on waste occurrence. Spontaneous efforts to remove waste almost always follow. For example, in the footwear company, a team that had been established to specify and implement a new company-wide IT system adopted Process Activity Mapping. This was then used as the basis for understanding current information handling procedures and specifying new procedures, with less waste, that were to be encapsulated in the new information management system.

Development of a ‘model’ lean value stream

The major element of improvement is the rapid development of a model lean value stream. Once data has been collected from all the maps related to the target value stream, the Lean Champion, together with representatives from the various operational groups along the chain, come together to review the overall situational analysis which has been developed. The critical areas of waste are highlighted, root causes identified and an action plan developed to streamline this value stream. Proposed improvements should work on the premise of requiring little or no investment (Womack and Jones, 1996). A much leaner supply chain can invariably be achieved with existing systems. If it is ultimately felt that major improvements can be achieved through investment in new equipment this can come later, after staff have learnt to work in a lean fashion within the existing environment. The key objective here is to make the sample value stream work more efficiently and effectively and to do it quickly. It then becomes a working model, not in some far-off Japanese factory, or a different industry sector, but right there in the company; a working demonstration that lean value streams can be achieved. This model is vital in persuading other groups not directly involved in the initial exercise that the lean approach is viable and valuable. It is also gives senior management confidence to authorize more widespread adoption of the lean principles.

Chapter 10 describes the Lean Vision and implementation plan that was developed and adopted by one automotive component manufacturer in the LEAP project.

INITIATIVE SIX: EVOLUTION OF SUPPLY CHAIN STRATEGY

It has long been recognized that different customers or customer groups have different service requirements and there is a need to segment customers in terms of the service provision (Christopher, 1992). This requirement is addressed implicitly within the value stream management approach.

Once the lean approach has been shown to work for a key product flow to a key customer, it can be rolled out to other customers or customer segments and other products or product groups. The first step is to identify the basic principles and operating characteristics of the model system. These might, for example, include small batch production, rapid machine changeover times, production to order, demand smoothing, minimal stock held at only one strategic point within the chain, on-line order capture and processing, or just-in-time delivery. Customers and products that can be well serviced by the same or a similar system are identified and the model system applied to them. However, other customer segments and/or product families may require different service levels or supply systems.

A vision of the lean supply chain thus evolves. The vision can then be formalized into a supply chain strategy statement, with a phased roll-out of the lean model to encompass all major product families and customer segments. The senior management team, which have been aware of development process to date, now become directly involved in the scoping, authorization and resourcing of the strategy. An important aspect of this approach is that the strategy will be achieved through incremental steps, value stream by value stream. It can thus be easily modified in response to changing operating or commercial circumstances.

In the footware company, two parallel value stream mapping and improvement projects were developed. The first was for the main product group supplied to the major domestic customer. Improvements were made that led to quantifiable improvements in quality, cost and delivery. Once these changes were implemented, the methods and ideas were used as a model for improved supply to other Category A domestic customers. The second value stream studied and improved was to the key account in China. A lean model was developed for SE Asian supply, which at the time of writing was being applied and monitored. Once it is tested and refined it will be applied to other customers in that region.

CONCLUSION

Four particular obstacles have been identified as a barrier to supply chain improvement, namely:

- time lags: the extended time period required for improvement processes due to the complexity of supply chains;
- functional silos: the need to break down functional barriers both within and between companies;

- hierarchical structure: the need to involve staff at all levels within the company;
- lack of understanding: of the scope and benefits of lean approaches to supply chain management within companies.

The Parallel Incremental Transformation Strategy attempts to overcome these obstacles by applying the techniques of value stream management in such a way as to create self-sustaining, incremental improvements.

Each of the six initiatives of the PITS approach is designed to do three things:

- educate the staff involved and thereby equip them for subsequent self-generating improvement activity;
- result in immediate outputs which can lead to identifiable operational benefits;
- contribute to a wider value stream improvement process both for the initially targeted value stream and for subsequent value streams.

PITS is designed to avoid the time lags associated with the more traditional approaches to planning and development, which usually proceed sequentially through data collection, data analysis, strategy formulation and operational improvement. In complex supply chain systems these time lags can be significant and there is a danger of losing the commitment and enthusiasm of staff who are not directly involved in the planning process, which may then jeopardize the success of implementation.

From the commencement of the initial educational seminars and waste analysis exercises, staff are involved in a series of parallel activities which have both immediate and longer-term benefits. As the initiative proceeds, more staff are involved as each manager or group spreads the techniques and knowledge to others.

Staff at all levels in the company are involved in data collection, analysis and improvement initiatives and also have an input into policy formulation. The senior management team are involved from the start and are hence more receptive to suggestions for improvement that they may ultimately have to authorize. The establishment of operational teams to address particular supply chain issues inevitably brings together cross-functional and/or cross-company teams which is a most effective way of putting into practice the ideas of horizontal process management and supply chain partnerships.

PITS is a disaggregated approach which applies lean philosophies to supply chain systems. Its incremental nature means that it is less traumatic and more flexible than many traditional approaches to supply chain improvement which often require large, expensive, aggregated solutions for the whole supply chain, e.g. new IT systems, new RDCs etc. The fundamental aim of PITS is to develop a structure and approach within a company or supply chain that will engender a spirit of continuous, incremental improvement.

NOTES

1 This is a revised version of a paper that was first published in the International Journal of Logistics Research and Applications Vol 2 No 3 November 1999.

REFERENCES

Christopher M. (1992) *The Customer Service Planner.* Oxford: Butterworth Heinemann.

Dimancescu D., Hines P. and Rich N. (1997) *The Lean Enterprise: Designing and Managing Strategic Processes for Customer Winning Performance*. New York: AMACOM.

Dimancescu D. (1992) *The Seamless Enterprise: Making Cross Functional Management Work*. New York: Harper Business.

Hines P., Rich N., Bicheno J., Taylor D., Butterworth C. and Sullivan J. (1998a) Value Stream Management. *International Journal of Logistics Management*, 9(2), forthcoming.

Hines P., Rich N. and Hittmeyer (1998b) Competing against ignorance: advantage through knowledge. *International Journal of Physical Distribution & Logistics Management*, 28(1), 18–43.

Morehouse J. (1993) Supply chain integration: a CEO's perspective. Council of Logistics Management Annual Conference, Washington, DC.

Shingo S. (1989) *A Study of the Toyota Production System from an Industrial Engineering Viewpoint*, Cambridge, MA: Productivity Press.

Taylor D.H. (1997) *Global Cases in Logistics and Supply Chain Management*. London: International Thompson Business Press.

Wass V.J. and Wells P.E. (1994) *Principles and Practices in Business Management Research*. Aldershot: Dartmouth.

Womack J. and Jones D. (1996) *Lean Thinking*. New York: Simon & Schuster.

Towards understanding supply chain dynamics: the Lean LEAP Logistics Game

17

Matthias Holweg and John Bicheno

INTRODUCTION

This paper describes how a participative simulation model is used to demonstrate supply chain dynamics and to model possible improvements to an entire supply chain.

Initial research in the LEAP project suggested that a lack of understanding of the core processes throughout the supply chain caused distortion and amplification of both demand and supply patterns. In consequence, this deficit of information is often replaced with inventory – resulting in increased lead times and pipeline cost.

At the start of the project there was relatively little collaboration in the supply network, and the 'Lean LEAP Logistics Game' was developed primarily to foster this collaboration. To achieve this, the game had to model reality, and was built on a series of mapping activities. Unexpectedly, it turned out that developing and running the game led to insights into scheduler behaviour, scheduling decision making, prioritizing improvement activities and into supply chain dynamics, especially the 'Forrester' or 'Bullwhip' effect (Forrester, 1961; Lee *et al.*, 1997).

THE BACKGROUND

Supply chain dynamics and their effects on the performance of the entire system have been discussed in the seminal work of Jay W. Forrester (1961) and John Burbidge (1983). Although explanatory research has been carried out, much of the research has focused on managing distribution networks and retail chains (Alber and Walker, 1997), Distribution Resource Planning (Martin, 1995) and the Quick Response initiative in the textile and food industries (Hunter, 1990; Kurt Salmon Associates, 1993). These generally assume that a standardized product unit exists. If this is not the case, the system complexity increases sharply, as every level of the chain not only represents a decision point, but also a standardized 'product' unit does not exist. Along the studied value streams, steel converts from an initial 20-tonne wide steel coil to slit coils, to blanked sheets, to press components, and finally to assembled automotive components in the 0.5–50 kg mass range. Unlike some supply chains, there is a problem of fixed batch sizes. Thus, a 20-tonne steel slab, made to customer specification, is in turn converted into coil, blanks, and pressings which are unlikely to correspond exactly with a demand period multiple. Yield and quality uncertainties, new product introductions, and varying capacity constraints further complicate supply chain dynamics.

LEAP is one of the very few studies to involve several manufacturing tiers in a supply network, as shown in Figure 1, providing a unique opportunity to study a multi-company three-tier supply chain.

In the initial project phase, intensive empirical research has been undertaken in the form of value stream mapping, using specific analysis tools (Hines and Rich, 1997, printed in Chapter 3) to enable a detailed understanding of the supply chain.

THE 'LEAN LEAP LOGISTICS GAME'

The use of games in management for educational purposes is well established, and JIT games had already been applied within the LEAP project at company level.

When the idea of using a more comprehensive supply chain game to demonstrate the dynamics in supply chains was first mooted, the possibility of using the MIT 'Beer Game' (Sterman, 1989) was discussed. However, several problems were encountered with the Beer Game:

- The Beer Game is a retail distribution game, and therefore does not take **product conversion** into account.
- Also, the Beer Game does not have the particular **characteristics** of set-up times, process reliability and quality problems found in a manufacturing-oriented supply chain.
- The Beer Game does not have enough **stages** to be representative to the steel supply chain, as at least the core processes (see Figure 1) needed to be modelled to create a representative simulation.
- The **capacity allocation problem** is not dealt with in the Beer Game, but proves to be a major constraint in the steel supply network.

Therefore, it was decided to model a new participative simulation of the all major processes, building on previously gained knowledge from the process mapping, called the 'Lean LEAP Logistics Game'. The game comprises two products (*RED* and *BLUE*) and six stages. The stages model the core processes in the supply chain:

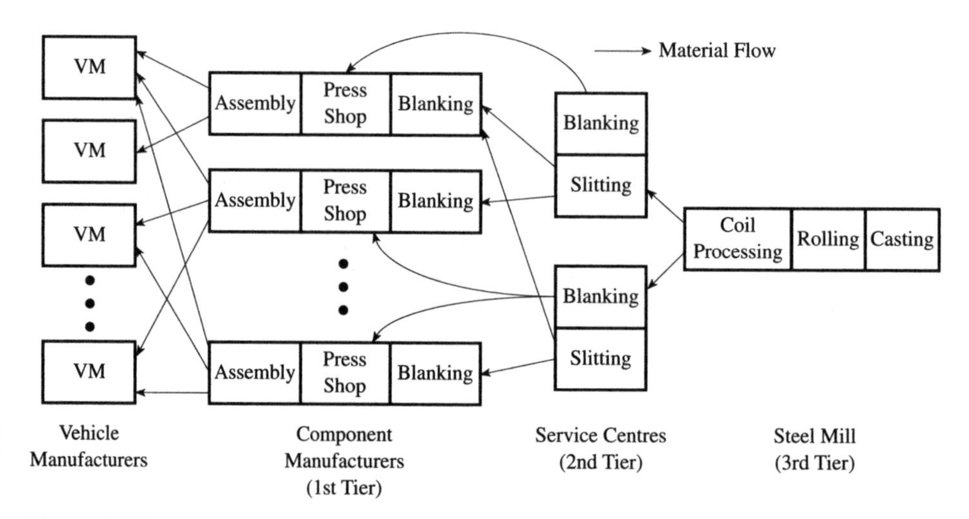

Figure 1 The automotive steel supply chain – core processes.

1. **The Final Customer** (i.e. the vehicle manufacturer, played by a game controller)
2. **Despatch**
3. **Final Assembly**
4. **Press Shop**
5. **Blanking Operations**
6. **Service Centre / Slitting**
7. **Steel Mill**

A typical gaming session comprises several rounds, subdivided into periods. One of the products (*BLUE*) has absolutely stable demand throughout most rounds. The other product (*RED*) is subject to some variation from period to period. This situation models the reality of vehicle manufacturers, some of which produce totally stable demand, while others do not. The products are simulated by the use of DUPLO and LEGO bricks that are gradually transformed along the supply chain, and these DUPLO and LEGO bricks are physically moved along the chain. Thus an eight-stud DUPLO brick, representing a steel slab, is transformed into two DUPLO four-stud bricks, representing a coil, which is transformed into two eight-stud LEGO bricks, representing a metal blank, which is finally transformed into two four-stud LEGO bricks, representing a product. Thus a 'slab' can make eight 'products'. (A real slab makes a few hundred to a few thousand products.) Each station has an individual 'factory', which is represented on a game sheet. A sample game sheet is shown in Figure 2.

Each station also keeps a record sheet for tracking orders, inventory and backlog of each colour for both supplier and customer. Each station plots a graph, each period. At the beginning of each period orders for the next stage in the chain are placed face down on the desk. Initial inventories are distributed to all stations. The game is synchronized by the use of a whistle: all layers must have completed all the steps before another period can be signalled.

Each period, each player or station must decide how much of each colour product to make (if allowed), and how much to order from the next player. During the first round, the only indication that players have the current overall demand pattern are the orders from the previous player. However, these orders are 'optimal'. Despatch and Final Assembly are given an indication of the demand pattern, which represents a fairly realistic order situation.

Several stations are subject to chance events, simulated by rolling a die. The probabilities built into the game reflect the current problems encountered in the actual supply chain. For instance, the roll of a die models the output from the steel mill to reflect yield and quality uncertainty, and in the press shop another die roll determines changeover difficulties.

A typical game session takes between 3.5 and 6 hours.

Unlike the Beer Game, no spike of increased demand is introduced during the first round. Nevertheless, demand amplification always occurs; only the extent of the oscillations is in question. It only takes one player to make one decision out of line with the others for the whole chain to experience a cycle of oscillation. For example, a change in safety stocks might be the cause for oscillation, as happens in the real case and was predicted by Forrester (1961).

Figure 3 shows an overview on the game flow, the game stations and the restrictions that apply to model real-life features of the processes.

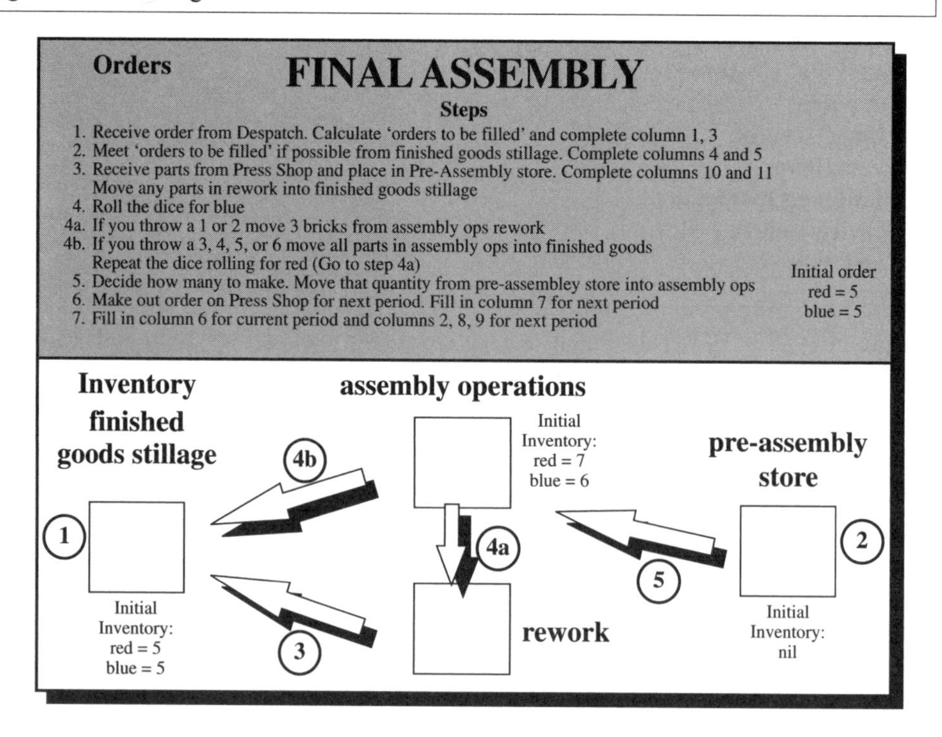

Figure 2 Game sheet example.

OBJECTIVES OF THE GAME

The first year of research in the supply chain highlighted a significant lack of communication, which caused severe disruptions to the information and supply flow in the chain. This lack of communication further results in a low level of understanding of the core processes in the chain, and finally, in unawareness of the

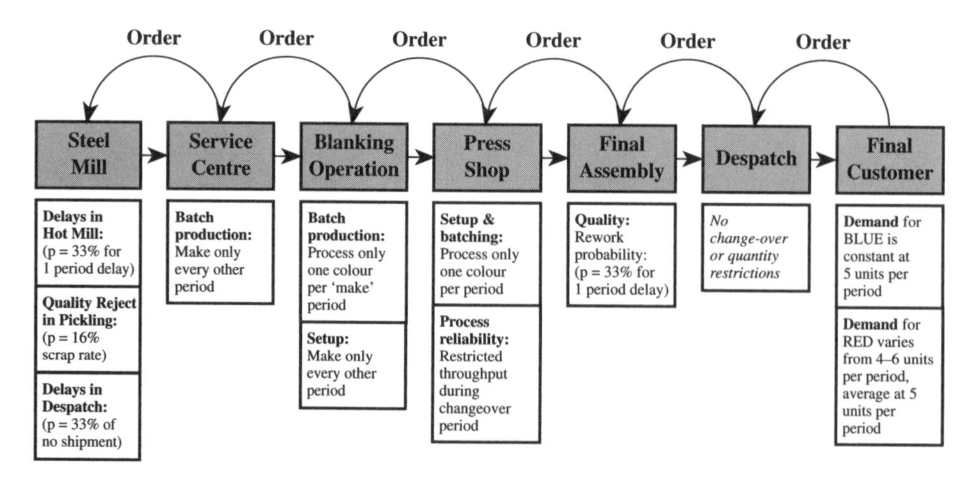

Figure 3 Game flow diagram and restrictions at the game positions.

cause and effect of the players' own behaviour on the whole system. The game was created to demonstrate this effect by simulating the entire supply chain, and also to trigger the understanding of where possible solutions could lie. In general, the game has two objectives: to create 'supply chain awareness', and to 'develop and validate improvements' to the current supply chain.

'Supply chain awareness' simply means to use the simulation to make the players understand the whole supply chain and its core processes and problems, as opposed to only understand and focus the own company's problems and neglecting the effects the own company's behaviour has on the entire system.

Therefore, the players are asked to play a position in the chain that is different from their own company. The intention hereby is to develop understanding of processes in the chain and understanding of the importance of communication in the supply chain, and to create awareness of the consequences of own decisions in the chain.

EXPERIENCES RUNNING THE GAME

The game has been played with various levels of management from companies along the same supply chain. The participants included directors, planners and schedulers, and entry-level staff. Where possible, the participants were drawn from the full range of companies along the supply chain.

Feedback from director-level participants has indicated an improved level of appreciation of the complexity of the supply chain. A typical reaction from directors is that although they feel that they have considerable knowledge about their own node in the chain, the difficulties of other nodes were not appreciated. In particular, the impact on other participants of policies adopted by their own company was not appreciated. Quantity discounts is a classic case in point. Directors who were also members of the research project steering committee felt that they were able to direct the project in a more effective way. For example, the relative merits of changeover reduction as against level scheduling were debated following the experience of playing the game.

In some of the supply chain links that were researched, participating companies have different philosophies to production control. These include JIT and kanban, MRP, and reorder-point spreadsheet-based systems. For scheduler-level participants, the implications of using a particular approach on other members of the supply chain are not always apparent. The game is able to simulate the effects of various ordering policies to some extent and the implications can be quickly seen and discussed amongst the schedulers.

At least two rounds of the game are played. During the first round no communication is allowed between players except the passing of order cards. After the first round players are asked to estimate the end-customer demand pattern. The estimates are invariably considerably inflated. Thereafter, the data and graphs produced by each player are displayed and discussed. This is followed by a short presentation on the theory of supply chain dynamics. In preparation for the second round, the players are asked to develop an overall control strategy. During the second round restricted communication between players is allowed to simulate the actual partnership communication. Once again data and graphs are kept on the record sheets. A third round may follow, since it is often found that it is difficult to avoid all supply chain instability despite a strategy and good communication, and the team may wish to undertake further refinements – which in fact is a lesson in itself.

Where a third round is played, the customer demands are made a little more variable. The third round and subsequent rounds may also be used to change the production characteristics – for instance, changeover times, run frequencies, and reject levels may be changed to simulate successful improvement activities.

In the following section the results of a typical game played with director-level managers are shown.

THE FIRST ROUND

Figures 4 and 5 show the demand patterns and inventory levels over ten periods. The graphs show the total demands for RED and BLUE products. The final customer demand pattern is constant at five units for BLUE and varies between four and six units for RED.

It may be seen that orders from the stages of final customer, despatch, and final assembly remain fairly stable throughout all periods of the game. However, orders from the press shop amplify sharply. The primary reason is that the press shop works on a four period cycle (representing four-week batches). One would expect order quantities in the order of 20 units, which in fact happens in early rounds, but these quickly rise to over 40 units. The reason is the demand amplification effect.

The real demand is not known, so blanking and service centre players forecast on the basis of the first two press-shop orders (22 and 20 units). Additionally, all game positions try to build up certain safety buffers. Therefore, the orders amplify up to 44 units per period which equals 8.8 times of an average period's demand and 2.2 times of an average press batch. The steel mill as the last member in the chain

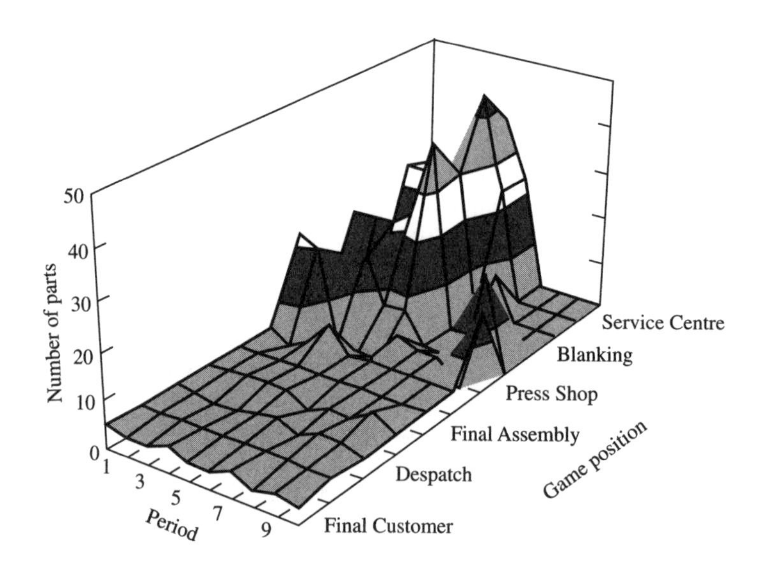

Figure 4 First round – demand patterns over time.

faces these amplified orders and is not able to deliver the required quantities. Consequently, the steel mill builds up a supply backlog to the service centre, and the service centre itself soon experiences a stock-out and builds up a backlog to the blanking operation.

This amplified ordering practice continues until period 5, where all orders from blanking and service centre suddenly collapse. The reason is that continuous over-ordering increases the inventory levels at the blanking operation to unrealistic levels (see Figure 5).

In Figure 5, as the inventory levels rise and actual demand works out to be far less than what has been ordered, the upstream game positions (blanking and service centre) simply stop ordering from period 8 onwards. During periods 8–10, the steel mill can catch up with the backlog and delivers the outstanding material to the service centre (order cancellation is not allowed), which then delivers to the blanking operation according to the backlog. Therefore, the inventory levels do not decrease, although the upstream positions stop ordering completely. Finally, the maximum inventory level is reached in period 8 at the blanking operation (96 units, equalling > 19 periods' supply).

THE LEARNING POINTS

The results of the first round clearly show the effects of distorted demand patterns. The order patterns, especially those from the press shop to the service centre, show highly amplified demand. Reasons are the lack of demand visibility and the incentive to build up safety stock to cover the 'own' game position against any demand or supply uncertainty. This leads to highly increased orders – the maximal order in the first round goes up to 8.8 times the average demand / period. The continuous over-ordering results finally in an inventory of greater than 19 periods' supply at the blanking operation.

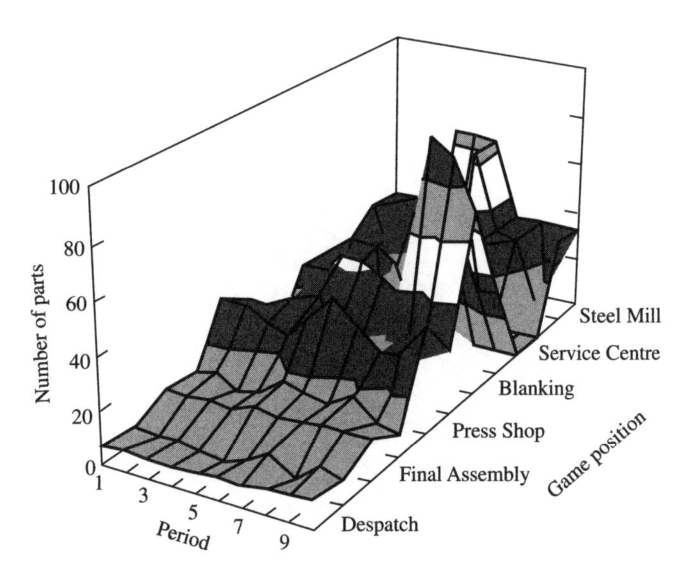

Figure 5 First round – inventories over time.

The first round also almost invariably illustrates an effect that was found during the mapping of demand amplification. This is the effect of the 'double wave' or 'supply amplification', whereby players in the middle of the chain (Blanking and Press Shop) get hit from both sides often at once. Shortages emanate upstream from the steel mill and downstream as a result of scheduling practices.

The follow-up discussion revealed both lack of demand visibility as well as process reliability and quality issues as main problems resulting in unstable demand and supply pattern.

ROUND 2 – SYNCHRONIZING DEMAND AND SUPPLY

In the second round, pre-game planning and strategy 'meetings' amongst players were encouraged. 'Synchronization' of the processes (and consequently the demand and supply patterns) was not encouraged explicitly, but the suggestion was made to players to forecast the demand and to determine the exact time periods when material or products were needed. This becomes crucial at press shop, blanking, service centre and steel mill positions, as these operate on different batch cycles or production restrictions. If this were neglected, redundant inventories would occur.

Synchronization was followed almost perfectly in the second round – accompanied by extensive communication amongst all players. This is reflected in the graphs shown in Figures 6 and 7. It is also very interesting to note that although detailed figures are not provided, the overall delivery performance to the final customer increased from 60 per cent (both products, Round 1) to almost 100 per cent (both products, Round 2).

Figures 6 and 7 are equivalent to Figures 4 and 5, but show the outcomes of the second round. Note that the scale of the two y-axes has been halved.

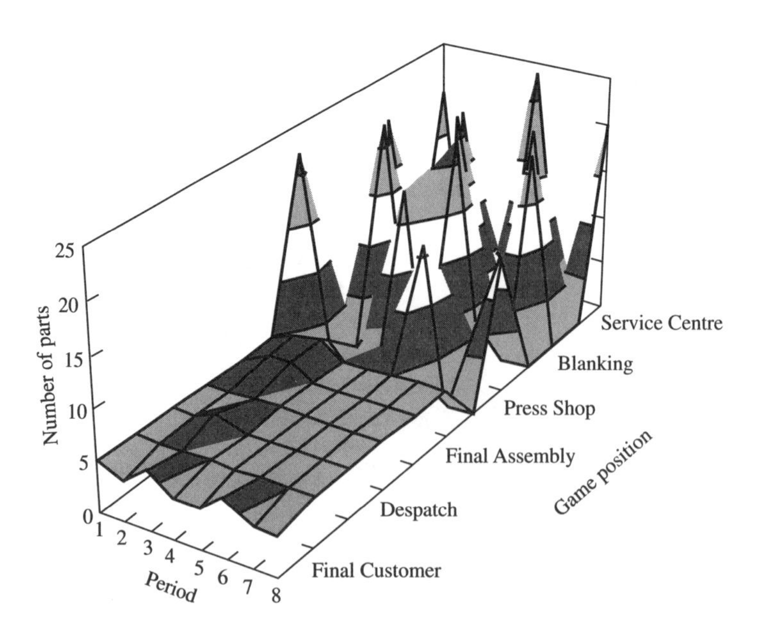

Figure 6 Round 2 – demand patterns over time.

Concerning the demand patterns in the second round, the orders are stable (+ 20 %, – 0 %) from the Final Customer to the Press Shop position. These orders are nearly identical with the real demand from the Final Customer, as there is no need to build up additional safety stock to cover any demand uncertainty. Only Final Assembly slightly increased orders in periods 1 and 2 to build up a small safety buffer. This buffer covers possible rework in the assembly process and ensured nearly 100 per cent delivery performance from Final Assembly to Despatch. The order patterns at the Press Shop, Blanking and Service Centre positions show characteristic peaks. These peaks reflect the four-period press batch cycles. The inventory levels in the second round (Figure 7) show similar stability up to press shop level, and even after the batch production is introduced, the maximum inventory level does not exceed 45 parts, i.e. eight periods' requirements.

In Round 2, synchronized demand and supply patterns throughout the chain could be achieved by transmitting the overall demand information directly along the whole supply chain, resulting in:

- stability in both demand and supply patterns;
- low inventory levels at all game positions;
- no amplification in demand patterns, although the batch operations still oppose a distortion factor.

SUPPLY CHAIN SYNCHRONIZATION – A NEW IDEA?

Synchronization in multi-tier information feedback systems, such as supply networks, distribution chains or retail channels, has been widely promoted by the systems dynamics research, and its impact could be clearly demonstrated in the game.

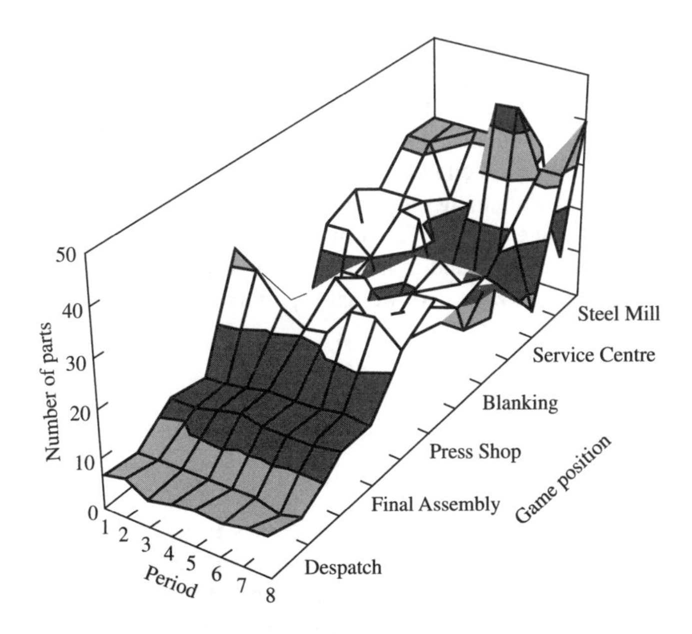

Figure 7 Round 2 – inventories over time.

The benefits of known demand along a supply chain have been shown by Lee *et al.* (1997). They further argue that 'double forecasting' can be a driver for the 'bullwhip effect' (as they call demand amplification). Double forecasting means that the incoming forecasts are adjusted at every decision point within the chain and then submitted to the preceding level, where the same process occurs and the information is adjusted again. In the synchronized chain, no double forecasting is happening, as all levels use the final demand to forecast their production.

Forrester (1961) demonstrated the benefits of visible demand. He showed, through a limited system simulation, that in the case of a +10 per cent increase in final demand the production peak at manufacturing level can be reduced from +45 per cent to +26 per cent by transmitting the information directly from the customer to the manufacturer.

Furthermore, Lee *et al.* showed for retail distribution chains that balanced and 'perfectly synchronized' retailer ordering can be achieved. Under that scenario (and only then), the variability of demand experienced by the supplier and the retailers is identical, and the 'bullwhip effect' disappears.

Synchronization improves the overall supply chain performance, as the demand visibility erases demand amplification. It also facilitates inventory reduction where safety stocks were necessary to cover demand and supply uncertainty, and improves the quality of forecasting and long-term planning in all supply chain parties.

Apart from the system dynamics research, 'synchronization' has also been demanded in the context of supply chain integration (Tetu, 1998). Tetu postulates that companies have to see themselves as part of the whole chain and accept the need to achieve a global optimum instead of striving for more and more sophisticated local optima within every single company or participant in the supply chain. Womack and Jones (1994) describe this goal as the 'Lean Enterprise', which would be formed by a group of individuals, functions, and legally separate but operationally synchronized companies.

Examples of existing supply chain scheduling concepts are the Quick Response Programmes (QRM or QRP) in the textile sector (e.g. Hunter, 1990), and the Efficient Consumer Response Initiative (ECR) in the fast-moving consumer goods (FMCG) industry (Kurt Salmon Associates, 1993).

Synchronized supply chain scheduling has to integrate a whole supply chain into one scheduling concept and to deploy demand visibility to all levels in order to achieve a 'zero waste' value stream. The aim is to include all the different scheduling systems found in a supply chain and to integrate as many parts as possible, even those that provide less stable and repetitive demand patterns. Essential factors to achieve synchronization are:

1. **Demand visibility**, providing every supply chain player with the actual overall demand that drives the network. This could be achieved via EDI (Electronic Data Interchange), Internet or other means, similar to the EPOS (Electronic Point of Sale) system in retailing.
2. **Process visibility**, where all production processes in the supply chain are timed against each other – or 'synchronized'. This requires further communication between supplier and customer to determine the point in time where the material is needed, i.e. the production process is scheduled to start.

3. **Appropriate time buffers**, to buffer the system against any kind of uncertainty or process unreliability. Remember, any kind of quality rejects or time delays forces the subsequent tiers in the chain to reschedule, and creates in return distorted demand information – which is one of the key drivers of the 'Forrester effect' or demand amplification.

The benefits clearly are reduction in inventory, increased delivery performances, lead-time reduction and last but not least – cost reduction, which increases the overall competitiveness of the whole network against other supply sources.

A synchronized supply chain would have to introduce 'time buffers' instead of safety stock at all points where unreliable processes might destroy the synchronized flow (Goldratt and Fox, 1986). Time buffers are not safety stock – safety stock is an agreed amount of inventory in front of a process step which will always be filled up to the agreed level. Safety stock covers against demand and process uncertainty, whereas time buffers will provide 'safety time' for unreliable processes. Time buffers are not stationary inventory, they are the 'right' product in the 'right' quantity and quality, just ahead of time. Time buffers are introduced ahead of all problematic processes to give more time to process the products. If more time (for unplanned set-ups for example) is required, the time buffers give the possibility to take more time than scheduled. Once the products are processed, the buffer will not be replenished. The long-term objective of course is to decrease unreliability and hence reduce the time buffers.

A special case is the steel mill, as a synchronized scheduling approach would have to rely on a slab stock as an intermediate buffer to cover processing lead times. The reason is that the caster schedule is driven mainly by technical issues and is not flexible enough to respond adequately to the demand, although known and stable, without this buffer stock. This slab stock also serves as safety stock to cover casting process reliability problems.

CONCLUSION

A realistic supply chain game, which models reality, has far greater impact on credibility than relying on a standard game, such as the Beer Game, however good this might be. To build a realistic supply chain requires a complete understanding of the supply chain. Full understanding of the supply chain was facilitated by the use of mapping tools (Hines and Rich, 1997).

Building the game itself proved to be a learning experience for the researchers, and one that the participants from the companies themselves could benefit from. Running the game with participation from managers from companies along the supply chain has proven an excellent facilitation device. Moreover, the physical simulation of various alternatives gives impetus to cooperative work – perhaps even as far as synchronizing demand and supply.

The game results clearly demonstrate the impact of the synchronized supply chain scheduling concept, although the simulation model is simplified and therefore not sufficient to demonstrate the general applicability of the concept. More sophisticated mathematical models will be necessary to verify the concept under the conditions of varying coil sizes and product ranges that include several hundreds parts, some of them showing less stable demand patterns.

The game can also show the impact of batch sizes and unreliable processes. Halving the press batch immediately halves the supply chain inventory for the particular tier and the supplying tiers of the chain. This is a point of vital understanding, which often conflicts with the idea of Economic Batch Quantities (EBQ), which is derived from unsuitable performance measures and tends to drive companies towards a local instead of a global optimum.

Concerning process reliability, the game clearly demonstrates that unless resolved, unreliable processes can cause disruptions and demand amplification by inducing short-term rescheduling decisions which rebound along the entire chain.

Supply chain simulations are a powerful tool to allow participants to see and understand their own supply network. Apart from this educational aspect, a game can also be used as a supply chain engineering tool to deploy, discuss and validate changes to real-world supply networks.

REFERENCES

Alber, K. and Walker, W. (1997) Supply chain management: a practitioner's approach. *APICS Conference Proceedings*.

Burbridge, J.L. (1983) Five golden rules to avoid bankruptcy. *Production Engineer*, 62(10).

Forrester, J. W. (1961) *Industrial Dynamics*, New York: MIT Press and John Wiley & Sons.

Goldratt, E. and Fox, R. E. (1986) *The Race*. Croton-on-Hudson, NY: North River Press.

Hines, P. and Rich, N. (1997) The seven Value Stream Mapping Tools. *International Journal of Operations & Production Management*, 17(1), 46–64.

Hunter, A. (1990) *Quick Response in Apparel Manufacturing: A Survey of the American Scene*. Manchester: The Textile Institute.

Kurt Salmon Associates (1993) *Efficient Consumer Response: Enhancing Consumer Value in the Grocery Industry*. Washington, DC: Food Marketing Institute.

Lee, H.L., Padmanabhan, V. and Whang, S. (1997) Information distortion in a supply chain: the bullwhip effect. *Management Science*, 43(4), 551.

Martin, A. J. (1995) *Distribution Resource Planning*, revised edition. New York: John Wiley &Sons Inc.

Sterman, J. D. (1989) Modeling managerial behavior: misperceptions of feedback in a dynamic decision making experiment. *Management Science*, 35(3), 321–399.

Tetu, L. (1998) Supply chain planning & synchronisation. *APICS – The Performance Advantage*, 8(6).

Womack, J. and Jones, D. (1994) From lean production to lean enterprise. *Harvard Business Review*, 74(2), March, p. 93–103.

Developing a lean improvement initiative in a major company: a case study of British Steel Strip Products

18

Chris Butterworth

INTRODUCTION

This chapter will describe the experiences of the LEAP project with relation to British Steel Strip Products (BSSP). At the time of the programme launch BSSP was part of British Steel plc and employed just over ten thousand people across several sites in the UK. Two large fully integrated works located at Llanwern and Port Talbot in South Wales made up the largest proportion of BSSP output.

It may seem an obvious statement but a large integrated steel company is very different from a component manufacturer in many ways. Firstly, from a customer perspective, rather than having a relatively small number of customers with very similar needs BSSP supplied a wide range a products into several very different market sectors. The main market sectors were construction, automotive and packaging and these three combined accounted for over half the total sales. All these very different customer needs had to be met from the same pieces of capital equipment. An upstream production line might on one shift be producing steel which will ultimately be used on the side of a steel-clad building and on the next shift be producing steel which will make the bonnet for a top of the range sports car.

This capital equipment is on an enormous scale. When a height extension was required to one of the main galvanizing lines at Llanwern it was to become so high that discussions had to be held with the local Royal Air Force representatives in order to ensure that there was no risk to pilots carrying out low-flying practice runs. Distances on sites can be measured in kilometres rather than metres and the reaction of first-time visitors is always the same: 'I can't believe how big it is.' The equipment is also extremely capital intensive. In 1998 the company invested in a new Continuous Annealing Line in order to reduce lead times and produce higher strength steel demanded by the marketplace. This single investment alone was over £120 million. This means that there is inevitably financial pressure to ensure that units produce sufficient volume to cover the capital investment.

In addition, the demands of the markets are changing. When the plants were originally built relatively basic grades of steel were needed in high volume. But the market now needs a massive range of different grades to meet a wide variety of performance criteria. In the five years from 1990 to 1995, over 50 per cent of the grades supplied to make a typical automobile were newly developed steels. Steel is now supplied lighter and stronger than ever before. It is supplied coated in a wide variety of alloy coatings which give extended corrosion resistance and it is supplied as a finished painted product used to produce household appliances from washing machines to videos. All of this variety was supplied by BSSP and in total over six million tonnes of steel were produced annually.

The above facts try to paint a picture of the enormously complex organization which the LEAP team had to try to understand in the context of Lean thinking. The development of Lean thinking in any large organization can present quite a challenge both to the organization and to the individuals who champion the initiative. The principles of lean are straightforward and few can disagree with the logic of what they are trying to achieve. But in any organization, agreeing on the principles is a long way from working out how to implement the changes that are required to achieve them. This chapter will seek to describe the experience of the LEAP team and share some of the lessons learned.

APPROACH

A consistent approach taken throughout the project was that of awareness raising, education and implementation. In every organization the importance of the awareness and education stages cannot be understated. In a large organization this represents a significant challenge. While it is important to start at the top it is not enough to raise awareness only at a senior level. Buy-in at all levels of the organization is important and the LEAP team decided that one way to do this was through the implementation of practical improvement projects. This learning by doing is very powerful and can quickly convert a healthy scepticism into enthusiastic support.

The original intention was to take the value streams that had been mapped at the component manufacturers and steel service centres and follow these value streams through all their processes at the steel mill. The sheer size of the facility provided challenges for the application of some of the tools. For example, the site mapped is over 3 miles long and walking the process to collect data for the process activity map would have taken several days. The plan was to map in detail a major value stream for the supply chain being researched, following the process flow backwards from the despatch of material to the customer to the start of the planning and order recruitment process.

The enormous size and complexity of the operation meant that these tasks were very time consuming. It quickly became clear that there was a danger of resources being expended on data collection that could take several months to gather and even longer to analyse. Instead it was decided to focus on two areas: the information flow from receipt of customer order and the detailed physical flow in one of the despatch bays. The despatch bay was chosen as it was the nearest point to the customer and the next link in the chain. A large proportion of the products supplied to LEAP customers were channelled through this particular despatch bay. Again, to try to give some idea of scale there are four despatch bays each of which could absorb several football pitches. Summary information was then collected for the downstream processes to give the overall picture.

MAPPING DATA

It proved extremely difficult to map an individual value stream as physical flows tended to be generic to particular product types. Some grades of a particular specification could follow a set route every time they were produced but in many cases a variety of routing options existed. The information flow was generic to all orders. An important lesson learned was that careful consideration need to be given as to how the Value Stream Management mapping tools are used in any given situation. For example, the process activity map of the despatch bay was very useful in quantifying muliple handling, distance travelled and delays but a detailed process activity map exploring the length of the hot mill would have provided limited useful information for this project. It would have been extremely useful for a detailed targeted improvement initiative focused in this area if the top-level mapping had identified the need.

FINDINGS

Once the data had been gathered and analysed a presentation of the main findings was given to a group of directors and senior managers. The key points from the presentation were:

1. demand amplification was a major factor affecting delivery to time;
2. opportunities existed for improved information flow in the order fulfillment process;
3. a mismatch in the balancing process between sales and capacity forecasts contributed towards delivery to time issues;
4. opportunities existed for improved product flow in the despatch bay.

As the team had limited resources available, it was agreed that projects would be pursued in the areas of demand amplification and product flow in the despatch bays.

THE DESPATCH BAY PROJECT

The mapping of the despatch bay showed opportunities for improved reliability for delivery to the customer, product flow, stock reduction and turnaround times. A project team was established to embed a clear process with detailed roles and responsibilities in the despatch bay. Locations were marked out more clearly and bar-code labelling and readers introduced. The bay was segregated by areas for road and rail transport and coils were coded by mode of transport at the point of packing so that they could be correctly transported. Fixed time slots were introduced for the loading and turnaround of trains and road hauliers and revised maintenance schedules implemented for the cranes. A new perpetual inventory process was introduced and stock accuracy designated as a Key Performance Indicator (KPI) for despatch bay operatives and management. The results are impressive:

- Stock accuracy has improved to 98 per cent.
- The average number of times a coil is handled has been cut by over 50 per cent.
- There are clear roles and responsibilities for everyone involved in the process.
- Time slots have reduced average turnaround time by over 20 per cent.

This work was piloted in one despatch bay and is being progressively implemented across other sites.

DEMAND AMPLIFICATION

It was possible to construct a demand amplification map for specific value streams and these proved extremely useful in identifying opportunities for improvements in demand planning. For example, it was possible to show that a demand that started out fairly level at the customer was received distorted by the mill. The demand amplification maps showed a significant distortion in the demand pattern as information moved back up the supply chain and into the mill scheduling systems. Even where demand started out fairly level it was distorted beyond recognition in the demand profile seen by the mill manufacturing operations. This meant that delivery to time information had little resemblance to the end customer's real requirements and highlighted a significant opportunity for further work. This is illustrated in the Demand Amplification Map shown in Figure 1.

This chart shows a customer demand at a relatively stable level of approximately 50 tonnes per week. The demand was then tracked through the order entry and purchase order systems at the service centre and the orders eventually placed on the mill overlaid onto the original demand. The elapsed time is sufficient to overcome any discrepancies which could be explained by lead-time differences. There are many reasons for this which are discussed in more detail in Chapter 9. One of the consequences is that the mill sees no pattern to the demand profile. What is actually a stable customer demand appears to be a totally random requirement.

When the information is received at the steel mill it is aggregated into all the orders received for a particular delivery week. In order to produce the steel required in this case the material must pass through multiple production units on multiple sites. Gathering information at this level proved to be extremely laborious and extremely difficult to get at. As such it was decided to record the actual orders coming in against orders despatched as this information was readily available and its accuracy could be verified through other systems, e.g. invoicing data. As Figure 2 shows, there is a major discrepancy between the two figures.

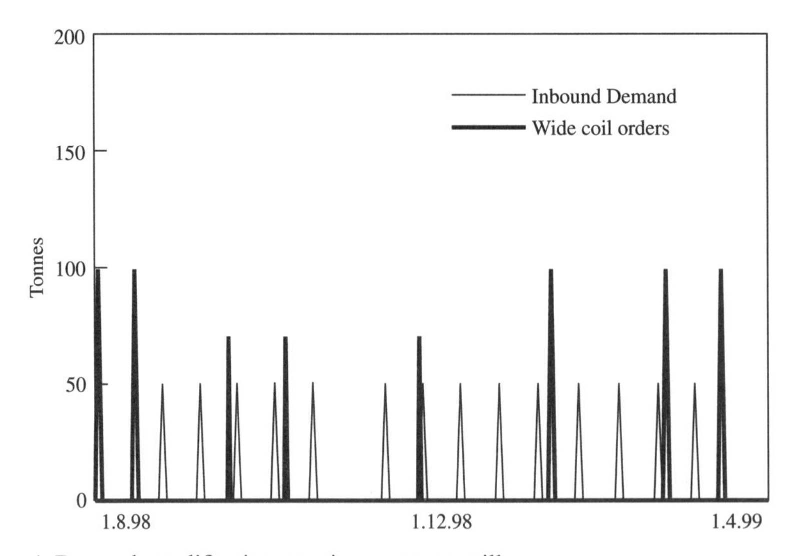

Figure 1 Demand amplification, service centre to mill.

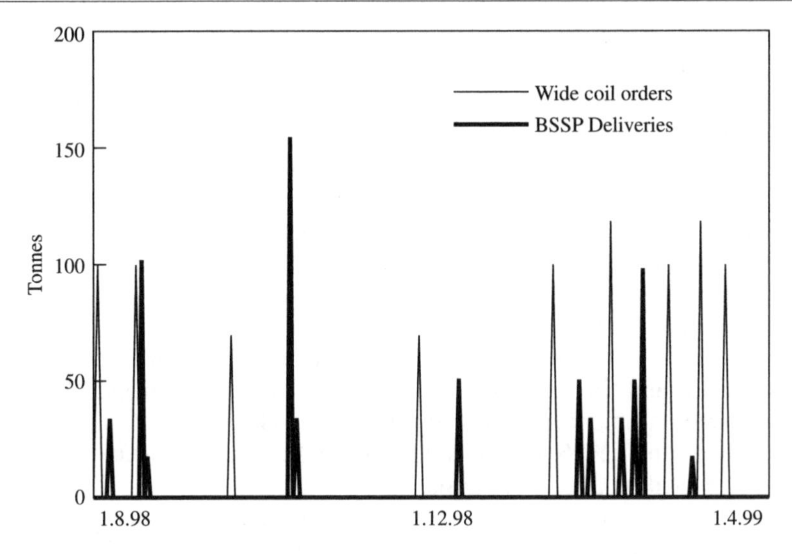

Figure 2 Steel mill delivery pattern.

Several underlying reasons were identified for this difference. These included:

- no clearly visible pattern of demand;
- no use of repeat demand information;
- no information link between true customer demand and mill production;
- a mismatch between capacity forecasts and actual achievement.

One lesson learned was that in a very large multi-product organization is easy to assume that there is no pattern to customer demand. Yet if a pattern can be identified and harnessed for a significant part of the order book, it provides tremendous opportunities to introduce more stability into production planning, improve delivery to time and reduce inventory.

DEMAND AMPLIFICATION PILOT PROJECT

A Pareto profile showed that five parts accounted for 45 per cent of the total value supplied to one of the major component manufacturers and 10 per cent of the total value through one of the service centres. As such it was agreed that a pilot activity on these parts could show significant benefits. A team with representatives from each part of the chain worked together to determine the true customer demand for the top five parts by analysing the schedules from the OEMs and then converting the part quantities into press shop and blanking shop batches based on the chosen batch cycle. An example profile is shown below:

OEM demand = 1200 parts per day
Press shop batch = 6000 parts (a one-week batch cycle)
Blanking shop batch = 6000 parts (a one-week batch cycle)
Tonnes equivalent = 50 tonnes per week
Minimum mill batch for this grade = 100 tonnes

This analysis was conducted for all five parts and buffer stock levels agreed and set at each point in the chain. In the example above, it was agreed that the supply chain would work to the fixed cycle of 50 tonnes per week and the mill would manufacture on a fixed two-week cycle. Buffer stock ahead of blanking equivalent to one day would be held at the component manufacturer using an internal pull system which would trigger raw material call in 24 hours ahead of when the blanking was to take place. The service would hold an average of three weeks' total stock consisting of a buffer of two weeks and a an average pipeline stock of one week. The mill would produce every two weeks and ship to the service centre on a fixed day every other week. This arrangement was essentially the same for all parts with slight variations in quantities.

A feedback loop was established whereby the component manufacturer advised the service centre and the mill of any changes in customer demand. These were formally reviewed on a monthly basis and any required changes to the pipeline agreed jointly by all three parties. Supply chain stocks and any performance issues where also formally revised on a monthly basis.

Measurements taken before the pilot showed that average raw material inventory in the whole supply chain was over six weeks. This consisted of:

- 1 week at the component manufacturer;
- 4 weeks at the service centre;
- 1 week at the mill.

After operating the pilot for several months the stock profile had changed to:

- 1 day at the component manufacturer;
- 3 weeks at the service centre;
- 2 days at the mill.

In other words, total raw material supply chain inventory had reduced from 6 weeks to just over 3.5 weeks – a reduction of over 30 per cent. At the same time, on-time delivery to the component manufacturer improved to 100 per cent.

This pilot was manually very intensive to operate at the mill as the parts had to be monitored and controlled outside of the normal systems. However, as a result of the work it was agreed that a new computer program would be fully specified so that the findings could be rolled out to a large number of customers. This 'pipelines and schedules', as it is called, will allow much more automatic control of repeat demand patterns with the users setting parameters and the system reporting exceptions that need intervention. At the time of writing, this system has now been specified and an implementation plan is being drawn up.

FOLLOW-ON PROJECTS

The success of these projects led to requests for the team to support work in other areas. One of these was in New Product Development (see Chapter 19) and the other was in support of a major programme working with suppliers, details of which can be found in Chapter 20. These projects provided practical benefits and also provided useful opportunities for expanding the application of the value stream management tools.

One issue that the team constantly faced was that of resource. Having identified a series of issues was one thing but clearly it was not possible to resource work on them all in such a large organization. It was decided to focus on the projects on demand amplification and despatch bays as these could be addressed with the resources available, could be used to demonstrate successful improvements and could be clearly linked into the other companies in the LEAP programme. The work in New Product Development and the project with suppliers were undertaken because they provided opportunities to explore new aspects of applying the VSM tools and because the requests for involvement were the result of customer pull from the senior people in those areas.

COORDINATING CHANGE IN A LARGE ORGANIZATION

At the same time a large number of improvement initiatives were under way across the business and it was clear to the directors that a coordinated approach was required which would pull the various initiatives together. As a result a major business-wide improvement programme was launched supported by the board and driven by strategic objectives.

This programme pulled together all existing initiatives into one coherent business improvement plan. A board-level steering group oversaw the direction and resourcing of the programme. Key processes were identified, such as sales acquisition and order fulfilment. Dedicated full-time project teams were established, focused on each key process and led by a senior-level champion. The order fulfilment process was targeted as a particular stream of work led by the Logistics Director. Members of the LEAP team joined this project. The data gathered and ideas for improvement coming from the LEAP work were fed into the team. The complete process from receipt of customer order to arrival of the product at the customer was analysed in depth and improvement opportunities quantified in terms of customer service and financial benefits. The project team was made up of a truly multifunction team with senior-level representatives from all areas. This was a great strength as all functions were represented in the change initiative and the proposals much stronger as a result.

After some weeks of detailed data gathering and analysis of proposals, an holistic work programme was designed consisting of several interdependent projects to address all the issues identified within the key process areas. In the order fulfilment process these included, for example, major pieces of work on the balancing of sales and capacity forecasts, understanding and using true underlying customer demand and improvements in information flow. In addition, a project team was proposed to create a pilot for pull scheduling and examine how pull could be implemented in a wider context. Existing projects connected to the order fulfilment process such as the work in the despatch bays was also incorporated into the overall project plan. The key objective of this work was to achieve a step change in customer performance levels.

BUY-IN

To get to this stage took a very large amount of time, effort and commitment. The order fulfilment team alone had ten full-time members at senior management level and was led by a line director who came 'off line' for eight weeks to lead the project. This representated a major commitment to change by the organization but also gave a clear signal that it the initiative represented a serious desire to see that change would happen. Indeed, it is unlikely that significant change will be achieved in any organization without a substantial resource commitment.

The organization fully bought into the improvement programme proposals, which represented a major investment in time and resources. This buy-in was achieved by involving people at every level in the change proposals. A lot of involvement was on a part-time basis and senior departmental heads had to sign up to the feasibility and willingness to implement the proposals. This process meant that a lot of resource was expended in communication and coordination but this is essential if the proposals are going to win the necessary support.

THE MERGER

The organization was in the process of preparing to launch teams to implement the agreed work programme when it was announced that at a plc level a major merger with a European competitor had been agreed. This merger would create a new global metals company. The new company, named Corus, would have a significantly different make-up of assets and customers and the merger had major implications for the work planned, especially in the area of order fulfilment. Partly as a result of this and partly due the human resource requirements of the merger, the decision was reluctantly made to put on hold the major part of the work until the implications of the merger could be more fully understood. Projects that could be continued at a local level, such as the despatch bay process improvements were allowed to continue so that the identified benefits could be realized. At the time of writing this case study, plans are in the process of being finalized to launch a jointly devised programme of work in the newly merged organization.

FURTHER LESSONS LEARNED

It takes a lot of time, effort and resources to achieve change in a large organization. Full support from the top is clearly essential but it is not enough in its own right. It is necessary to engage key players at all levels in order to ensure that the vision is carried through. It is important not to underestimate the resources needed to effectively communicate to the wider organization.

An important aspect of the communication is that the initiative should not be perceived as being driven by a particular functional area with its own agenda. Implementing any lean improvement requires the organization to focus on the key processes and these invariably cut across several artificially created departmental boundaries. In some large organizations, entrenched departmental views may need to be broken down. One way of doing this is by using cross-functional teams with strong representation from each area that will be affected by any change.

It is important not to underestimate the importance of the awareness and education process. When it comes to implementing changes this can only really be successfully achieved by the people in the line jobs buying into and believing in the change. If implementation is seen as a project issue then it is unlikely to succeed. Instead the line managers must own the change as their own improvement initiative which can be facilitated by project teams.

In order to convince the organization of the need to change and the potential benefits, it may be necessary to undertake projects in specific areas to demonstrate what is possible. This is fine as long as they are in support of the overall strategy and can be tied into an overall coordinated programme. If they are isolated islands of success without any hard link into an overall programme they are at best unlikely to be sustainable and at worst could waste valuable resources.

CONCLUSION

In any organization, obtaining buy-in at the right level and with the right people is critical to the success of the change programme. In a large organization it takes a lot of time and effort to persuade people of the need to change and keep everyone on board with the work. The approach taken by the LEAP team was to illustrate improvements through the implementation of practical projects focused on improved customer satisfaction and reduced supply chain costs. The success of these projects was very useful in providing a level of credibility for the adoption of new ideas.

A coordinated strategy-driven approach to improvement initiatives is essential for long-term success but at the same time it is often necessary to demonstrate the art of the possible through focused projects addressing specific processes or areas for improvement. These help to achieve the buy-in required for a larger initiative and provide a groundswell of support for what will undoubtedly be a major programme of work. It is important not to underestimate the resource commitment that will be required.

In the fast-moving pace of today's business world, even when this support is fully achieved and aligned outside factors such as a merger can have a major impact on the timing of the planned improvement programme. It is not, after all, a destination but a never-ending journey.

ACKNOWLEDGEMENTS

Data for demand amplification charts was provided by James Sullivan.

Section 6
Lean Applications in Other Situations

Applying lean tools in new product introduction 19

Ann Esain

INTRODUCTION

How New Product Introduction links with LEAP

Lean Thinking contends that Lean Production is 'Lean' because it uses less of everything compared with mass production (Womack and Jones, 1996 [1]). If using Lean tools and techniques in manufacturing enables such improvement then New Product Introduction should also be able to benefit from the Lean approach.

The Lean Enterprise Research Centre (LERC) had been exploring the potential benefits of applying Lean principles of Lean in New Product Introduction (NPi) whilst working on a previous research programme 'The Supply Chain Development Programme' and had produced a theoretical approach to study the current state of the process (Hines *et al,* 2000 [2]). LEAP offered the opportunity to test the theory and validate the use of diagnostics to achieve the objectives stated in Chapter 1.

It was during the LEAP project time frame that a link was formed between another related research project, Cogent (Nakamura and Milburn, 1997 [3]). Cogent was an Innovation in Manufacturing Initiative (IMI) project (as is LEAP) that was managed by Cranfield University and sponsored by Nissan. The main purpose of the project was to focus on the New Product Introduction (NPi) process and more particularly the improvement of this process along the supply chain.

Phases of the sub-project within LEAP

The three-phase approach, which is explained fully elsewhere in this book, adopted for the LEAP project and also used to explore order fulfilment, formed the basis of the review of NPi.

- Phase 1 was initiated in the last quarter of 1997. It included meetings to explain the logic, ideas and gain approval to conduct the research. This was completed within a short time frame.
- Phase 2 focused on the diagnostics process being tested, with data collection, waste and analysis, the reviews of strategy versus process (a further development of the original theoretical work), and proposals for improvement action, took place from October 1997, with an interim review in January 1998, from which some early action for change was initiated.

- Phase 3 was the initiation of the change programmes selected. These were particularly monitored against the projected savings – the use of less of everything. The key research was undertaken in the early part of 1998.

British Steel Strip products and New Product Introduction

At the time of this research British Steel was ranked third in their marketplace, with Nippon Steel being the leading world supplier and the Korean company Posco second. Improving this position was part of the British Steel plan. Fifty per cent of the production of steel was for overseas markets. In South Wales alone, British Steel employed ten thousand employees. Wales contributes to over 45 per cent of total UK steel production (Pathway to Prosperity, Welsh Office, 1998 [4]).

New Products were becoming increasingly more important to British Steel Strip Products (BSSP) whose main marketplaces were Automotive and Construction. The Automotive sector was proactive and it was believed that the retention of accounts with the lead players in this market would keep the rest of the business ahead of the technological game.

The type of product development undertaken for the automotive sector was broken into three segments:

- Customer Specific;
- Market Led;
- Business Innovation.

The vast majority of development was 'Customer Specific' and there was a strong desire to cross-fertilize these technological advances across the ranges and sectors of the strip products business. The key driver of the organization was profitability and that was applied equally to all areas of British Steel.

The customer pressure, from the automotive original equipment manufacturers (OEMs), of year on year improvements was a theme for change, as was under-performance relative to the competition. British Steel Strip Products were involved in a number of initiatives to enhance working practices, organization and technology to realize their goals of providing the customer with value propositions in advance of the competition.

A group of senior technical employees had already been recruited to 'make a difference' and the Managing Director of the business had placed an emphasis on New Products. Staff had been reorganized to be 'customer facing' and an account management approach had been adopted to determine customers' needs.

Method

Introduction

The structure, which was employed during this work, was designed for the product introduction process along with the organizational strategies, which effect that process (Clarke and Fujimoto, 1991 [5]). The steps involved a review of the strategies at different levels of the organization and the detailed activities of the new product introduction processes, particularly related to time and waste. The outcome was to establish a detailed plan of action for continuous improvement at all levels of the firm's organization.

At the outset a quantitative study of the process was proposed for diagnosis. The use of surveys as a method of analysis was considered and rejected because depth not breadth of knowledge was required. Depth, particularly relating to data, to highlight issues of complexity and constraints was considered necessary to obtain continuous improvement. This depth was required particularly in personal behaviour of teams (microcosms of the organization) and between different functions and levels within the organization. A case study approach was selected, with structured observation (quantitative) and interview/brainstorm (qualitative) (Saunders *et al.*, 1997 [6]).

Theoretical framework

If Stalk and Hout's (1988, [7]) proposition is correct, that time is a powerful source of competitive advantage, then the focus for improvement in this process must include time-based methods. Time alone will not result in the success of product development. The strategy of the organization is also crucial to the success of the process. Clark and Fujimoto (1991, [5]) claim that a strong strategy or concept can provide a thematic framework; a direction that clarifies, simplifies and integrates the many detailed decisions and actions that define, and are embodied in, the new product. New product strategies are important determinants of long-running company success (Copper, 1984, [8]). If, however, the whole portfolio of organizational strategies is in contradiction with the new product strategy or concept, there may be a direction that confuses, complicates and decomposes the many detailed actions of the new product process.

The motivations for new products are demanding customers, ever-shorter product life cycles, better IT, competition, a desire for differentiation in the marketplace etc. (Kanter, 1989, [9]). This also creates the need to fully understand what happens during the process and the issues, which eat into the time for product development. With this in mind and using the earlier research from the Supply Chain Development Programme (Hines *et al.*, 2000, [2]) an approach to evaluate processes was devised (see Figure 1).

One of the missing elements of the original approach, described in the above paragraph, was that the diagnostics used had no clear linkage to a company organizational strategy. A simplistic use of quality functional deployment (QFD) was proposed (Clausing, 1994, [10]). The employment of this tool would be in a slightly different environment than its intended use; however, the intended use had originally been in the development of new products.

The aim was to deploy the 'mileage chart' feature and use this to identify conflicts between the objectives of the organization, as detailed in the strategy documentation (Esain, 1999, [11]). In Figure 2 the conflicts are illustrated through the use of a cross, while arrows depict that one objective leads to another or is mutually supportive. The combination of diagnostics and evaluation was then tested at BSSP, an enthusiastic company looking for new ways to 'keep ahead of the game'.

To ensure that the project was owned by British Steel a single point of contact was established. This person was also the champion of the project within British Steel. A simple project plan was established with milestones and review points agreed. A product was selected which could be followed through the process and a team was nominated based on their involvement with the development of this product.

Diagnostics of New Product Introduction

Value Stream Mapping Tools Adapted from Hines and Rich, 1997 [13]	Applicable to New Product Introduction		
	Yes	No	Yes
1. Process Activity Mapping	Y		
2. Quality Filter	Y		
3. Product in Process Funnel	Y		
4. Skills Constraints Funnel	Y		
5. Value Added Time Profile	Y		
6. Decision Point Analysis		N	
7. Human Structure	Y		
8. NPD Responsiveness	Y		
9. Demand Amplification	Y		
10. Committed Design Cost Curve	Y		
11. Design Relationship Mapping	Y		

Figure 1 An approach to evaluate processes. Adapted from Esain, 1999 [11].

Strategic Alignment

1. Long-term profitability
2. Maximization of Assets
3. Improved Product and Market Development
4. Cost Competitiveness

↑ A leads to B
↓ B leads to A
X Negative

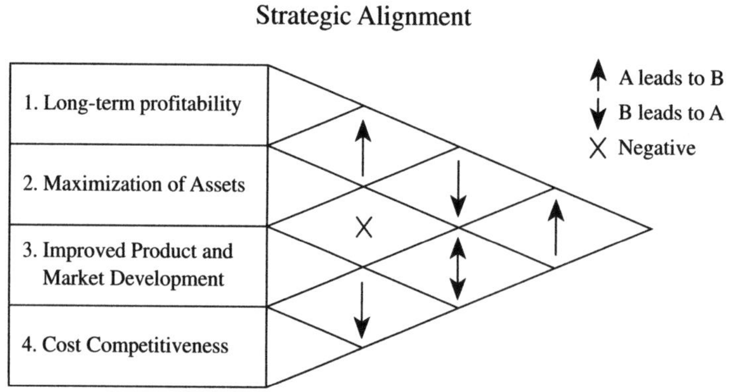

Figure 2 An approach to evaluate strategy. Adapted from Esain, 1999 [11].

CASE STUDY FRAMEWORK

Detailed below is the case study framework which was used to evaluate process and strategy. The framework has four steps:

- Education and Overview of Process
- Structured Observation
- Review
- Conflict Resolution and Formulation of System

These steps are a reflection of Deming's (Walton, 1991, [12]) plan, do, check, act, cycle (Figure 3). The case study framework was designed to establish the current state of the process at various levels of the organization and in conjunction with company strategy.

Diagnosis of the process relies on the selection of current or recent products, which are typical within the organization. Once the product selection has taken place, a top-level review of the process is undertaken with the senior management. This activity is designed to:

1. gain an understanding of the process through the senior management team's eyes;
2. provide a route map of who in the organization should be interviewed; and
3. gain approval for further interviews of participants in the process.

Building on the value stream analysis tools – VALSAT (Hines *et al.*, 2000 [2] and Hines and Rich, 1997 [13]), the types of wastes that exist within the process are identified through brainstorming by a cross-functional group. These are then assessed against the series of tools and techniques shown in Figure 1 above. The wastes are then evaluated and ranked in order of applicability to the process. These tools and techniques are both quantitative and qualitative and enable validation of perceptions, and will theoretically help organizations visualize their process across all the participating functions.

The framework also reflects the blocks and barriers to change and innovation, and how this affected the case study. Figure 4 shows how, generically, these elements fit together over time and in relation to effort. The case study framework will focus on the issues surrounding New Product Introductions (NPi) and in particular the approach taken to reduce lead time.

Additionally, the issue of complexity is addressed. Figure 5 illustrates the types of processes that exist in the delivery of a new product to market. The process, that is

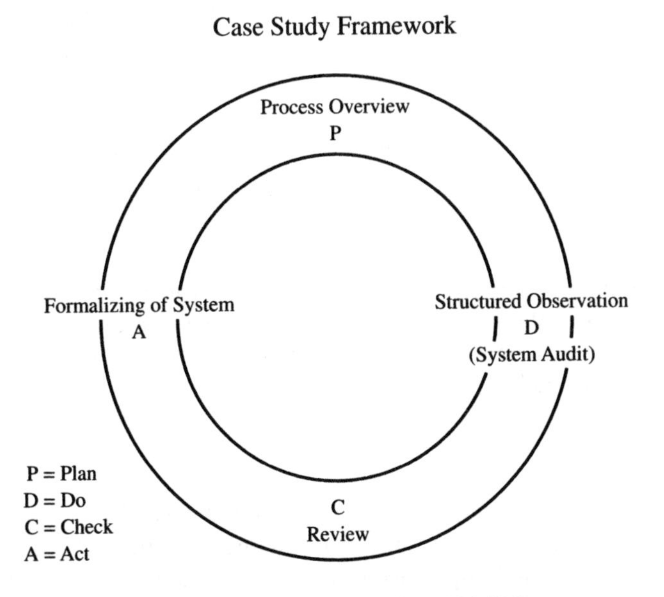

Figure 3 Case study framework, adapted from Esain, 1999 [11].

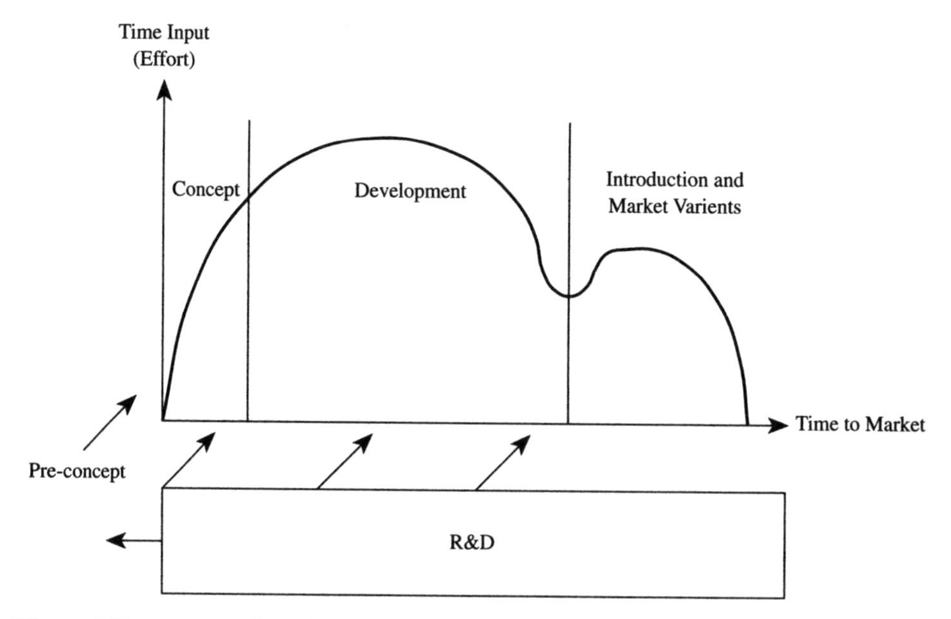

Figure 4 Key aspects of product development, adapted from Hines *et al.*, 2000 [2].

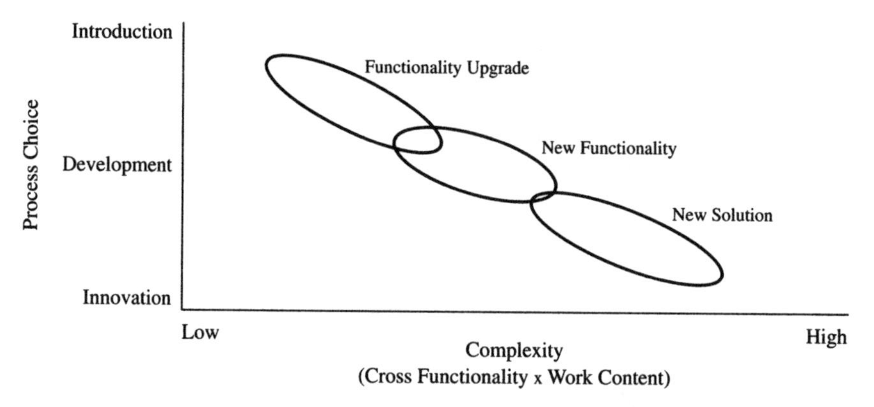

Figure 5 Process choice vs. complexity.m, adapted from Hill (1995 [23]).

applied often through experience will be in correlation to complexity. It is seen in some organizations that the same rules of engagement for new products are employed regardless of the process complexity. This is a constraint on time to market and needs to be part of the case study approach.

EDUCATION

The principles of Lean Thinking (Womack and Jones, 1996 [14]) are often considered to be counter intuitive, therefore at the initiation of this research an element of the background logic needs to be explained to the participants. To reinforce the key principles, cross-functional exercises were devised to create buy-in at all levels of the organization. Three particular elements, detailed below, were deemed to be crucial.

Visualization

What is a process? LERC have been using the following definition for processes to help firms understand the focus.

> Processes are patterns of interconnected value adding relationships designed to meet the business goals and objectives. Processes are in a continual state of change, adapting to both expected and unexpected events. The way in which information is communicated and knowledge is managed affect the quality of decision making and innovation within a process (Dimencescu *et al.*, 1997 [15]).

In order to help visualize the process and address complexity (as detailed in Figure 5), the key aspects of bringing new products to market were separated (Hines *et al.*, 2000 [2]); these classifications were:

- Innovation (Research and Development including Concept) – the process of creating technology seeds and innovations which may not be practical today but will be in the future;
- New Product Development – the step change in products or processes which converts the technology seed into products that customers want; and
- New Product Introduction – the creation of variants to the New Product for target markets.

Waste

Additionally, the use of wastes as a means of time reduction requires firstly, definition and secondly the use of customized language. The terms used should not only relate to the process under examination, but also comply with the unique language of the organization.

The seven wastes, which are listed below, therefore needed to be translated into BSSP's unique language to facilitate understanding.

- Defects – in products;
- Overproduction – of goods not needed;
- Unnecessary inventory – of goods awaiting further processing or consumption;
- Inappropriate processes;
- Unnecessary motions – of people;
- Transporting – of goods;
- Waiting – by employees for process equipment to finish its work or upstream activity.

The list above was specifically designed for the production of goods. Taiichi Ohno (1988 [16]), originally developed these waste definitions (Bicheno, 1994 [17]). Hence the use in a different process also necessitated the translation into terms more appropriate to BSSP.

In the Supply Chain Development Programme, the firm, RS Components, translated the terms for process activity mapping (called in their case 'Information

Flow Mapping') into waiting, communicating, operation and decision. These terms were used in order to capture the language of the organization (Hines *et al.*, 2000 [2]). In the LEAP research this principle was adopted at the stages of education and overview of the process, rather than the detailed data collection stage. This was a conscious decision to validate understanding within the team and reduce the risk of barriers to change.

Value

The final idea to be shared with the organization is 'Value'. All waste is related to what the ultimate customer values, and hence buys. An exercise to understand the written and unwritten requirements is undertaken. This can then be used to evaluate the process steps.

What is value? In the book *Lean Thinking, Banish Waste and Create Wealth in your Corporation* (Womack and Jones, 1996 [14]) value is defined as 'a capability provided to a customer at the right time at an appropriate price, as defined in each case by the customer', whereas process steps can be categorized against the following definitions:

- Value Adding – where the customer's requirements both written and unwritten are provided by the activity, e.g. where a product is changed from steel to a screw, where the screw was the item valued by the customer;
- Necessary but Non-Value Adding – where the activity must take place in order that the product or service required can be produced but adds nothing to the customer's requirement, e.g.. the computer system must generate an order to initiate production but the ultimate customer gains nothing from this activity; and
- Non-Value Adding Steps that add no value to the customer and can usually be easily removed from the process, e.g. waiting to authorize a quotation.

Once these definitions are translated for the organization processes then the team is able to quantify activity. Once quantified, decisions and assessments can be made regarding appropriate removal of activity to achieve customer focused processes.

OVERVIEW OF THE PROCESS

Tool 1 – Brown paper model or cartoon

To initiate understanding the process is visualized by using the first tool which is known as a brown paper model or cartoon (Figure 6). This is the top-level picture of a process, which may be available in the procedures of the organization, and is collected as quantitative data. A brown paper map is a graphical tool, which traces the flow of work and activities between departments in the sequence from initiation to conclusion. (Bichemo, 1994 [17]).

During the brown paper model exercise the commitment to structured observation/interview is sought. There will be detailed interviews at the next stage to gain individual involvement. The brown paper model provides a suitable platform to inform and involve, as well as allay any fears prior to interview, while primarily being the planning forum for the project.

Responsibility	Customer Need	Business Acceptance	Feasibility	Planning PDT1	Approval	Scheduling	PDT1	Validate	PDT2	Product Manual Listing	SOP
Account Manager/ Commercial									XXX		
Customer	XXX								XXX		
Account Team	XXX		XXX						XXX		
Product Dev Manager (CYS)				XXX					XXX		
Commercial		XXX							XXX	XXX	
Works Tech						XXX	XXX	XXX	XXX	XXX	
Works							XXX	XXX	XXX	XXX	
Logistics					XXX				XXX	XXX	
WTC					XXX	XXX			XXX		
Product Engineer (CTS/C)					XXX		XXX	XXX	XXX		

Figure 6 An example of a brown paper model.

This top-level picture is the level against which most senior managers of the organization operate. During the structured interview a picture of the true activities taking place emerge, often subtly different from that which was required, and enables qualitative data to be recorded. This parallel management, where managers believe one thing is happening and the workers are doing something else, misses the opportunity for problem resolution. Hence this stage of activity is crucial to facilitate problem elimination (Bicheno, 1994 [17]).

The brown paper mapping exercise is used to record individual perceptions of barriers to change within the organization or supply chain. This will help validate the strategic conflicts identified during the strategic review.

The brown paper exercise was undertaken to understand the process. Also, a four-fields map (generated out of the Welsh Development Agency programme 'Time to Market' in which BSSP had previously participated), which had already been constructed by the automotive team, was handed to the research group as a means of familiarization to the process (Hines, 1994 [18]).

One of the key issues raised by this map was the degree of effort being expended at the latter stages of the process. Best practice suggests that early involvement results in more successes for the new products. As a result BSSP created a graph of the type of profile which would reflect how their business could operate. This is reflected in Figure 7.

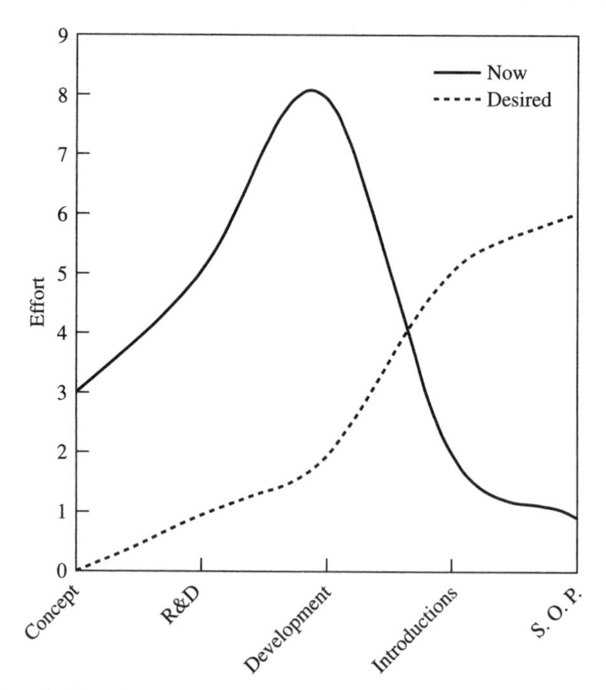

Figure 7 Profile of effort time for each stage of new product introduction.

Tool 2 – The Value Stream Analysis Tool (VALSAT)

As indicated earlier, the nominated individuals are asked to rank the wastes. Ranking is against the greatest probability that the waste exists in the process, i.e. if the process is the supply of stationery, the individual may consider that waiting is the most common waste and will rank this highly (up to a maximum of 10 points), whereas the individual may consider that overproduction will never occur and hence this will be given a ranking of zero. In order to retain an indication of scale the individual is allocated 35 points, which are awarded against the seven wastes listed in Figure 8.

In Figure 8 the waste of overproduction is ranked highest. The averaging process is used to ensure that no one functional area can prejudice the outcome; in this case nine individuals, from different levels in the organization and from five different functional areas, were asked to rank the wastes. The focusing exercise is also used to record individual perceptions of barriers to change within the organization or supply chain.

This ranking of the wastes forms the basis of the value stream analysis tool – VALSAT (Hines *et al.*, 2000 [2]). The relationship between the wastes and the corresponding maps is numerically valued. So, dependent on which wastes the group select, the tool suggests the most relevant maps to use. In this example the maps selected are shown in Figure 9.

Waste perception for BSSP New Product Introduction

Prior to the scoring exercise, staff at all levels in NPi were involved in a brainstorming to ascertain their views and perceptions of the waste in the NPi

VALSAT

Waste	1	2	3	4	5	6	7	8	9	Average
1. Overproduction	5	10	8	2	10	8	10	0	7	6.67
2. Waiting	6	8	7	5	4	10	5	10	3	6.44
3. Transportation	8	4	2	10	3	9	4	0	5	5.00
4. Inappropriate Processing	6	4	2	0	10	2	3	10	2	4.33
5. Unnecessary Inventory	3	3	6	10	3	3	2	7	3	4.44
6. Unnecessary Motion	5	1	2	0	3	2	3	6	9	3.44
7. Defects	2	5	8	8	2	1	8	2	6	4.67
Totals	35	35	35	35	35	35	35	35	35	35.00

Figure 8 An example of ranking the wastes within a process. (Esain, 2000 [24]).

VALSAT

	Process Activity Mapping	Supply Chain Response Matrix	Product Variety Funnel	Quality Filter Mapping	Forrester Effect Mapping	Decision Point Analysis	Physical Structure
	1	2	3	4	5	6	7
1. Overproduction	6.67	20.00	0.00	6.67	20.00	20.00	0.00
2. Waiting	58.00	58.00	6.44	0.00	19.33	19.33	0.00
3. Transportation	45.00	0.00	0.00	0.00	0.00	0.00	5.00
4. Inappropriate Processing	39.00	0.00	13.00	4.33	0.00	4.33	0.00
5. Unnecessary Inventory	13.33	40.00	13.33	0.00	40.00	13.33	4.44
6. Unnecessary Motion	31.00	3.44	0.00	0.00	0.00	0.00	0.00
7. Defects	4..67	0.00	0.00	42.00	0.00	0.00	0.00
Totals	197.67	121.44	32.78	53.00	79.33	57.00	9.44
Rank	1	2	6	5	3	4	7

Figure 9 An example of the selection of tools to validate perception. (Esain, 2000 [24]).

process. A summary of the key comments arising from the interviews is given below, categorized in terms of the seven wastes (Bicheno, 1994 [17]).

To understand the outcome of this exercise, an overview of the steps of the New Product Introduction process is required. In more simple terms than the brown paper model the NPi process at BSSP consisted of five steps:

1. Idea generation
2. Feasibility assessment

3. Development
4. Piloting
5. Launch

The brainstorm with the nominated participants reflected these different aspects. Particularly during the steps of development and piloting, production resources, in the form of people and assets, were often being used to achieve new product introduction output.

Defects

The causes of defects were considered to be:

- trial design – bringing products to market before being ready;
- scheduling errors, poor planning – particularly due to lack of understanding by the scheduling group (a production function) as to what were the unique requirement of NPi – one outcome was the misappropriation of product either to satisfy production shortfalls or due to no real process to separate a development or piloting product;
- unnecessary handling – in all ways across the process;
- lack of risk analysis – seen as not enough work at feasibility stage;
- process not adhered to – samples to use for future feasibility stage would go missing.

Senior management also saw defects as a major waste, particularly when defects (or errors in data) were not spotted at source and when they were identified it was usually due to a catastrophic failure. It was felt that this centred on process capability.

Overproduction

The causes of overproduction were considered to be:

- width and product changes, lack of modelling – seen as a result of not enough work undertaken at feasibility stage and the constraints around this part of the process;
- large trial tonnages, production record attempts, machine limitations, campaigns (laying down stock to run minimum batch quantities), just-in-case mentality and make to stock – all resulting in excess product and much of which was effected by the order fulfilment process;
- customer change of mind, order input errors resulting in incorrect information.

Senior management considered that overproduction was due to being uncertain that enough consistently good product would be the output of the process.

Unnecessary inventory

The causes of unnecessary inventory were considered to be:

- series development and piloting system;
- badly defined objectives, which relates to poor or incorrect information as detailed as part of 'defects' above;

- poor customer interface – resulting in incorrect information, incorrect materials etc;
- lack of laboratory simulation facilities as well as size and appropriate use of sample bank – resulting in the product being tested on production equipment and poor use of the sample bank.

Inappropriate processes
The causes of inappropriate process were considered to be:

- poor visibility of requirements and poor definition of requirements, which manifested themselves in costly and inappropriate methods used;
- just-in-case inspection – resulting in damage caused by inspection, this being the outcome of not enough/correct feasibility work;
- unnecessary or inappropriate activity simply to 'appear to be doing something' to the customer;
- over-elaborate approval procedures and lack of authority to technical managers – tasks undertaken at inappropriate level of authority.

Unnecessary motions
The causes of unnecessary motions were considered to be:

- unnecessary distribution of paperwork, re-approval of documentation – lots of walking and/or overuse of paper and printing.

Transporting
The causes of transporting were considered to be:

- doglegging of material, layout of equipment – this relates to the development and piloting stages and results in timely delays of materials;
- sample evaluation procedure – longwinded and unnecessary piloting approval systems and data collection for customer;
- travel between sites (as a multi-site environment), waiting to travel between sites and non-productive meetings.

Waiting
The causes of waiting were considered to be:

- poor scheduling (see 'Defects' above);
- incorrect information, communication errors;
- elaborate approval system, lack of empowerment, and overemphasis on sampling – resulting in delayed feedback, long R&D lead-times and delays at specific machine test release;
- customer trailing process seen as overemphasis on trials.

The perceptions collected above are quite typical. The quantity of the issues noted can be one of the barriers to improvement and change. These issues were subsequently reflected in the action plan drawn up at the end of the structured observation phase of the research project.

The next part of the exercise was to collect facts using the nominated cross-functional team (Dimancescu, 1994 [18]).

STRUCTURED OBSERVATION

Introduction

This section will discuss the tools tested from the original theoretical framework shown in Figure 1. In this case study the maps listed in Figure 1 were discussed by the nominated team and some were rejected as not applicable to this process at BSSP. These were:

- Products in Process {3} and NPD Responsiveness {8} – where data collection proved to be an issue;
- Value Added Time Profile {5} and Committed Cost Curve {10}– once again this was due to availability of data;
- Human Structure {7} – where, although produced for this case study, the map was rejected as it provided no useful content, partly due to the way in which indirect and direct staff were allocated;
- Demand Amplification {9} – amplification did exist due to batch size constraints and 'just in case' scheduling but the team felt that this map would only show an effect and not the cause, hence it was agreed not to use it.

Additionally, there were two tools that were added to reflect the unique nature of product introduction. The first was the generic slip rate of projects. This tool was designed to compare with the benchmark of best in class product development. It is said that Japanese companies are not faster in the development process but they manage the slip rate better (Carter, 1992 [20]). The benchmark of less than 15 per cent slip rate was given. Although this tool was not applied fully at the time, subsequent work has enabled the measurement and management of this element of product introduction.

The second tool was the generic understanding of customer needs – this is discussed more fully below.

Process Activity Mapping {1}

The pivotal technique applied during the data collection stage is Process Activity Mapping (Hines and Rich, 1997 [13]). This map is taken from the industrial engineering discipline and applied in a different manner. It focuses on fact and detail (see Figure 9), which provides the information against which improvements can be made. The critical path is mapped. It also allows for further perception of the process to be collected on a one-on-one basis.

The use of the Process Activity Map differs from its application in an order fulfilment process. In the NPi process the map is following mainly information flow. In the case study at BSSP the five stages described above were separated. Hence, where the information flow was the critical path this was tracked, i.e. at the idea generation or feasibility stages. Where physical product being produced was the critical path, in the main, this was tracked. Often there is a grey area when the data resulting from the production is key, not the product itself. In this instance it is imperative that during the interview the objective of the activity is established. These stages are then knitted together to form a coherent route map.

A selection of the observations collected while mapping the process at BSSP were reported back at an interim review with senior management at the end of January 1998, along with a Gantt chart and process activity map similar to that in Figure 10.

Activity No	Step	Area	Dist (m)	Time (min)	People	OP	TRAN	INSP	DELAY	COMMENT
1	Sales Inform Project Team of an opportunity	QEP Office	100	10	2	1	0	0	0	xxx was a well received tender for L
2	Delay	QEP Office	0	240	0	0	0	0	1	Estimated at 4 hours
3	Decision Team to go to Review meeting	QEP Office	10	5	6	1	0	0	0	Used Team Structure Identical to xxx
4	Delay	QEP Office	0	480	0	0	0	0	1	Estimated time 1 day
5	Travel to Review Meeting	Car		60	6	0	1	0	0	
6	Delay	Cust	0	20	0	0	0	0	1	
7	Meeting with LR and other runners	Cust	100	60	6	1	0	0	0	Companies Competing Meeting on 9/
8	Delay	Cust	0	20	0	0	0	0	1	
9	Travel to Lunch Review Meeting	Cust		10	6	0	1	0	0	Provided Cad plots and tapes at the
10	Delay	Cust	0	10	0	0	0	0	1	
11	Review Opportunity	Cust	50	120	6	1	0	0	0	
12	Delay	Cust	0	10	0	0	0	0	1	
13	Return to Office	Car		60	6	0	1	0	0	
14	Model Generation	QEP Office	0	2160	2	1	0	0	0	Estimated at 1 week less 1/2 day of
15	Delay	QEP Office	0	240	0	0	0	0	1	
16	Produce questions (specification clarification)	QEP Office	400	120	6	0	0	1	0	

Figure 10 An example of a Process Activity Map.

Different departments usually saw the root of their problems in other departments. Such things as 'plan from "x" too late to provide capacity'. This constraint had never been discussed prior to the mapping exercise and was immediately rectified enabling capacity to be scheduled.

Other interesting observations were related to solutions without justification. Such as, a dedicated process resource focused on continuous improvement would make a difference. This was further supported by a view that New Product Introduction should benefit from dedicated resources.

Other observations could be justified. For example, a skills shortfall in one area against the projected volume of work to be performed in that area. Additionally, the accounting methods meant that often management information was clouded by the 'need to lose overhead'.

This interim review of the process map also quickly highlighted the immense period of time when the new product introduction was in a waiting (delay) state. Much discussion ensued and ideas for simple changes were introduced by BSSP.

A benchmark for the best process for the production of goods is 10 per cent of the elapsed time is on operations (Hines *et al.*, 2000 [2]). In product development it was estimated that this figure would be less, between 5 per cent and 7 per cent. The process activity map provides a different view of the process and presents the opportunity to improve on the effectiveness of the remaining 90 per cent of the time. This makes process activity mapping so powerful.

Research across a wide range of UK firms has shown that the majority of processes have less than 1 per cent of value-adding operations (Hines *et al.*, 2000 [2]). The process activity map revealed that of the total time to bring a variant of a product to market 4.8 per cent of the activities were operations (see Figure 11). The nominated team on reviewing the data felt that 4.5 per cent of the total time was value adding (see Table 1). All indications are that BSSP has a better than average process.

The process activity map enables evolutionary change, i.e. the removal of steps that add no value. Also, the process activity map highlights radical areas for improvement.

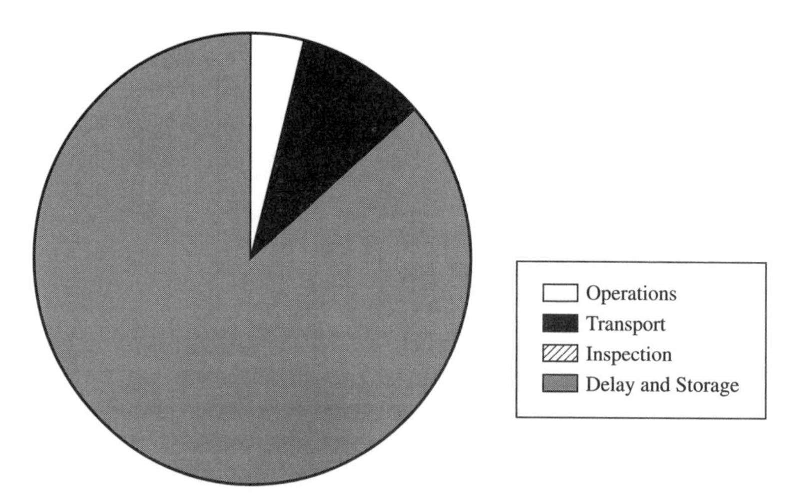

Figure 11 Summary of Process Activity Map.

Table 1 Total value added in the NPi process determined by the nominated team

Activity	*Number of steps*	*Time (100 = Current total time)*
Total process	138	100
Total operation	40	4.8%
Total value added	22	4.5%

The product being brought to market is a component incorporated into other products and hence time compression alone would not provide competitive advantage. The prime contractor is driving the timescales and is dependent on many other components being available. However, the opportunity to provide a variant rapidly may well become a competitive advantage in the future and BSSP has initiated change to be prepared for any future developments.

During the interview process qualitative issues are being recorded to help the interviewer to determine which quantitative techniques in the tool kit are applicable to the organization and/or supply chain. Hence the tool kit is designed to triangulate data gathered against perception, which has also been identified from the team, through interviews.

Quality Filter Map {2}

The Quality Filter map is used to establish if and where quality issues exist within the NPi process. These quality issues may be infrequent but when experienced will have serious consequences or may occur regularly and be considered an annoyance (often something which has to be lived with). Evaluation of these issues is undertaken in relation to three parts of the process: firstly, at the point the activity in the process originates, the second when the quality issue is apparent at a later stage in the process and the last is related to the supporting activity (see Figure 12).

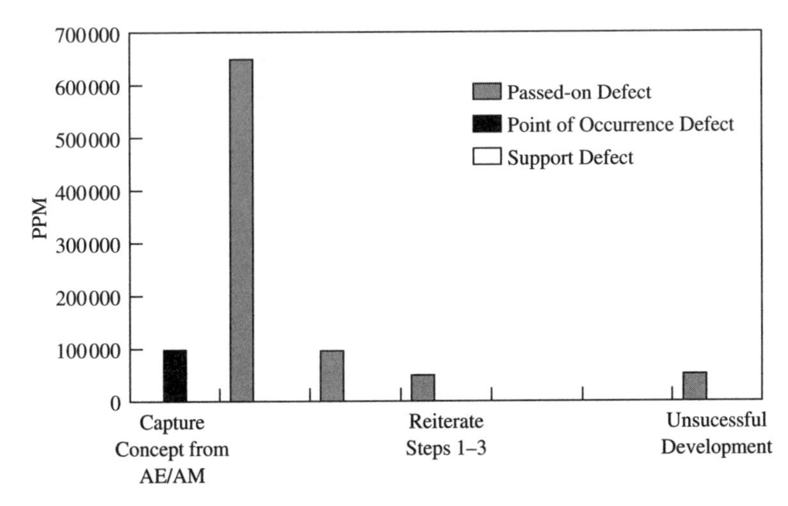

Figure 12 An example of a Quality Filter map for product development.

Using triangulation several other issues were validated. The quality filter map suggested that after feasibility, time was being used to validate the customer requirements (passed-on defects), and 65 per cent rework was being experienced as a result. The capture of concept requirements was taking place too late in the activity, i.e. not when the customer is initially being asked to detail the requirements. This was also shown in the process activity map. Finally this was also shown in the map used to understand customer needs. This new map suggested that only 60 per cent of requirements were captured at early phase of the process and that there was room for improvement. This is also validated by the perceptions caught during the waste brainstorm.

Additionally, a number of product developments went through the complete process but were not successful. The root cause of this was considered to be lack of understanding of customer requirements at the commencement of the project. This situation has been significantly changed and this type of issue should not occur in the future.

Skills Constraint Funnel {4}

Operations management literature suggests that the bottleneck of a process (Goldratt, 1990 [19]) will be the limiting factor. The volume that, can be produced from that bottleneck will be the maximum process volume. In product development the bottleneck generally is related to the availability of skills. In the structured observation, data is collected for a range of skills required to achieve successful product development. A cumulative graph is produced to establish the shortfall (see Figures 13 and 14). This shortfall when read in conjunction with the other maps may provide a clue to the root cause of waste.

The bottleneck of the process was identified through the skills constraint funnel. Once established and read in conjunction with the quality filter map it was established

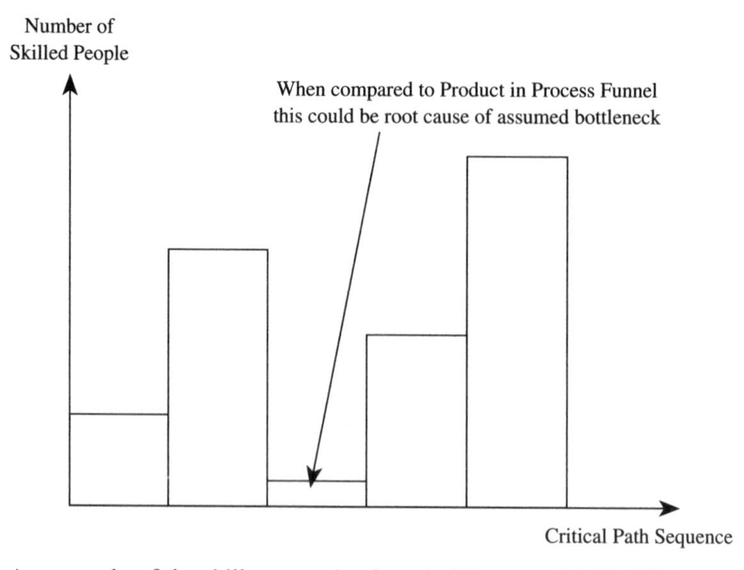

Figure 13 An example of the skills constraint funnel. (Hines *et al,* 2000 [2].)

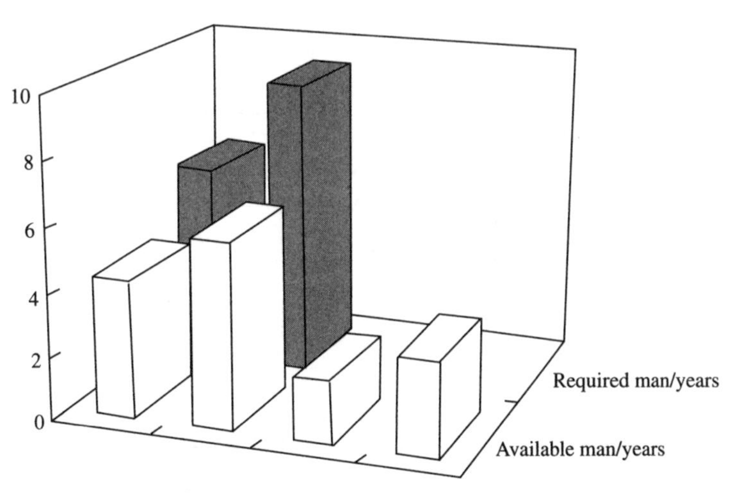

Figure 14 Skills constraint funnel.

that work undertaken by the bottleneck skills could be avoided if the definition of customer requirements was more clearly defined at the initiation of the development process. This would hence increase the capacity of the process, enabling more enquiries to be addressed or uncovering a new bottleneck in the process to be managed.

This was not the only root cause. Another outcome of the process was the analysis of work to ensure the skill bottleneck is utilized for strategic work only.

Product Development Relationship Map

The Product Development Relationship Map, which reflects the work of Alber and Walker (1997, [21]), provided a simple illustration of the movement of the product being introduced across departments in BSSP.

As is shown in the example given in Figure 15, this map simply illustrated how many interfaces there were in the process and as a result how the customer requirements were filtered to these departments. It was a good medium to show the senior management how complex the process has become when this should have been the least complex of the product development processes (see Figure 6).

Understanding Customer Needs

This map was in addition to the original theoretical framework (Figure 1) discussed earlier in this chapter. There is deemed to be a correlation between understanding customer needs and wants at the beginning of a programme and the successful launch of those products. This being the case, it was agreed to produce a tool to reflect this issue (Dimancescu *et al.*, 1994 [21] and Dimancescu and Dwenger, 1998 [22]).

At the outset of product development, early understanding of value, as defined by the customer, enables a successful development process. Part of the structured observation is to evaluate the degree to which early understanding is part of the process. This additional tool (see Figure 16) was developed to complement the original set of tools identified by Hines *et al.* (2000 [2]).

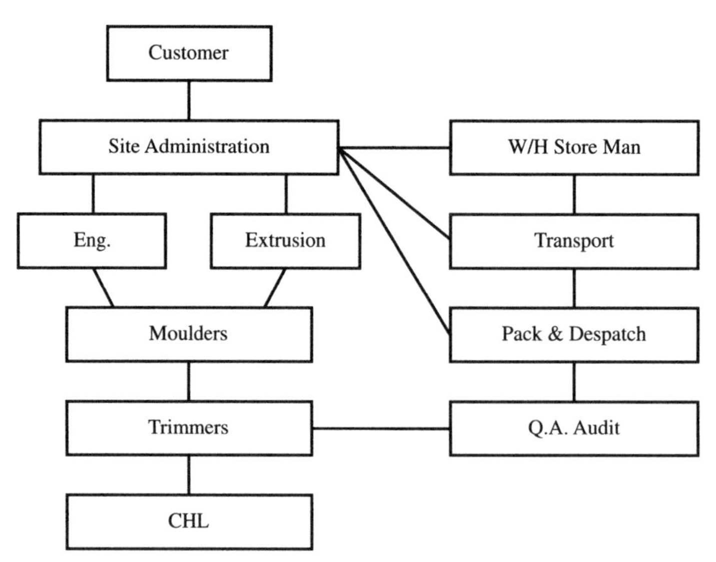

Figure 15 An example of a relationship map.

Checklist for Understanding of Customer and User Needs – At Concept Stage		
Method	How Used	Applied 0% – 100%
Customer Needs	Automotive Engineering Group	
Survey and Questionnaire	General (Not Targeted)	
QFD	NO	
Customer Prototype Evaluation	Material Selection to Late	
User Prototype Evaluation	Trailing Just In Time	
Customer Complaints	Unsure of formal routine	
Regulatory Issues	Awareness	
Market Segments Defined	Looking at Innovators or followers	
Learning – Previous Products	Yes	
Competitive Analysis	Considered Good	
Key Performance Indicators	NO	
Priority of Requirements	NO	
Order Qualifiers	NO	
Order Winners	NO	
Voice of the Customer	AEG (Recently)	
Embeddedness	Booklet Pilot with Customer	

Figure 16 Understanding customer needs.

For BSSP the ways in which the customers needs and wants were collected were varied. A list of the most popular and recognized mechanisms for collecting this information was listed and the degree to which they were applied was recorded. This was based on perception but even in these circumstances there seemed to be a need for improvement. One of the conclusions drawn was regarding measures, particularly that if there were a measure in the organization to record the success of collecting customer needs against successful launch this would result in focusing on this part of the process.

Additionally, the degree of effort in the early stages of the process could also be an indication of the level of customer needs being collected. It had already been seen in the brown paper model that people involvement was more cross-functional at the later part of the process (see Figure 6). Similarly, Figure 7 shows pictorially that effort in delivering the process was, at the time of mapping, at the latter stages of elapsed time for a product introduction. Hence it could be concluded that more effort at this stage of the process could result in more successful product introductions.

Strategic Objectives

Research and development, product development and product introduction rely on cross-functional groups to bring complex issues into an understandable form. If, however, the group are being directed to achieve aims that conflict with product introduction then confusion and misinterpretations occur. Hence, to achieve competitive advantage through time, it is also essential to review new product introduction in the context of the organizational strategy. The strategy itself may be the constraint (Dimancescu *et al.* 1997 [15]).

The key therefore is the strategy, tactics and measures used to focus the organization. In Figure 17, the example given shows strategic objectives that are potentially conflicting. At a working level they could cause confusion and personal conflicts, particularly where individuals or teams are carrying out specific roles.

When the maps were read in conjunction with the company strategy, a further set of constraints were identified. In particular, the maximization of assets was interpreted by the operations staff to mean long batch runs. Products being developed also needed to be trialled on production equipment; these were short batch runs and hence the aims of the two groups were in conflict.

These issues were affirmed by the process team, but were not seen as clearly by the senior management (Figure 18), and the highlighting of constraints enabled a set of measures to be produced, which balanced the corporate objectives across the functions.

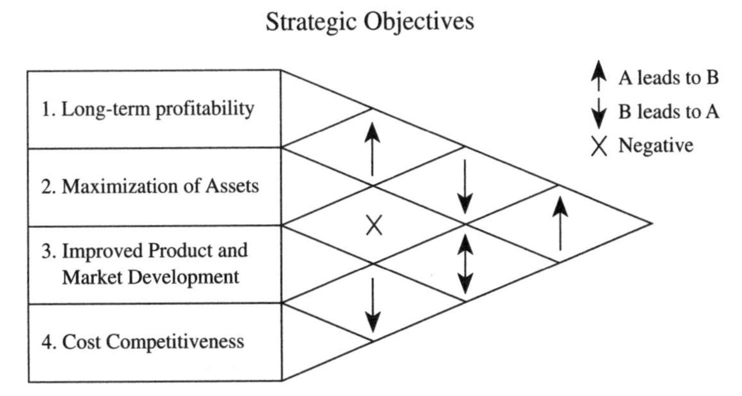

Figure 17 An example of potential conflicting strategic objectives. (Esain, 1999 [11]).

Theoretical Framework Tested

As can be seen from the work above, the theoretical framework and the method for evaluating the new product introduction process as shown in Figure 1 were tested during the structured observation phase of the research programme. The outcome of testing the method using the case study provides a basis for the diagnostics, which could be used in other companies. It is contended that these tested maps would be a mechanism to achieve improvement in their new product introduction process.

REVIEW

The process team reviews the data collected during the structured observation and any perceptions that are in conflict with the data. As the data collected refers to one point in time it is possible that it reflects non-typical issues.

At this point the variation between the brown paper model and process activity map (Figure 18) will be reviewed and, where appropriate, problem-solving techniques (Bicheno, 1994 [17]) will used to identify the root cause(s) and generate plans to counter the problem(s), i.e.

- Pareto analysis;
- Histogram and Measles Charts;
- Ishikawa Diagram (Cause and Effect).

Finally, the group is asked to visualize a vision of 'How good could the process be' by the use of facilitated brainstorming. The participants are encouraged to use the information collected from the Process Activity Map and any other charts produced from the interviews to reconstruct the system. This is used to encourage ownership and create a standard process.

FORMALIZING THE SYSTEM

The feedback from the team formed part of a presentation to the senior management, required to gain a mandate for the revision of the system. Similarly, strategic conflicts (Figure 17) are discussed and barriers to change highlighted.

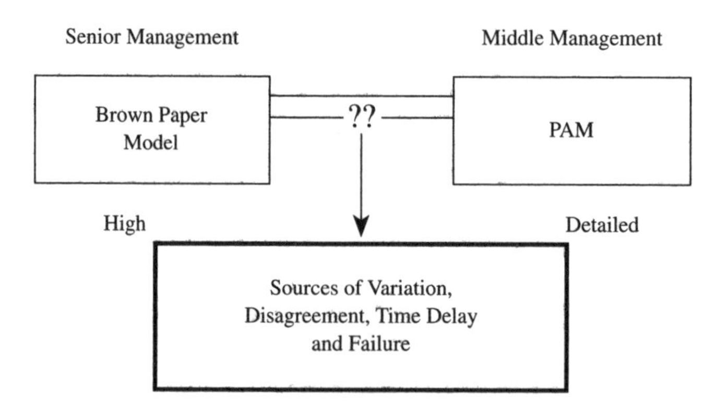

Figure 18 method of identifying conflict between maps. (Esain, 1999 [11]),

Additionally, a factual, step-by-step action plan is presented. As the time (Stalk and Hout, 1990 [7]) taken to deliver new product introduction is the key measure under review, the plan looks at how time can be removed from the process.

There are usually three stages to an action plan. The first looks at quick and easy changes. These are generally related to non-value adding activities, which relate to the wastes identified through the maps. The next stage is the medium-term changes, which will also relate to non-value added activity. In addition, they will also include some necessary but non-value added activity, that is, activity which is required today but could be overcome tomorrow with technology or investment or change in strategy etc.

Finally, long-term changes, the third stage to the action plan. This will detail activities related to some sort of step change. This could be changing the whole approach to New Product Introduction or could relate to significant capital investment.

Each stage to the plan of action will have different time-scales. There is no standard time frames for the action plan as it is dependent on the level of complexity found in the organization or supply chain being analysed. The time for improvement is also related to the complexity of the processes under scrutiny. The methodology described above and the case study framework is designed to be a cycle (Figure 3) with each turn of the cycle bringing variation and standard time down in the formalized system. The approach facilitates cross-departmental communication as well as achieving the quantified competitive benefit of time compression.

The work undertaken on this project resulted in the reduction of time in the process (Figure 19) by a substantial amount.

CONCLUSION

The case study framework (Figure 3) discussed enables the visualization of process complexity through the process overview and structured observation and hence the process is simplified.

Time is used as the common unit of measure and the quantified action plan is used to provide value and eliminate waste. The case study framework applied for the process of new product introduction and waste elimination highlighted the sources of variation across an organization.

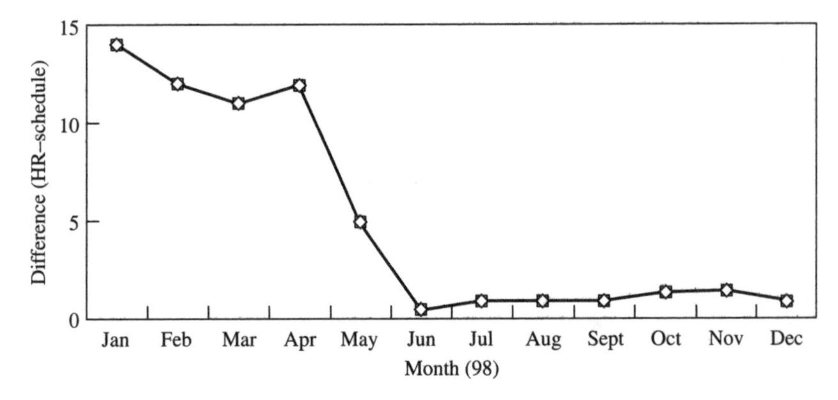

Figure 19 Improvement in New Product Introduction process over time.

If the change programme were focused on the process alone, the potential for identification of sources of variation would be significantly limited. Additionally, the understanding of the strategy of an organization can ensure that any proposed changes are in line with the overall direction of the company. This particularly creates a case for buy-in.

The continuous improvement and innovation in the product development process becomes aligned to the strategy and enables delivery of customer value through this framework.

The case study indicates that this approach can lead to significant improvements in time to market.

ACKNOWLEDGEMENTS

Thanks particularly go to the following staff of Corus (formally BSSP): Bernard Hewitt, Terry Goodwin and Keith Bartlett for sponsoring the project and the subsequent implementation work, James Sullivan from BSSP, Donna Samuel and Peter Hines from the Lean Enterprise Research Centre for their help with theoretical perspectives and support.

REFERENCES

[1] Womack, J., Jones, D. and Roos, D., (1990) *The Machine that Changed the World.* New York: Macmillan.

[2] Hines, P., Esain, A., Francis, M. and Jones, O. (2000) Managing New Product Introduction & New Product Development. In Hines, P., Lamming, R. Jones, D., Cousins, P., and Rich, N. (2000) *Value Stream Management: Strategy and Excellence in the Supply Chain*, pp 401–434. ISBN 0273642022. Harlow: Financial Times, Prentice Hall.

[3] Nakamura,Y. and Milburn, I (1997). Global design and development in the European automotive industry. The ImechE 1997 James Clayton Memorial Lecture, London, 12 February 1997.

[4] Pathway to Prosperity (1997) Welsh Office.

[5] Clark, K. and Fujimoto, T. (1991) *Product Development Performance: Strategy, Organisation and Management in the World Auto Industry.* Boston, MA: Harvard Business School Press.

[6] Saunders, M., Lewis, P. and Thornhill, A. (1997) *Research Methods for Business Students.* London: Pitman Publishing.

[7] Stalk, G. and Hout, T. (1990) *Competing Against Time.* The Free Press, New York.

[8] Copper, R. G. (1984) The strategy – performance links in product innovation. *R&D Management*, 14(4), 247–259.

[9] Kanter, R. (1989) *When Giants Learn to Dance, master the challenge of Strategy, Management, and Careers in the 1990's.* New York: Simon and Schuster.

[10] Clausing, D. (1994) *Total Quality Deployment: A step by step Guide to World-Class Concurrent Engineering.* New York: American Society of Mechanical Engineers Press.

[11] Esain, A. (1999) *New Product Introduction* – the triangulation tools for time compression and competitive advantage – a process industry case study. *Proceedings of the 5th International Conference on Concurrent Enterprising*, The Hague, The Netherlands, 15–17 March, pp. 473–479.

[12] Walton, M. (1991) *The Deming Management Method.* London: Mercury Business Books.

[13] Hines, P. and Rich, N. (1997) The Seven Value Stream Mapping Tools. *International Journal of Production and Operations Management*, 17 (1), 46–64.

[14] Womack, J. and Jones, D. (1996) *Lean Thinking, Banish Waste and Create Wealth in your Corporation.* New York: Simon & Schuster.

[15] Dimancescu, D., Hines, P. and Rich, N., (1997). *The Lean Enterprise, Designing and Managing Strategic Processes for Customer-Winning Performance.* New York: American Management Association.

[16] Ohno, T. (1988). *The Toyota Production System: Beyond Large Scale Production.* Portland, Oregon: Productivity Press.

[17] Bicheno, J. (1994). *The Quality 60, A Guide to Gurus, Tools, Wastes, Techniques and Systems.* Buckingham: Picsie Books.

[18] Dimancescu, D. (1994). *The Seamless Enterprise: Making Cross-Functional Management Work.* New York: Wiley.

[19] Goldratt, E. (1990) *The Goal.* Great Barrington: North River Press.
[20] Carter, J.(1994) Japanese product development, myths and realities. International Association for Electronic Product Development.
[21] Alber, K. and Walker, W. (1997) *Supply Chain Management: Practitioners Notes*. Fall Church, VA: APICS Educational & Research Foundation, October.
[22] Dimancescu, D. and Dwenger, K. (1998) World-Class New Product Development, Benchmarking Best Practice of Agile Manufacturers American Management Association, New York.
[23] Hill, T. (1995) *Manufacturing Strategy, Text and Cases.* Macmillan Press, London.
[24] Esain, A. (2000), Networks, Benchmarking and Development of the Strategic Supply Base: A Case Study, *International Journal of Logistics: Research and Applications*, Vol. 3, No. 2, pp 157–171.

20 Lean supplier development: the LEAP Paint Project

D. Brunt

INTRODUCTION

Over recent years a major issue for professional managers has been how to cope with ever-increasing change, complexity and competition (1). In the search for long-term competitive advantage, it has been suggested that there is a triangular linkage between the company, its customers and its competitors (2). The source of competitive advantage is found firstly in the ability of the organization to differentiate itself in the eyes of the customer from its competition and secondly by operating at a lower cost and hence at greater profit (3).

The search for competitive advantage has caused fundamental changes both internally and along organizations' supply chains (4). There is an increased awareness of the benefits effective purchasing can bring to an organization through technological innovation and strategic collaboration (5), mutual obligation and partnership sourcing (6) and the integration of key internal processes with those of their suppliers (7).

The chapter is divided into five parts. Firstly, the strategic agenda set by lean thinking is explored in terms of the management of the supply base. Secondly, an outline of the process used to work with suppliers is described. This leads to the third part of the chapter in which the case study from the LEAP programme will be discussed. Finally, conclusions will be drawn from the case. The research sought to understand whether supplier associations could be developed with suppliers that competed with one another and if a structured framework of 'learning to see' maps and value stream maps aided the awareness and education phases of developing buyer–supplier relationships.

STRATEGIC AGENDA SET BY LEAN THINKING

The strategic agenda set by lean production and lean thinking has huge implications for the purchasing operation and for a company in terms of how it manages its relationships with its suppliers. Part of the success of Japanese manufacturing companies has been caused by the thorough and speedy implementation of their manufacturing systems throughout the complete supplier network. This is done through integration of their key internal processes with those of their suppliers and the rapid development of the latter community of firms (8).

Reviewing the key internal processes that have allowed the lean assemblers to gain their pre-eminent position is an important step in understanding the new strategic role for the purchaser within lean production. In a lean organization, these processes run cross-functionally, rather than functionally, through the business. This results in employees focusing on key company goals rather than traditional narrow functional targets. Benefits of this approach are the avoidance of functional optimization at the expense of company optimization and cross-functional buy-in to decision making and implementation. By adopting this cross-functional approach, the processes of quality, delivery and cost become visibly important across organizations.

The search for improvement to gain competitive advantage cannot be confined to an individual company and must be done throughout all the firms involved in their respective supplier networks that add value to the final product (9). As value added activities occur in the supplier networks, purchasing has a role to play in the achievement of adopting lean production by externalizing these processes in these firms.

The strategic agenda for companies wishing to adopt lean production involves making major changes that lead to improvement, as well as continuing kaizen (constant improvement) activities. While purchasing can aid a company to double its productivity, it has the potential to be the leader of change to help suppliers double their productivity. For a company wishing to follow this route, it needs to eliminate waste in the supply base. The waste can be categorized into two areas – intracompany and intercompany waste. The operationalization of this logic is depicted in Hines' Network Sourcing (8). The Network Sourcing model is shown in Figure 1.

Intracompany waste arises due to the inability of suppliers to be more efficient in their own internal processes. Customers at each level individually developing their suppliers can tackle this waste. In the Network Sourcing model supplier development involves the activities carried out by the customer to help improve the strategies, tools and techniques employed by suppliers to improve their competitive advantage. For example, disseminating customer strategies, improving factory layout and assisting firms to implement lean tools and techniques such as set-up time reduction and kanban operations.

The second waste is intercompany waste, or waste due to the inability of the different value stream members to share strategies and integrate internal processes. This is operationalized in the Network Sourcing model as a high level of supplier coordination by the customer company at each level of the tiered supply structure. Supplier coordination involves those activities made by a customer to mould their suppliers into a common way of working so that competitive advantage can be gained, particularly by the removal of this waste. For example, developing common quality standards and using the same systems.

The mechanism used to carry out supplier development and coordination utilizes the Kyoryoku Kai or Supplier Association. A 'Supplier Association' is a mutually benefiting group of a company's most important suppliers. This group is brought together on a regular basis to coordinate activities and develop cooperative attitudes, as well as to assist all members who can benefit from an exploration of concepts and techniques such as just-in-time (JIT), statistical process control (SPC), continuous improvement and kanban.

1. A tiered supply structure with a heavy reliance on small firms
2. A small number of direct suppliers with individual part numbers sourced from one supplier but within a competitive dual sourcing environment
3. High degrees of asset specificity among suppliers and risk sharing between customer and supplier alike
4. A maximum buy strategy by each company within the semi-permanent supplier network, but a maximum make strategy within these trusted networks
5. A high degree of bilateral design employing the skills and knowledge of both customers and suppliers alike
6. A high degree of supplier innovation in both new products and processes
7. Close, long-term relations between network members involving a high level of trust, openness and profit sharing
8. The use of rigorous supplier grading systems increasingly giving way to supplier self-certification
9. A high level of supplier coordination by the customer by the customer company at each level of the tiered supply structure
10. A significant effort made by customers at each level individually to develop their suppliers

Figure 1 The Network Sourcing model overview.

THE LEAP PAINT CASE PROCESS

The Leap paint case study began as the pilot for 'Phase 3' of Corus' suppliers initiative. The company had previously rationalized its supply base and wanted to develop a framework to achieve the following objectives:

- Improve the skill level in both its own and its suppliers' operations in terms of lean manufacturing.
- Develop a cost reduction programme by mapping processes (rather than by negotiation alone).
- Improve working relationship with the selected suppliers.
- Develop an effective forum for managing quality, delivery, cost, development and management elements of the supply relationship.

It was decided that in order to meet the project objectives, value streams stretching back to each of the suppliers should be mapped so that opportunities for improvement could be highlighted. In addition, one site which received deliveries from each of the suppliers should also undertake 'Value Stream mapping' in order to highlight both intercompany and intracompany waste, so that efforts could be put in place to reduce it. Senior management from each of the suppliers were asked to attend a meeting in July 1998 to obtain feedback on the idea and express if they were willing to proceed with a pilot programme. Buy-in to the project was achieved at the meeting.

To improve the skill level in both Corus' and its suppliers' operations in terms of lean thinking, a two-day workshop was organized at a Corus site in November 1998. The activity involved cross-functional personnel from Corus and each of the suppliers. Attendees represented both technical and commercial areas of each business. In addition, personnel directly involved both in the customer–supplier relationship and in the operational areas of each business were invited to the event.

The two-day activity at the Corus site was designed to meet the overall project objectives. In addition, the session aimed to raise the awareness of the group to the logic of lean thinking and introduce a number of lean tools and techniques, particularly those used to map value streams. The morning of the first day introduced the team to lean thinking and the key mapping tools to be used. This was done using interactive learning and simulation. The tools to be used in the project were the 'Learning to See' value stream maps (10) and three of the seven value stream maps highlighted in Chapter 3: the process activity map, demand amplification map and quality filter map.

The afternoon session involved the whole team using a process activity map to map a product from each of the suppliers through the Corus site. By opening the customer's site to such tools, suppliers could begin to understand the flow of their products and the issues the customer company faced. In addition, it was felt that working together on a common process could improve the working relationship with each supplier.

On day two, members from each of the companies divided into cross-company teams to use each of the remaining maps. In the afternoon, the group re-formed to discuss their findings from the mapping sessions. Finally, a structured agenda developed in order to map each of the suppliers was proposed and dates agreed for Corus personnel to visit each of the suppliers. Each supplier visit was completed by the end of November 1998.

Following mapping in each of the suppliers, a review meeting was held in December 1998. This included both senior representatives (who had signed off the pilot activity in July 1998) and operational management involved in the mapping process. Presentations were developed showing the results of value stream mapping at each supplier and the customer company. The aim of the meeting was to highlight the common themes that had become apparent in mapping each of the firms so that teams could be developed between the companies to tackle specific improvement opportunities. These teams could then report progress at regular review meetings as part of a tiered meeting structure whereby senior management were reviewing and setting the strategy for the project and operational activities were taking place aligned in the same direction.

THE LEAP PAINT CASE STUDY

The mapping process both at Corus and at each of the suppliers highlighted a number of areas for potential improvement. Figure 2 shows a 'Learning to See' value stream map for the Corus site. A description of how to construct these maps is developed in Chapter 7.

By analysing both the maps developed during the workshop at Corus and the maps generated at each of the supplier sites a number of key issues to be addressed were identified. The issues were addressed using the five lean thinking principles as a framework. These are shown below:

- Specify what creates value from the customer's perspective – not individual firms, functions and departments.
- Identify all the steps necessary to design, order and produce the product across the whole value stream to highlight non-value adding waste.

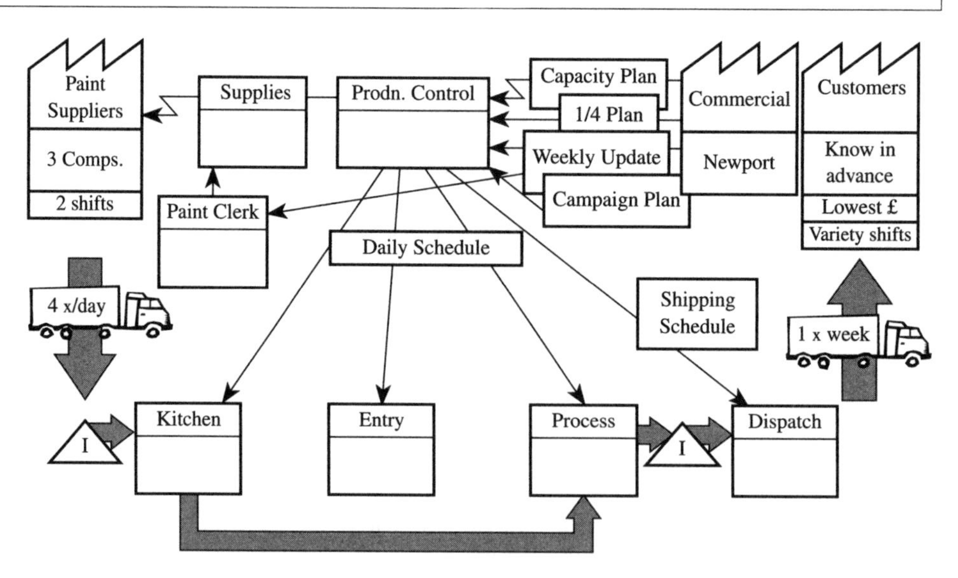

Figure 2 'Learning to See' value stream map for painted products.

- Make those actions that create value flow without interruption, detours, waiting or scrap.
- Only make what is pulled by the customer just in time.
- Strive for perfection by removing successive layers of waste, as they are uncovered.

The first principle relates to value. For the coated products value stream a simple definition was suggested. This is shown in Figure 3 and utilizes the 'seven R's' framework of 'ensuring the availability of the right product, in the right quantity and the right condition, at the right place, at the right time, for the right customer, at the right cost' (11). These seven R's can be further divided into three main areas: Quality (right product, condition and longevity), delivery (right quantity, place, time and customer) and cost (right cost for the market.) In addition to the fulfilment of an order, value may be expressed in terms of the dynamics affecting businesses and relationships. For example, the ability for a firm to respond to its customers' needs and the level of improvement a company shows in order to compete in the marketplace.

When all the maps had been generated for each of the firms involved, it became apparent that value was not directly measured between buyer and supplier firms. Opportunities existed to improve the measurement system in terms of both quality and delivery monitoring so that shortfalls in performance could be made visible and relevant action plans put into place.

All the activities currently required to produce and deliver a product were analysed. Therefore, it was possible to assess the proportion of time that value was added to a product (a value added ratio) compared to the total time the product was on a site (from raw material storage through to finished goods storage.) A value-added ratio of less than 4 per cent was found at each of the sites. This means that, typically, value was being added for less than 4 per cent of the total time that a product was on any individual site.

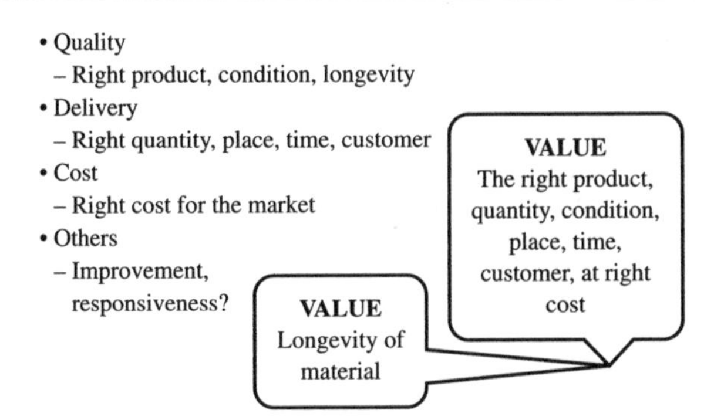

Figure 3 Value in the coated products value stream.

If value is being added for only a small proportion of the total time that a product is on a site then the product is not 'flowing' without interruption, detours, waiting or scrap (lean thinking principle three). Each map highlighted that paint production is a batch-driven system where economic and process considerations have resulted in the philosophy that 'the bigger the batch, the better the product.' In addition, variability in raw material consistency resulted in the need to tint each batch in order to produce a colour to the desired specification. Outside the paint industry, such a process could be thought of as rework that at present is a necessary non-value added activity.

Each of the maps also revealed that all the firms involved 'pushed' material through their plants. Each process along the value stream operates as an isolated island, producing and pushing product forward regardless of the needs of the customer. Demand amplification maps were particularly visual in displaying this issue to the team. In addition, the stock generated that subsequently became obsolete (as paint has a shelf life) enabled cash sums to be put against this 'push' mentality across the supply chain.

As a result of the analysis, the team developed a number of recommendations and actions. These are shown in Table 1 along with the next steps and responsibilities for completing the projects. Three projects were identified as being key to improving the performance of the painted products value stream. The first of these involved the creation of a clear, visual measurement system that could then be used to highlight and monitor further improvement opportunities. Evidence of the need for such a system was highlighted through quality filter maps (QFM) and demand amplification maps (DAM). This measurement system needed to take into account 'value' in terms of quality and delivery to the customer.

The second project aimed to address the issue of demand alignment. All of the demand amplification maps had provided evidence of the need to make demand information transparent and align that information between the customer and supplier firms. In addition, this project would start to improve visibility so that overproduction or producing too soon could be reduced. Initially it had been thought that demand was not regular across the range of colours the painted products value stream produced. However, analysis of the demand patterns showed that paint colours could be categorized into runners (regular demand every week), repeaters (regular demand but over a monthly time frame) and strangers (irregular demand patterns).

Table 1 Recommendations and actions from the mapping activity

	Mapping evidence	*Next steps*	*Who/When*
Measurement system	• QFM – BS • QFM – Suppliers • Service quality few measures • DAM – Double stocks • No baseline to measure improvement	• Prepare measurement system • Discuss for best practice • Review after 3 months	Purchasing Purchasing All
Demand alignment	• DAM – Mix/volume • DAM – Inventory • VSM – Lead time to value added ratio • Obsolescence	• Identify runners etc. • Identify joint buffers and agree safety level	Scheduling, Commercial, Purchasing, Manufacturing Suppliers
Quality management – right 1st time	• QFM – issues @ each supplier	• 1 to 1 workshops with suppliers	QA, and Supplier Quality

The third project concerned the management of the physical transformation of products and, in particular, the management of quality to achieve right first time production. Without such a project, the work on demand alignment would yield marginal improvements in stock reduction as the ability to produce good quality output would be variable and therefore stock would be required along the value stream in order to ensure continuity of supply.

Having identified the potential projects, the team needed to allocate resource to each of the improvement areas. As the process of value stream mapping enabled both inter and intracompany waste to be identified the projects fell into four categories:

1. Projects to be carried out by Corus. For example, preparation of the measurement system.
2. Projects to be carried out by suppliers. For example, internal quality improvement activities.
3. Projects to be carried out on a one-to-one basis. For example, quality improvement activities (not carried out in [2] above) as the suppliers are competitors affected by commercial issues regarding the management of projects and confidentiality.
4. Projects coordinated by Corus involving all suppliers. For example demand management issues.

The research sought to understand whether supplier associations could be developed with suppliers that competed with one another and if a structured framework of 'Learning to See' maps and value stream maps aided the awareness and education phases of developing buyer–supplier relationships. In attempting to do this a number of benefits and key learning points have developed.

In terms of the benefits of such an approach, the visual nature of both the 'Learning to See' maps and value stream maps enabled participants to gain a shared understanding of each other's processes. This allowed them to jointly identify and remove waste such as the removal of duplicate buffer stocks, obsolescence reduction and improvements in quality performance. A framework illustrating how improvements can be of mutual benefit is shown in Figure 4. Thus it was possible

	Cost of total (%)	
Waste category	Corus	Supplier
Obsolescence	w	a
Inventory	x	b
Non right 1st time	y	c
Total	z	d

- Realistic target for both firms = 3–5%
- Only waste (non value added) attacked in first year

Figure 4 Mutual benefits framework.

to identify real cost reductions (not price/margin reductions.) Developing the improvements through the use of a quantifiable measurements system allowed process improvements to be focused where they would have most benefit. In addition, the process helped develop the relationship between the customer and supplier firms and internally between departments in firms as it is both cross-company and cross-functional in approach. Finally, the supplier association route makes a more effective use of limited key resources and enhances the knowledge base of those resources. This research found that even though competitors were involved, the costs of education and training (and the time in carrying out these phases of the project) are reduced.

However there are a number of key learning points from the research. The first of these is that the project could have been managed more effectively if competitors had not been involved or had been part of a larger group of non-competing firms. The underlying issues of competitive advantage meant that much of the work was carried out on a one-to-one basis and therefore the opportunity to carry out more awareness and education projects involving suppliers on supplier sites was not possible. The second key learning point is that the need to raise awareness in terms of what lean thinking is and what the project is attempting to do is important. It is necessary to stress that this is not an overnight solution.

SUMMARY AND CONCLUSIONS

The research presented in this chapter has investigated the use of lean tools and techniques within the painted products part of the steel supply chain. The case study shows how 'Learning to See' maps and value stream mapping tools can be applied to other process industries other than Corus' mainstream steel-making activities. In addition, the chapter highlights that the benefits gained from lean thinking will be realized when a total value stream view is developed. This value stream approach needs to involve not only suppliers (as shown in this chapter) but also customers for the value stream concerned. The research piloted a method for the firm to initiate supplier development within a structured framework of a 'Supplier Association' using competitors as the members of the association. While this methodology displayed benefits over a traditional buyer–supplier relationship, it was felt that the project could have been managed more effectively if competitors had not been involved or had been part of a larger group of non-competing firms.

REFERENCES

1. Pettigrew, A. and Whipp, R. (1991) *Managing Change for Competitive Success*. Oxford: Blackwell.
2. Ohmae, K. (1982) *The mind of the strategist, the art of Japanese business*. New York: McGraw-Hill.
3. Christopher, M. (1992) *Logistics and Supply Chain Management, Strategies for reducing costs and improving services*. London: Pitman.
4. Steele, P. and Court, B. (1996) *Profitable Purchasing Strategies, A manager's guide for improving organisational competitiveness through the skills of purchasing*. Maidenhead: McGraw-Hill.
5. Lamming, R. (1993) *Beyond Partnership: Strategies for Innovation and Lean Supply*. Hemel Hempstead: Prentice Hall.
6. Macbeth, D. and Ferguson, N. (1994). *Partnership Sourcing, an integrated supply chain approach*. London: Pitman.
7. Hines, P. (1996) Purchasing for lean production: the new strategic agenda. *International Journal of Purchasing and Materials Management*, Winter, 2–10.
8. Hines, P. (1994) *Creating World Class Suppliers. Unlocking Mutual Competitive Advantage*. London: Pitman.
9. Womack, J. P. and Jones, D. T. (1996) *Lean Thinking, Banish waste and create wealth in your corporation*. New York: Simon & Schuster.
10. Rother, M. and Shook, J. (1998) *Learning to See, Value Stream Mapping to add value and Eliminate muda*. Brookline: The Lean Enterprise Institute.
11. Shapiro, R.D. and Heskett, J.L. (1985) *Logistics Strategy: Cases and Concepts*. St. Paul, Minnesota: West Publishing.
12. Womack, J.P., Jones, D.T. and Roos, D. (1990) *The Machine that Changed the World*. New York: Macmillan.

Integration of the supply chain for total through-cost reduction

21

Brian Daniels

INTRODUCTION

Corus Engineering Steels (CES) is one of Europe's largest special steels producers. From its two major plants in Stocksbridge and Rotherham, 1.5 million tonnes of steel are manufactured and processed for a wide variety of applications, ranging from aerospace to mining. By far the largest end-user sector is the automotive industry (see Figure 1).

Fifty-six per cent of all products find their way into chassis and power train components for cars, commercial vehicles and off-highway equipment. Typical components include crankshafts and connecting rods for engines, gears, shafts and constant velocity joints for transmissions, and springs for suspensions systems.

The future of the European automotive sector is vital to the success of CES, but critically, less than 1% of products go to the vehicle manufacturer directly. By far the largest proportion of products are via component producers. The forging and

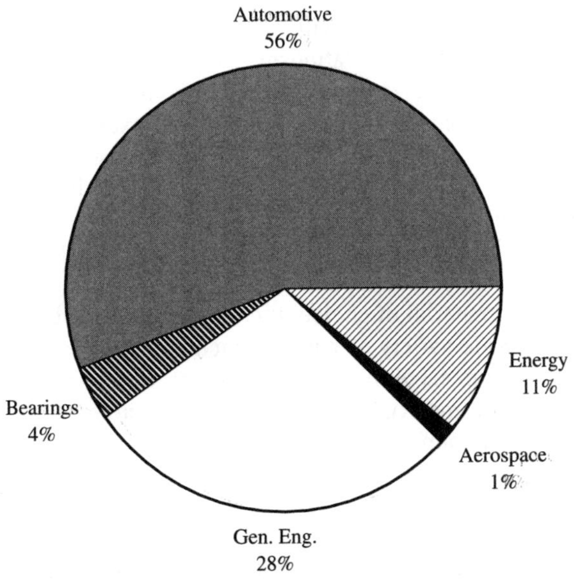

Figure 1 CES sales by end-use sectors, 1999.

Table 1 CES supply chains: engineering steels for automotive

	Total (Kt)	*Share (%)*
Direct to VMs	34	5
To tier 1	70	10
To tier 1/2 forgers	333	50
To tier 2 and below	238	35
Total	675	100

machining sectors are the major immediate users of engineering steels, and even they may not be first-tier suppliers they are often two or three stages back in the supply chain (Table 1).

CES has invested heavily to achieve the highest levels of product quality and systems standards to meet the ever-increasing demands of the automotive sector, yet it has traditionally had little direct control over the derived purchasing of its products by the automotive companies.

Supporting and managing the supply chain has therefore become a focus for CES's marketing activities, and the achievements over recent years have been considerable.

PRODUCT DEVELOPMENT

The decisions related to type and quality of steel for automotive components have generally not been outsourced by the vehicle manufacturers to tier 1 suppliers. Most vehicle manufacturers maintain their own core competencies in this area, which has led to strong technical relationships with the steelmaker.

CES has maintained a high level of skill in this particular area. With highly trained and experienced product technologists, and extensive research and development capabilities, CES has worked closely with the vehicle manufacturer to achieve weight and cost solutions. The working relationships and network of activities are world-wide, with successful projects completed and ongoing as far apart as Tokyo and Detroit. CES has proven that it can not just match but exceed steel quality standards anywhere in the world.

During November 1997 Corus established its Automotive Engineering Group. Based in Coventry, this extension to its research and development facilities has introduced the capability to add automotive engineering and design to its already established product development capabilities.

Steel is not the only material of choice. It competes with cast iron, sintered powder and aluminium for a wide variety of parts, particularly in the engine and suspension systems. Using this design capability, it is CES's intention to prove to the automotive designer that steel is the material of choice, at a decision point very early in new vehicle development programmes.

THE SUPPLY CHAIN

Despite CES's product development activities, if left unattended they could not guarantee commercial success. The competitiveness of CES's customers, and often beyond this the whole supply chain – is critical. One break in one link of the chain and the business is lost.

CES has therefore been keen to develop its products with its customers, with the vision that we combine all of our competencies to provide the ultimate, most competitive solutions to the end-user. It is on the basis of this vision that new methods have been employed to pursue through-cost solutions. This paper describes one of these activities, in detail, with a major European vehicle builder – a company with whom CES could have no direct steel supplies yet who use large quantities indirectly, particularly from producers of forged steel components.

THE OLD RELATIONSHIP

A close and significant technical working relationship existed with this particular vehicle manufacturer for many years, yet CES had no exclusivity of contract. There were other steel producers, each with their own product/technical specialities. It was therefore not possible to achieve such differentiation through those activities that would ensure commercial success. Projects were diverse, and although the vehicle manufacturer had his own solutions in mind there was no commercial focus for CES. There were occasional commercial meetings. Steel prices are notoriously cyclical, and short-term fluctuations in price result from significant changes in world prices for essential elements such as chromium or nickel levied as a surcharge on steel users. This was the basis of the vehicle manufacturer's interest – a need to be aware of their suppliers' exposure to changes in raw material costs. It was also in CES's interest to understand the changes in vehicle building programmes, as this affects significantly derived demand for engineering steels.

At this point, although relationships were good, there was no long-term plan for either party.

COMMERCIAL AND PRODUCT FOCUS

Through discussion, the following issues revealed a way forward for greater cooperation.

1. There was potential to reduce significantly raw material costs for specific transmission components through the development of a new steel type.
2. An analysis of the supply chain revealed a highly fragmented raw material supply base on these components. Despite their management of a limited number of first-tier suppliers, it was the first tier who chose the steel manufacturer.

Raw material cost reduction

The vehicle manufacturer specified to their suppliers a steel for gears and shafts with a particularly high and expensive alloy content. There was an assumed reluctance to change this, as their engineers were facing difficult processing problems which resulted in high levels of rework on their machining lines. They did not want additional complications. However, their competitors were using far less expensive steels for similar applications.

It was agreed to pursue a programme of CES support to develop a lower cost grade. The forecast financial benefits would be significant, with a 15% reduction in the basic steel price and, due to the lower alloy steel content, potential for a further 7% on effective steel prices due to the elimination of short-term changes in alloy surcharges (Figure 2).

A further and longer term programme of work was also agreed to investigate methods of improving fatigue life of gears consolidating and concentrating effort on this single component area.

Raw material supply base

The analysis of the supply chain revealed an alarming number of variants in steel supplies due to the uncontrolled purchasing of raw materials by suppliers. Within 25000 tonnes of a single steel type there were 50 different size combinations, ranging from one to over 1000 tonnes per annum from 11 different steel suppliers, of which CES was one.

As volume invariably has an effect on steel price, consolidation in this area offered potential for further cost benefit. By encouraging economies of scale with a single steelmaker, costs can be reduced.

It was also acknowledged that the variation in product, causing the expensive rework in the assembly plant, could be due to the variation in raw material caused by the differing processes employed by each of the 11 steel suppliers, i.e. greater control and less variability could ensue through mono-sourcing across the whole supply chain. This was a contentious point, not to be taken as a reliable factor for cost reduction at this stage.

SUPPLY CHAIN INTEGRATION

As the vehicle manufacturer shared more of its strategic intentions with CES it became necessary to develop further plans for improved competitiveness.

They had considerable experience from their global operations to give the capability to benchmark component costs across continents, and it was disclosed that forged steel products could be obtained from India and the Far East which were as high in quality, but 40% lower in cost than in Europe. This threatened not only their

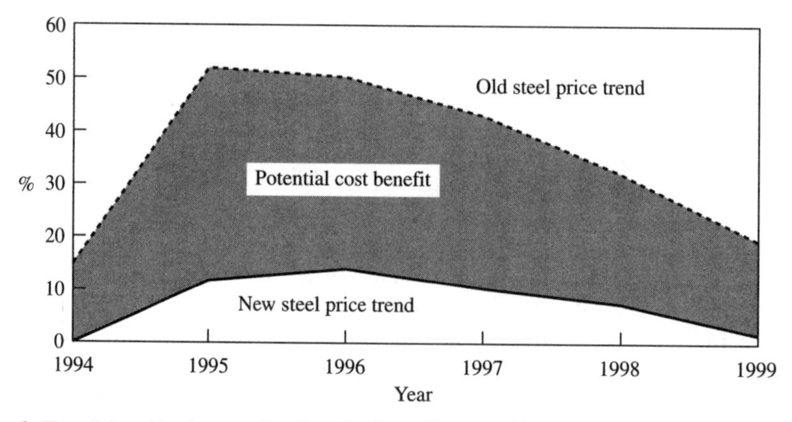

Figure 2 Trend in effective steel prices higher alloy steel/old steel compared with new steel.

European supply base but also the very existence of their own assembly operations. They had the capability to localize component and assembly production in these areas and supply complete systems to their vehicle assembly plants in Europe and North America.

Outsourcing was also becoming a reality within the business, as was the reduction in number of first-tier suppliers by as much as one-third over a period of 5 years. There was a danger that supplier reduction would not be in favour of CES, i.e. the survivors of the rationalization programme may not have been CES customers.

At this point it was agreed that further cost reduction should be sought by involving selected first-tier suppliers in the project, effectively grouping the whole supply chain from steelmaking to final delivery of transmissions to the assembly line.

The vehicle manufacturer and Corus had experience working with the Cardiff Business School Lean Enterprise Research Centre. Their successful application of lean thinking into European business, based on extensive research into Japanese organizations, could add a new and significant contribution to our cost reduction activities. Discussions with the Lean Enterprise Unit took place in order to establish the suitability of lean management techniques to this particular process. The paper 'Outsourcing competitive advantage' (Hines and Rich, 1998), describes the way in which Toyota has achieved competitive advantage over its UK rivals based on its effective use of its supplier network. This leverage of advantage from suppliers is not limited to the first tier, but has cascaded down the chain, culminating in the alignment of thousands of direct and indirect component supplies. It goes on to describe successfully created European suppliers associations of varying types. The possibility of creating a supplier association in our own small area was attractive.

A high-level meeting took place between CES, the vehicle manufacturer and each of three major suppliers of forged components. Each supplier was a major customer of CES. There was agreement in principle to proceed. High-level objectives were set and the emphasis was on aggressive time-scales. This was not to be a paper exercise but a challenging project, with clearly defined goals. For each company, in addition to achieving a favourable result for the vehicle manufacturer (for all of us the ultimate customer), here was an opportunity to develop overall competitive advantage which could be applied across all of our businesses.

The alliance was unique. In normal everyday circumstances each of the first-tier suppliers compete, yet through participation in the project they were prepared to be open about their manufacturing processes and systems in fine detail.

VALUE STREAM ANALYSIS

The value stream management process (see Figure 3) began with a senior management awareness session, the purpose of which was to assign management sponsors and a dedicated resource. Three value streams were selected, each a unique product flow to the vehicle manufacturer. The broad process was identified and a programme of work established to map each flow in detail, applying the concept of the value stream, which involves identification of the specific parts of the process responsible for adding value and waste, the elimination of waste being the primary target.

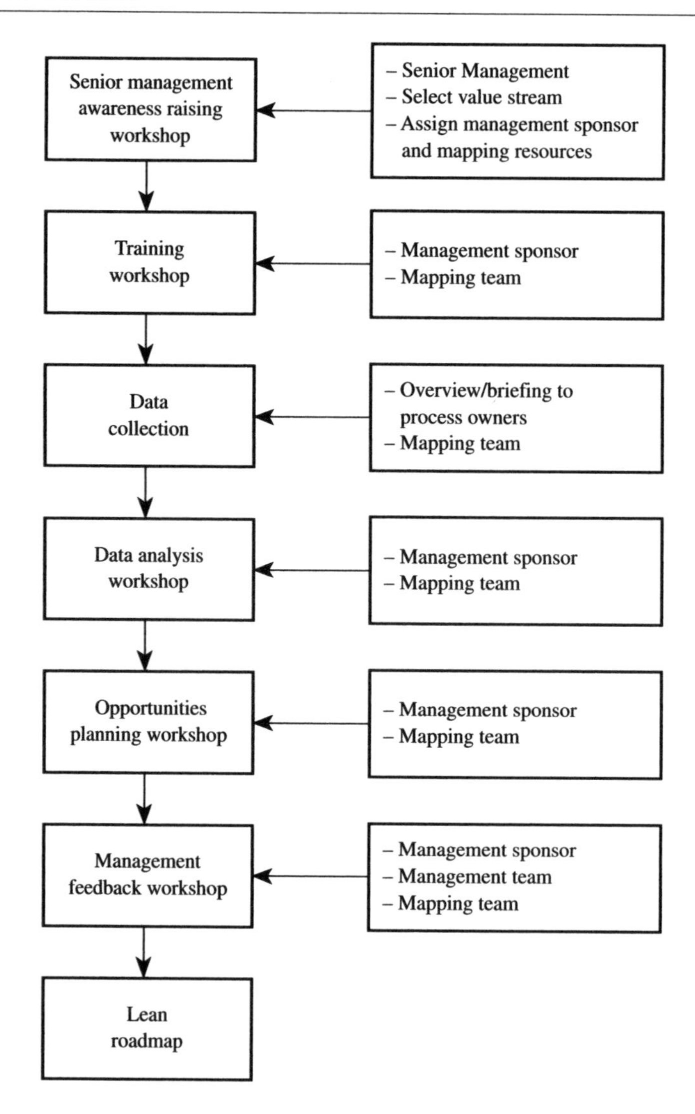

Figure 3 Implementing lean value stream management: process overview.

Wastes were to be identified within seven generic areas: over-production, waiting, transportation, inappropriate processing, unnecessary inventory, unnecessary motion and defects. Following this, the senior management groups appointed representatives who would work together on the mapping teams at each plant. Working as a team at each of the eight manufacturing sites, they would develop an in-depth knowledge of each other's processes, adding further to the benefits of supply chain integration through an understanding of their suppliers and customers' needs.

Prior to the mapping activities, training sessions in 'lean thinking' were arranged for both the management sponsors and the mapping team. The agenda was to develop an understanding of the main elements of lean management and to learn the practical aspects of its implementation particularly related to data collection. Table 2

Table 2 Implementing lean: the value stream management tool kit

Waste	*Process activity mapping*	*Supply chain response matrix*	*Product variety funnel*	*Quality filter mapping*	*Demand amplification mapping*	*Decision point analysis*	*Physical structure volume value*
Overproduction	L	M		L	M	M	
Waiting	H	H	L		M	M	
Transportation	H						L
Inappropriate processing	H		M	L		L	
Unnecessary inventory	M	H	M		H	M	L
Unnecessary motion	H	L					
Defects	L		H	H			

Source: S.A. Porter, Lean Enterprise 1998.

summarizes the value stream mapping toolkit, with the relative importance of each tool against each classification of waste. Process activity mapping and demand amplification mapping were chosen as the main tools for this exercise.

Within 3 months both data collection and analysis was complete. Maps had been created for each process activity throughout the whole supply chain for each of the three parts identified. Every step had been identified as either operation, transport, inspection or delay. Each step was also quantified by distance, time (in minutes) and people (number of operations). During the data collection process key causes of waste were documented and early ideas for elimination of waste proposed. Table 3 shows a typical activity map for one process of one of the companies involved. Table 4 combines all the maps for a total process and indicates (not untypically) the enormous proportion of non-value adding activities involved, with only 1.40% of time within the 'operation'.

OPPORTUNITIES FOR IMPLEMENTATION OF LEAN

The output from the mapping gave immediate opportunities for improvement, and in line with the 'just do it' approach (which was agreed at the outset) these were implemented. The forward plan for significant opportunities was agreed in the planning workshop and management feedback sessions. Each company developed its own process improvement plans, which in some cases were extensive. The inter-company project plan was established with a quantified cost rationalization target. The majority of benefits were at the company interfaces within the supply chain and included the following six major elements:

1. steady scheduling/reduced planning horizons/supply chain pull/EDI;
2. review of inter-company documentation goods inwards/test certificates/acknowledgements/bar coding/quality checks/weigh checks;
3. co-design new product development;
4. outsourcing opportunities;
5. process elimination opportunities;
6. logistics integration.

Table 3 Process activity map process activity: goods receiving

Number	*Step*	*Flow*	*Area*	*Distance (m)*	*Time (min)*	*People*	*Comment*	*O*	*T*	*I*	*D*	*O*	*T*	*I*	*D*
								1	1	1	1	0	0	0	0
1	Lorry arrives							1	1	1	1	0	0	0	0
2	Slinger goes to driver for paperwork	T	Goods received	45	1	1	Delivery note and certificate	0	1	0	0	0	1	0	0
3	Take paperwork to steel stores office	T	Goods received	86	2	1		0	1	0	0	0	2	0	0
4	Take a copy of delivery note, sign driver's copy	T	Goods received		0.5	1	Pass back to slinger	0	1	0	0	0	0.5	0	0
5	Compare against purchase deliver status	I	Goods received		0.5	1	Status for 1 month. Check Pt No. Enter quantity	0	0	1	0	0	0	0.5	0
6	Slinger informs crane driver	D	Goods received	40	5	1		0	0	0	1	0	0	0	5
7	Slinger goes to lorry	T	Goods received	60	1.2	2	Whilst walking checks for spaces	0	1	0	0	0	1.2	0	0
8	Start unloading by check weighing	O	Goods received	4	2	3	Lifts of 4/5 tons	1	0	0	0	2	0	0	0
9	Mark bars with cast and part number	O	Goods received		1	1	Scales broken	1	0	0	0	1	0	0	0
10	Crane travels to steel stores space	T	Goods received	45	2	3	5/6 runs per lorry	0	1	0	0	0	2	0	0
11	Slinger writes combined weight on top lift	O	Goods received		1.5	1	Amend paperwork if wrong	1	0	0	0	1.5	0	0	0
12	Return paperwork to office	T	Goods received	86	2	1		0	1	0	0	0	2	0	0
13	Enter information on to card system	T	Goods received		1	1	Double check	0	1	0	0	0	1	0	0
14	Stamp and sign delivery note	O	Goods received		0.5	1		1	0	0	0	0.5	0	0	0
15	Enter on to computer system	O	Goods received		1.5	1	PO No, Part No, Weight, Cast, Location, Find Lot No	1	0	0	0	1.5	0	0	0
16	Place near computer	D	Goods received		150	1	Empty 3 times per day	0	0	0	1	0	0	0	150
17	Print off receipt sheet	O	Goods received		0.5	1	This shows lot No	1	0	0	0	0.5	0	0	0
18	Mark bars with lot number	O	Goods received	46	10	1	Traceability	1	0	0	0	10	0	0	0
19	Place delivery note and certificate in purchasing mailbox	T	Goods received	159	5	1	Do not always receive certificate with delivery	0	1	0	0	0	5	0	0
20	Return to steel stores office	T	Goods received	159	5	1	Extra door, auto release	0	1	0	0	0	5	0	0

21 Purchasing check mail box	T	Goods received	56	90	1	Check 5 times per day	0	1	0	0	0	90	0	0
22 Check Pt No. against paperwork	I	Purchase		5	1	To get lot number delivery status report	0	0	1	0	0	0	5	0
23 Enter into system	I	Purchase		5	1	Check certificate against standard	0	0	1	0	0	0	5	0
24 Delay for receipt of certificate	D	Purchase		120		On system	0	0	0	1	0	0	0	120
25 Stamp paperwork	D	Purchase		2	1		1	0	0	0	2	0	0	0
26 File delivery note	D	Purchase	1	0.5	1	1 for Tro 1 for accounts	0	0	0	1	0	0	0	0.5
27 Mark off delivery	0	Purchase		0.5	1	Against supplier delivery report	1	0	0	0	0.5	0	0	0
28 Place copy of delivery note in Tro and accounts mail box	T	Purchase	56	2	1	Placed in main box once per day	0	1	0	0	0	2	0	0
29 Delay for and loading lorry	D	Goods received		30			0	0	0	1	0	0	0	30
30 Delay for steel storage	D	Goods received		7200			0	0	0	1	0	0	0	7200
31							1	1	1	1	0	0	0	0
32							1	1	1	1	0	0	0	0
33							1	1	1	1	0	0	0	0
end														
Total		Activity	843	7647.2	31	Comment	14	16	8	11	19.5	111.7	10.5	7505.5

Notes: O = Operation; T = Transport; I = Inspection; D = Delay.
Key points:

Causes:	Effect:
Certificate not received with delivery	Contact supplier for copy 120 min delay
Layout between steel stores and main office	318 metres of travel
Manual input of data	Delay in processing of receipt information

Major wastes:

Causes:	Effects:
Defect	Delay
Inappropriate process	Transport
Inappropriate process	Delay

Opportunities: construct extra entrance door opposite side door of main office; ask supplier to bar code advice notes and material certificates to enable auto entry and auto cert checks; steel delivery does not leave supplier without certificates; review the use of bar coding to verify weight; review the use of pre-delivery advice EDI.

Table 4 Total process data

	Operation	*Transport*	*Inspection*	*Delay*	*Total time (min)*	*Distance (m)*	*People*
Logistics	15.10	496.00	15.00	2685.00	3211.10	600.00	14
Goods receipt	2.50	10.80	2.20	20.60	36.10	920.00	32
Storage	0.00	4.00	0.00	5977.00	5981.00	780.00	4
Soft machining	8.18	6.08	0.00	30694.00	30708.26	224.00	13
Hardening	1095.50	70.20	10.00	24296.20	25471.90	4196.00	25
Machining after hardening	2.20	6.00	0.00	6706.10	6714.30	223.0	11
Assembly storage	3.20	15.10	1.10	3685.30	3704.70	485.00	18
Assembly	29.00	28.00	0.00	8502.00	8559.00	256.00	20
	1155.68	636.18	28.3	82566.2	84386.36	7684	137
Total	0.01	0.01	0.00	0.98	1.00		
	Operation	Transport	Inspection	Delay			
		Total hours = 1406.44					
		Total days (24 hours/day) = 58.60					

Each element was assigned a cost down potential of high (>5%), medium (3–5%) or low (<3%), and an overall target of 15% agreed.

The deadlines for the achievement of target cost improvements were short with most to be completed during 1999. Beyond this each member of the project would have achieved further significant improvements of their own capable of practising lean management within their own businesses, their suppliers and all of their customers.

The group is now well on the way to achieving these targets.

BENEFITS ACHIEVED

The group has met the majority of its objectives. Here are some of the benefits achieved:-

1. The end-user has exceeded cost targets as a result of the elimination of waste and focused product and process development without enforced price reductions on suppliers.
2. Improving planning interfaces through the supply chain has released time for more value adding activities at the sales/planning desk. Stock has been reduced and opportunities have been identified from greater planning integration.
3. New business has been obtained by CES from the end user. This business places CES as a tier 1 supplier, conducting downstream value adding processes previously not offered by CES.
4. The whole supply chain has been brought into the new business planning phase with the end-user. They now see working with the complete supply chain from an early stage as a vital means to achieve onerous new product strategic targets, i.e.:-
   - Each new product must have 15% lower cost than that which it replaces.
   - Each new concept must be more fuel efficient, lighter and safer.

This type of involvement is a major change for CES, and very significant when considering that from design to end of product life, components are produced and stocked for 15 years.

The application of lessons learned in this project to other supply chains has been a success for all parties. CES has used value stream mapping proactively with other clients. In one specific case, 20% increased uptime was identified in the machining of hydraulic components. This presented an opportunity for CES to increase volumes to this company where a 100% market share position already existed.

Some of the technical work of a non-exclusive nature which achieved breakthrough cost reduction has also been used to satisfy similar cost targets with another key end-user.

Lastly behind all opportunities there are threats. CES competitors have recognised the power of integrated supply chains and the use of value stream mapping to streamline them. Our nearest competitor in this sector has recently concluded value stream mapping with the same end-user, seeking similar advantage to those demonstrated within the original project. The message from this is not to be complacent and to continue to seek opportunities to enhance competitiveness.

CONCLUSION

Value Stream Mapping and the development of supply chain groups has become a core process within CES's service offering. The original project, which was to create value from the integration of the supply chain from raw material to the end-user, has become a blueprint, applied a number of times since. The needs and trends of the automotive industry have driven this process. However, the techniques and experience gained are now proving invaluable in other sectors, where CES is proactively achieving substantial cost savings for its customers.

REFERENCE

Hines, P. and Rich, N. (1998) Outsourcing competitive advantage: The use of supplier associations. *International Journal of Physical Distribution and Logistics Management,* 28(7), pp. 524–546.

ACKNOWLEDGEMENTS

A version of this chapter first appeared as a paper in Total Quality Management, Vol. 10, Nos 4&5, 1999, S481–S490 and is reproduced here with the kind permission of Taylor & Francis Ltd.

22 Automotive after-sales: service fulfilment – a true measure of performance?

John S Kiff and David Simons

The LEAP programme has focused on the upstream automotive supply chain. However, work ongoing at the Lean Enterprise Research Centre analyses the automotive supply chain downstream from the factory gates. This chapter looks at the automotive supply chain closest to the customer, looking at the interaction with the service provider.

This chapter proposes that the consumer requirement of the interaction with a service provider, at its simplest level, is to achieve the return of his or her car within the time that it has been explicitly promised or is implicitly expected; and, that the car has been fixed correctly on the first visit and that it will therefore not require a return visit. This can be expressed as:

'Right first time on time'

The ability of an after-sales provider to meet this consumer requirement will depend on the interaction of the workshop and the parts supply system. Working together they should aim to achieve 'right first time on time' for the consumer with the least waste for the system.

In almost all current after-sales providers the only measurement of whether consumer requirements have been met is the measurement of the 'Service Level' of the parts supply system. This measurement only measures the availability of one part at a certain point in the system – typically delivery from a national warehouse to franchised dealer's warehouse. Typically it is quoted as being around 90 per cent and is used as a proxy for the performance of the system in delivering customer satisfaction.

As this chapter will demonstrate, this measure is inconsistent – from one manufacturer's system to another – inaccurate and actually irrelevant to real consumers' requirements. The significant issue is that for any given consumer interaction with a service provider, more than one part is needed. In addition, the 'Service Level' measurement does not take into account the other variables in the system and the performance of the service provider's workshop.

The chapter proposes a 'Service Fulfilment Index' for all service operations based on the work of Kiff *et al.* (1998) and shows how a lean parts supply system can help to achieve the highest customer service fulfilment.

INTRODUCTION

The after-market parts supply system in the European motor industry has, until the past ten years, been characterized by infrequent deliveries of large batches between the various echelons in the system. In addition, the level of customer service has been poor and true demand has been obscured by promotional activity.

Over time, manufacturers have increased the number of different models in their ranges and the number of variants of each model. Cars themselves have increased feature content and therefore contain more individual parts. In addition, the life cycle of a model has reduced – encouraged by shorter new product development lead times. All this has vastly increased the number of discrete part numbers that are stocked to serve the demands of the after-market.

To date, business research has concentrated largely on the upstream activities, from design through to manufacture and from raw materials down the supply chain (Womack and Jones [1]). Such research has worked on making the manufacturing and supply systems lean but in isolation of the fact that factories are producing finished goods stock which can be measured in months of sales and which hitherto has relied on 'push' strategies to sell.

In the after-sales area of the industry there are also a number of distinctions between parts that are in current production and those that are not, which frequently results in two different upstream supply chains. In addition, the after-market demand for parts for cars that are in current production has to compete with the requirements of the production line itself and often the more 'visible' factory demand wins the battle for scarce component supply.

These issues have been part of an overall research programme into the car supply, servicing and repair processes as part of the International Car Distribution Programme (ICDP).

INTERNATIONAL CAR DISTRIBUTION PROGRAMME (ICDP)

ICDP is the world's leading cooperative research programme into the future of car distribution and retailing. ICDP was founded as a pilot programme in 1992. The first full programme began in January 1994. Such was its success that ICDP2 and ICDP3 followed in 1996 and 1998 for two years and ICDP4 is now part way through. ICDP is a non-profit-making organization, funded by over 30 sponsors ranging from car manufacturers to IT companies. This chapter discusses the work begun in ICDP2.

ICDP has a balanced team of industry experts and academics. The European research is mainly conducted in four markets, France, Germany, Italy and the UK. In each market there is a Business School at the core of the research effort – respectively Pole Universitaire Leonard de Vinci, Institut for Automobilwirtschaft Fachhochschule Nurtingen, University of Venice and Cardiff Business School, Lean Enterprise Research Centre.

ICDP2 had six research streams ranging from Consumer Requirements to Public Policy. John Kiff at Cardiff led the After-Sales stream, which was a two-year study of after-market demand trends and parts supply systems in four European markets.

Fundamental to this research was the investigation of distribution structures encompassing all decision and storage points along the supply chain from

---

The observations, views and conclusions expressed in this chapter are those of the authors and not those of ICDP.

Manufacturer Central Warehouse to Retail Dealers. The principal research tools were benchmarking and simulation for which data was gained using semi-structured interviews with quantitative questionnaires together with a mail-out questionnaire survey to actors at each echelon in the system.

CONSUMER CHOICE OF SERVICE PROVIDER

For a consumer, buying or acquiring a 'new' car is on balance a positive, if not an exciting, experience. Having that car maintained (serviced and repaired) is in contrast a 'distress purchase'. A consumer will tend to want to minimize both the cost and the inconvenience of running a car and carrying out maintenance.

The consumer's choice of a service-provider for after-sales maintenance will be based on perceived Value in terms of a combination of Quality, Cost and Delivery (QCD) factors unless the consumer is contractually bound by a leasing agreement or similar. This QCD combination is unlikely to be accurately evaluated, but will be based on a mixture of experience, values, information, hearsay, judgement and economic imperatives.

From the consumer's point of view Quality, Cost and Delivery will mean a combination of the provider's ability to get job done 'right first time on time' (Quality) and at the right 'Cost' – both monetary and in terms of convenience. The latter includes the location of the provider relative to the consumer. This is also part of the 'Delivery' element which itself includes the 'soft' factors in the after-sales experience – i.e. the way the consumer is 'looked after'. In the majority of cases, the consumer will be making trade-offs between these factors.

This QCD concept of customer value is closely related to that expressed by Reicheld and Sasser [2] in which:

$$\text{Customer Value} = \frac{\text{Results} + \text{Process Quality}}{\text{Price} + \text{Customer Access Costs}}$$

The UK after-sales market is highly competitive and highly fragmented. There are thousands of service-providers with different consumer offers in terms of QCD. For vehicle manufacturers' franchised dealers, the revenue and profitability of their after-sales operations is vital to their total business as the profitability of new car sales continues to come under pressure from over-supply in the market. However, their market is primarily in cars of up to three or four years old and the ongoing improvements in quality and reliability of cars means that the significant market demand for maintenance and mechanical repair is moving to older cars. [3]

This 'ageing after-market' means that franchised dealers face a loss of business to independent garages who have a large share of the market for cars over four years old. To maintain their vital profit stream, franchised dealers must provide a better value proposition to customers. In order to do so, they need to measure their current performance; i.e. how well they are currently satisfying consumers. A critical element of that performance is the performance of the parts supply system.

OBJECTIVES OF THE PARTS SUPPLY SYSTEMS

The most important viewpoint from which to examine the objectives of the parts supply system is that of the consumer. However, the consumer has no interest in the performance of the parts supply system itself. It is the authors' contention that the consumer is only interested in the following: the return of his or her car within the time which has been explicitly promised or is implicitly expected, and he or she also expects that the car has been fixed correctly on the first visit and will therefore not require a return visit. In addition, the consumer is usually not interested, nor aware in advance of the number of parts required to fix the car. This contention forms the core of the conceptual framework for this research [4].

In order to achieve the objective of satisfying every consumer on every visit, it would be necessary to ensure that the following had been achieved as part of the After-Sales function:

- The parts required are made available to complete the job by the time promised. To do this requires the correct and complete diagnosis of the job in advance.
- The workshop space and technician's time is available to complete the job by the time promised.
- The technician has the necessary skills and tools to complete the job correctly by the time promised.

In addition, for complete satisfaction, the consumer will expect the part(s) fitted to be reliable in service.

As well as satisfying their own workshop customers, dealers will also wish to wholesale parts to independent garages. This is generally a secondary consideration, but is one that the Parts Supply System has to take into account in terms of demand.

So, as a primary objective, dealers will wish to satisfy their consumers but they will naturally be seeking to make a profit from this aspect of their business, while at the same time minimizing the costs they incur from keeping stock. Dealers will also want to ensure that the Parts Supply System does not allow the workshop to be idle. Parts and Service Labour are a very important aspect of a dealer's business. They form, on average, 50 per cent of the gross profitability of a typical European dealer.

Manufacturers will also share the dealers' desire to satisfy the consumer. They will also be seeking to return a profit on their part of the parts supplied business. They will also want to minimize their stock (although the manufacturers recognize that they are the ultimate holders of certain types of Parts – especially those that are no longer in production).

Manufacturers will, to some extent, also recognize that the dealers need to make a profit from the parts supply system and that these profits are a critical part of the overall profitability and health of the distribution network.

SERVICE LEVEL

This measurement is the one most used in the Parts After-Market to describe the performance of the system in satisfying customers' needs for parts. It can be applied to the dealer's ability to satisfy the workshop customer, or, more usually, it can be applied to the National Sales Company (NSC) warehouse's ability to supply the dealer with a part. The averages of the figures quoted by dealers and NSCs in each market to the research team are shown in Figure 1.

The Service Level represents the ability of the system to supply any **ONE** part-line (part-number) from one echelon to another. In view of the fact that the majority of the jobs in a workshop require more than one part-line (if a part is needed at all), this

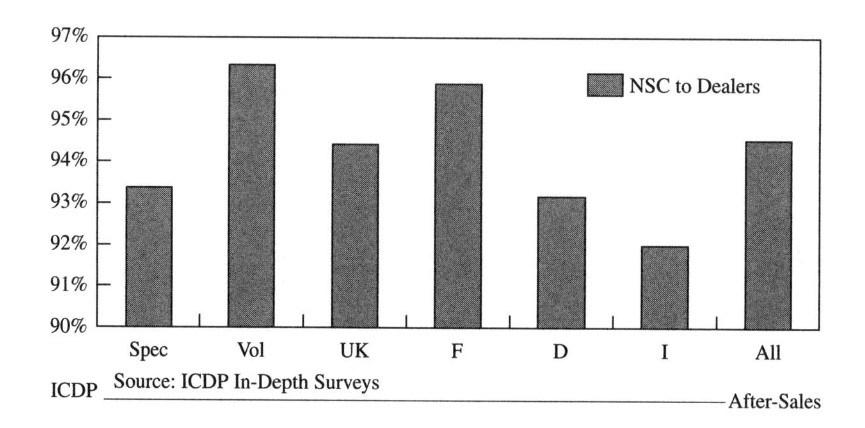

Figure 1 The oft-quoted 'Service Level' – 'Everyone defines it in a different way!'.

immediately makes the 'Service Level' an inadequate measurement as far as customer satisfaction is concerned.

In addition to this, there are a multitude of measurement methods! No two manufacturers' systems measure Service Level in the same way. Some manufacturers state that only if the full order-line is sent will it count towards achievement of service level, while others count it even if some parts were missing. Table 1 illustrates some of the measurement issues.

In the authors' opinion 'Service Level' is a meaningless comparison and worse still it does not measure true service level! It serves no purpose from the consumer's point of view since he or she is only interested in the return of his or her car within the promised time and it serves no purpose as an inter-franchise comparison, since the measurement criteria are different.

Table 1 Measurement issues

Some count if . . .	*Others count if . . .*
Full order-line sent	Part order-line sent
All arrive on time	Some arrive by some time
All come from NSC	Some come from elsewhere
They are original orders	They are back orders

SERVICE FULFILMENT INDEX

Table 2 illustrates the concept of the **Service Fulfilment Index**, taking as an example the current situation in a parts supply system of a volume manufacturer. It is calculated, in a relatively conservative way, as follows:

- The chance of *any one part* being in stock is taken as being the 'first time pick' level or 'service level' in this example, this is taken as 90% at the dealer parts store.

Table 2 Service Fulfilment Index – current

Chance of part being in stock	90%	
No. of parts required	**3**	
First time pick of parts basket		73%
Chance of job completed on time (where all parts available)	93%	
Sub-total		68%
Chance of these jobs being done right	92%	
Service Fulfilment Index		**63%**

- The average number of parts required for *all* workshop jobs is 3. This is derived from the research team's examination of 400 service invoices. For all jobs that require parts, the average is 4, but the more conservative position has been taken here for illustrative purposes.
- If the chance of any one part being in stock is 90%, the chance of all three parts required being in stock is 90% × 90% × 90%. This amounts to **73%.**
- The chance of a job being completed **on time** is taken as 93%. (The view of volume dealers was that only 7% of jobs were not completed on time.)
- This reduces the chance of the consumers expectations being fulfilled to **68%** (73% × 93%).
- The chance of a job being completed **correctly** is taken as 92%. (The view of dealers was that only 8% of service and repair job customers returned to the dealership.)
- This gives a **Service Fulfilment Index of 63%** (68% × 92%).

It is the principle of the Index rather than the absolute accuracy of the figures that is important. It illustrates clearly the extent to which consumers are being inconvenienced or dissatisfied at the moment and why 'Service Level' as a measure of the performance of parts supply systems is misleading.

It is accepted that many of the fast-moving parts may have much higher first-time pick rates and therefore the likelihood of picking the required basket of parts for more straightforward jobs will be much higher. However, the research team did not come across any franchise that measured the variation in first-time pick for the different movement rates of parts.

So from this analysis, we can see that unsatisfied customers are of the order of 37 per cent. Research on customers requiring repeat repairs indicates a figure of 29 per cent. The difference between the two may be explained as follows:

- some customers may wait (after the promised or expected time) for their repairs to be completed,
- some customers may be offered a courtesy car and may not perceive that they have had to 'return' for repeat repairs, while
- other customers may never return to the dealership and have their remedial work done elsewhere.

Clearly, there are many problems of definition in this area, but the principles are evident and the industry needs to provide itself with a tool to measure true customer satisfaction for the after-sales process.

This measurement index is analogous to the Lean Production method of measuring Overall Equipment Effectiveness, which takes into account machine downtime and quality etc.

SERVICE FULFILMENT INDEX – FUTURE, LEAN STATE

In order to improve the Service Fulfilment Index it is necessary to improve:

1. the logistics of supplying the correct parts to the workshop, and
2. the workshop time and skill/experience scheduling.

The former can be improved by changing the Control, Centralization (of ordering) Time and Structure elements of the Parts Supply System (Kiff *et al.*, 1997 [4]; Simons and Kiff, 1998 [5]).

The latter can be improved by significantly increasing the **pre-diagnosis** of the jobs coming into the workshop. This is done by contacting consumers in advance and ensuring that, in so far as is possible, the work required is understood in order that the correct parts are delivered by the system in advance and made available so that the job can be completed by the time the consumer expects.

Using this method, the number of unexpected parts shortages will be minimized, but even with a high level of pre-diagnosis, in a few cases it will not be possible to have **fully** diagnosed the requirements of job. In such circumstances, the revised, leaner structure of the supply chain will be able to deliver the vast majority of the parts required within the day from a local distribution centre where a much larger range of parts (say 25,000 lines covering over 95 per cent of all needs) will be available.

Using the computer simulation model developed by Simons as part of this research programme, it is possible to simulate the Service Fulfilment Index of a Lean parts supply system in which pre-diagnosis of jobs plays a major role.

The results of the simulation are shown in Table 3.

Table 3 Service Fulfilment Index – future

Chance of part being in stock	98%	
No. of parts required	**3**	
First time pick of parts basket		94%
Chance of job completed on time (where all parts available)	96%	
Sub-total		90%
Chance of these jobs being done right	94%	
Service Fulfilment Index		**85%**

CONCLUSIONS

The current after-sales activity within franchised car dealers is clearly unsatisfactory in terms of customer satisfaction performance. The current measures of system performance do not measure what is important to the customer. It raises the question: how many other systems are not truly measuring what is important to customers in terms of Quality, Cost and Delivery, yet are continuing to use inappropriate measures? Using a Service Fulfilment Index at every dealer for every customer would allow improvement activity to be targeted much more accurately than at present. Although dealers could institute better pre-diagnosis activities and SFI measurement immediately, it is incumbent on the manufacturer to institute improvements to the Parts Supply System to make it leaner.

ACKNOWLEDGEMENTS

The authors express their acknowledgements and thanks to Professor Dan Jones of LERC for his conceptual input into this research and his advice and encouragement.

REFERENCES

[1] Womack, J. and Jones, D. (1996), *Lean Thinking – How to Banish Waste and Create Wealth in Your Corporation,* New York: Simon & Schuster.

[2] Reicheld, F.F. and Sasser, W.E. (1990), Zero Defections: Quality Comes to Services, *Harvard Business Review,* Sept–Oct.

[3] Kiff, J.S. and Chieux, T. (1998), The Impact of Technology, Quality and Reliability on After-Sales Operations, *International Car Distribution Programme Research Report.* June 1998.

[4] Kiff, J.S., Chieux, T. and Simons, D. (1998), Parts Supply Systems in the Franchise After-Market in France, Germany, Italy, U.K., *International Car Distribution Programme Research Report 7/98.*

[5] Simons, D. and Kiff, J.S. (1998), Automotive After-Sales Distribution Analysed Within A Conceptual Framework For Supply Chain Improvement, Proceedings of the 2nd Annual Logistics Research Network Conference, Cranfield University.

Section 7
Towards the Future

23 Policy deployment: from strategy to action – from the firm to the supply chain

Nick Rich

INTRODUCTION

'Policy deployment' is probably one of the last remaining elements of the 'jigsaw puzzle' that supports both the lean manufacturing and supply chain management systems. It provides the skeletal structure through which improvement activities are generated and sustained in the world-class Japanese factories such as Toyota Motor Corporation (Akao, 1989). The policy deployment approach is not culturally specific to Japan and indeed the founding principles upon which this system is grounded are Western in origin. However, it is in Japan that this approach has been perfected and where the process forms the touchstone for all change management activities within the enterprise and beyond. It provides meaning to every action at every level of the business and reinforces all the values of the 'learning organization'. The birthplace of this 'modern' technique is, unmistakably, Total Quality Management (TQM) and the logical adaptation of these values within the enterprise 'system' and its interrelated parts or departments.

The policy deployment process also forms an infrastructure that values, integrates and nurtures the human resources within the business as the single most important input to long-term competitive advantage. In essence, policy deployment is a 'software' system and it is through the focusing of human effort that improvements become capable of commercial exploitation. At the heart of the system is a concern to marry the human resources with the business hardware and manufacturing process. This marriage is focused on growth and the improvement of the manufacturing process at a rate that is in line or quicker than predicted changes in customer and market requirements. In this respect, policy deployment is also the process through which waste is banished from the lean enterprise (Womack and Jones, 1996) and those people who will be responsible for its efficient and effective operation create the implementation of the new business system. The policy deployment is therefore the cement that bonds departments, individuals, customers, suppliers, and all improvement initiatives into a unified force for change that is forward focused, mutually reinforcing and synergistic. As a catalyst for change, the power of policy deployment comes through participation, integration, alignment, control and most of all focus on what needs to be changed in the short term to achieve medium-term improvements in customer service and superior business performance.

To a large extent, the policy deployment process is hidden, and less overt than mechanisms like kanban or standardized operations, yet it dovetails with these practices. These practices are the physical manifestation of 'how' a business improvement requirement has been solved. These mechanisms are therefore the result of the policy deployment process and the 'things that needed to be changed', not necessarily the logic and motivation for the change. As such, policy deployment is the process that embodies the business 'wants' whereas these mechanisms reveal only 'how' a business 'want' has been satisfied. Emulation of the policy deployment approach is not sufficient. The process is unique to the company and its market circumstances; however, the process is universal in application but requires logical localization to match with these unique circumstances and the market pressures that affect the business. Wholesale emulation is therefore not practical, not possible and will not identify correctly the 'things that must change'.

The hidden 'guiding hand' of the lean administrative and management system, provided by the policy deployment process, is 'hidden' in the respect that the system relies largely on meetings and group activities of employees. It is the result of these meetings and implementation plans that start the process of improvement and until this point there is little visible evidence of change. However, as implementation plans get under way, the system switches from the covert to the overt. It is perhaps little wonder that the process has not enjoyed the volumes of importation that other practices, such as quality circles, kanbans and set-up time reduction, have achieved in the UK. Also, it must be suspected that the 'less visible' nature of the policy deployment process has meant that aeroplanes full of managers on the industrial tourism circuits of Japan have missed this vital ingredient of world-class manufacturing and Japan's ability to switch gears (Hayes and Pisano, 1994). To compound matters, few attempts have been made to promote, through publications, the concept in the English language due to the unique localization of the approach for every business that employs the method. After decades of studying Japanese manufacturing organizations, and the mixed implementation success of kanban and other practices in the UK, the 'system' that supports world-class performance has been missed.

The objective of this chapter is simply to promote awareness of the policy deployment process and to provide an understanding of how, through the structuring of activities, a defined strategy becomes enacted in a cascade process of targeted improvements within and beyond the business. The latter section of this chapter will highlight the role of the purchaser as the focal point for the development of the supply chain and the externalization of the policy deployment approach to focus the improvement efforts of suppliers. It is the management of the business and its supply chain that has been the focus of the LEAP initiative and while it is not desirable to discuss the individual activities of sponsor companies the chapter will provide much of the basic logic[1] that has underpinned all improvement activities.

POLICY DEPLOYMENT: A DEFINITION

Hoshin kanri is the Japanese term for what has been coined 'policy deployment' in the West. *Kanri* is the Japanese word for 'control' or 'management' and *Hoshin* means 'shining needle' and is analogous to the needle on a navigational compass. So putting the two elements together we get 'a system of directional management'. This is the very essence of policy deployment. It involves the senior levels of the business setting the necessary direction for change and the identification of weaknesses in the current business system in relation to these future market requirements. The resolution of these weaknesses is central to policy deployment and the future competitive position of the business over the mid-term (three to five years). These forward-focused plans, derived from the mission and strategy of the business, are therefore translated into an 'enterprise-wide' challenge to improve and grow the business. In this manner, policy deployment touches every element of the business and integrates all employees in a process of ongoing improvement.

Watson in Akao (1989) provides the following definition of the approach: 'Hoshin kanri provides a step-by-step planning, implementation, and review process for managed change. Specifically, it is a systems approach to management of change in critical business processes. A system, in this sense, is a set of co-ordinated processes that accomplish the core objectives of the business.' Here we can see the emphasis of policy deployment on the *system* and the *management of processes* that create *customer service* and *shareholder value*. It is important to note that this definition places an emphasis also on the key business *processes* that will enable the strategy and goals of the business to be achieved. As such, re-engineering of business processes is implied by this definition but unlike the use of this technique in the West, the combination of re-engineering effort in a company or supply chain is guided by the deployment of strategy. The Western approach has, however, lacked this clear intent and focus, and, while benefits may accrue, these tend to be local and bear little significant relationship with the business or its improvement.

More recently, Womack and Jones (1996) proposed that policy deployment is 'a strategic decision-making tool for a firm's executive team that focuses resources on the critical initiatives necessary to accomplish the business objectives of the firm. By using visual matrix diagrams . . . three to five key objectives are selected while all others are clearly deselected. The objectives are translated into specific projects and deployed down to the implementation level of the firm. [Policy deployment] unifies and aligns resources and establishes clearly measurable targets against which progress toward the key objectives is measured on a regular basis' (p. 307). This definition, compatible with the previous, emphasizes the 'means' through which the goals or 'wants' of the business will be satisfied and in particular the internal processes of alignment and selection that create the common direction for all business departments, teams and employees. Here, the concentration on the TQM values of the 'vital few' actions that will improve the business is stressed. By implication, this means that resources will be directed towards initiatives that generate the highest 'market' value for the investment of time and money.

The key features of the policy deployment process can therefore be summarized as:

- A systems approach to managing the entire enterprise and its supply chain.
- A process of goal setting from the strategy selected by the senior management.
- A process that identifies business weaknesses in order to improve the robustness and stability of the future system.
- A system that emphasizes the core processes and achievement of the future goals needed to maintain the highest levels of value generation (for shareholders, customers and employees).
- A system that emphasizes growth in the enterprise in order to the meet future needs of the market or industry.

- A system that focuses on the 'vital few' initiatives that will enable increased business performance.
- A system that involves all employees in target setting, improvements and reviews.

Conversely, it is important to understand what policy deployment is not. It is not the strategy process itself. The strategy process is conducted before any policies can be deployed within the organization and the starting point of the policy deployment is following an internal revision process concerning the strategy itself. The strategy process, although not conducted in a vacuum, is therefore the parent process to *deployment*. Policy deployment is not a panacea and is not a wish list of superficial changes that can be passed to individual managers in the business. No manager can, as an individual, influence the key business processes to such an extent that the competitive position of the business is improved and therefore a consensus approach to management (using groups of managers who improve key processes by collective, directed and continuous effort) is required. These targets are therefore future-focused 'stretch goals' that focus on the weaknesses of the business, then reconcile how to improve these weaknesses for the benefit of the customer now and in the future. As such, they are not easy to achieve and success is a function of changing the business, not isolated islands within that business. Policy deployment is therefore a system that governs the renewal processes of the business as if it were one highly complex and highly integrated organism.

THE ORIGINS: A LINEAGE OF TQM

The origins of policy deployment are not difficult to trace; it is a derivative of the Total Quality Management (TQM) family. This is unsurprising given that policy deployment was developed and promoted across Japanese industry during the 1960s where it gained popularity with major corporations (such as Toyota and Komatsu) and followed in the wake of initiatives to promote total quality management – spearheaded by Deming himself. In effect, the activity can be seen to represent the application of TQM at the strategic level of the business in order to create one enterprise-wide system of quality control in its broadest sense. As such, the concept of policy deployment includes both the techniques of continuous incremental improvement (Kaizen) and breakthrough objectives or challenges that create 'stretch goals' as targets for company performance. It also reinforces the processes of planning for implementation as well as standardizing the results of any improvement to form the foundation for subsequent improvement initiatives.

To conform to the 'total' dimension of TQM, the policy deployment process is also the means of aligning and integrating all the many sub-systems of the factory that must be brought together in order to make major improvements (in the control items) of the business as a whole. These sub-systems themselves must also be, by design, microcosms of TQM and mutually reinforce the goals and direction of improvement required of the business as an entity. As such, the approach is based on consensus management that covers both the direction of change and the extent of changes needed to realize the goals of the business. The influence of the traditional quality gurus, of Deming and Juran (Bicheno, 1994), are easy to see and include the manifestation of such advice as:

- Create a constancy of purpose.
- Create pride in work that can only come from exercising empowerment.
- Eliminate slogans that push people forward without any real support or meaning.
- Exceed expectations of customers in order to retain them and exclude competitors.
- Break down departmental barriers that restrict information and incorrectly assign responsibilities in the factory.
- Create leadership to drive forward improvements.
- Adopt the Plan–Do–Check-Act (PDCA) then Standardize-Do–Check–Act (SDCA) cycles of change management.
- Continuously improve towards perfection and ultimate levels of value generation.
- Drive out fear and remove blame for under-performance as this can never be the fault of an individual in any factory but a collective shame.
- Cease doing business on price alone as this has no correlation with customer service which itself is based upon a complex evaluation of quality, cost and delivery issues.
- Eliminate quota thinking both for the shop floor employees and for management. This promotes local efficiency above the effectiveness of the business in satisfying and exceeding customer expectations.
- Focus on the vital few sources of problems and solutions in order to avoid deploying resources to activities with only marginal benefits to the company.
- Standardize practices so as to act as the foundation or lever upon which improvements can be sustained and improved upon by subsequent cycles of problem solving.

The Japanese quality revolution of the 1960s occurred during a period when Japanese manufacturing performance lagged far behind the rest of the industrialized world and therefore these fundamental principles, resonated with a manufacturing base that was hungry to improve and needed to export. As such, the environment was right for change and the TQM philosophy provided the catalyst for improvement as well as a fundamental system for business management. From this point, eminent Japanese scholars, such as Ishikawa (Bicheno, 1994), began to weave new tools and techniques for the mass application of Quality Control and, as such, the strategic advice became reinforced by concrete application of quality improvements at the team level. This completed the picture of company-wide quality improvement from the top-down and the bottom-up. This potent combination has eluded most Western manufacturers who have struggled to blend activities to the point that they form a true system whereby the end of the top-down process is indistinguishable to bottom-up actions.

Consensus within the business is central to the policy deployment approach and means that all levels in the business and all 'human resources' need to receive periodic and ongoing promotion of the 'business cause'. This promotional activity is also the means of incorporating and motivating all the 'key players' to improve their business sub-systems in order to improve the enterprise. Promotion of the total quality values and principles is therefore present in all actions and all levels of the business, not solely on the shop floor. It is the system of policy deployment and the parallel series of reviews that allow each element of the business to define its role, its 'worth', and the justification for improvement actions. Another result of the policy deployment process is the ability to 'sum' actions taken in local departments and to witness the 'domino effect' whereby improvements in one element of a key process releases the potential to improve other elements (internal customers and suppliers). As such, the promotional activities are important and communicate the results of 'quality' improvements across the business. In this respect, departments with similar processes can gain from this ability to transfer successful practices in a form of sharing and horizontal deployment. In this manner, even small changes and improvements to working practices can be very quickly incorporated within the factory and more recently, globally.

In order to create and sustain consensus, the business must be a learning business, both in terms of external learning from others and internally through learning about the business itself. This attention to learning also serves another purpose and reduces the variability of individual skill sets. Thus training is used as an investment to reinforce goals, values, knowledge and key techniques (especially a 'common approach to problem-solving') throughout the lifetime of work for the business. Each employee is therefore equipped with the necessary skills for their position at a level in the business (technical and interpersonal skills) and the regular planning and problem-solving activities with peers equips the individual with greater and greater detail of the business and its many sub-systems. For the Japanese, this system is reinforced by periodic job rotation (approximately every five years) to enhance the understanding of the business 'as a system' and the constraints within key business units. Thus middle managers are trained formally and informally in the 'business' and better understand the implications of their actions for the key business processes and their immediate customers and suppliers. It is the middle managers who therefore hold a great deal of tacit knowledge about the business and its systems. These individuals are, as recipients of investments made by the company in training, highly valuable. This investment in training, while not directly quantifiable as a financial investment, has a potential to make many paybacks every year over a working lifetime of each employee and it is the middle managers who can make the highest contributions.

The use of regular problem-solving meetings, for teams and managers, also ensures that the tools and techniques provided by the investment in training are used regularly and form common (business-wide) standards of analysis. It is these features that are missing from the UK situation. In the UK, training is sporadic, general and does not necessarily mean that the techniques learnt will ever be applied or will be applied infrequently. This last comment is perhaps an indictment that the UK has, to date, failed to sustain continuous improvements at the business level and failed to connect the top-down with the bottom-up. To illustrate the power of the policy deployment approach and the integration of all employees in the solution to business problems, Toyota Motor Corporation used its policy deployment approach to stress the need to reduce operational costs throughout 1993. The emphasis of the cost reduction followed awareness that high competition and adverse exchange rates would severely impact on the business unless costs were reduced. Over 40 years since the Toyota Production System was introduced (deemed the most powerful form of lean production by Womack *et al.*, 1990), the 150,000 employees in the global business saved $1.5 billion and the final accounts for 1994 led with that as its headline. This is perhaps testament to a business that has developed its policy deployment approach during the 1960s and sustained continuous improvements for over 40 years, setting many benchmarks from the 1980s onwards.

THE INGREDIENTS OF POLICY DEPLOYMENT

The ingredients of the policy deployment approach are effectively human resources at every level of the business, the appropriate training of these resources and the formation of groups at each level. However, before an attempt is made to discuss the process of policy deployment and how it works, it is wise to consider two key dimensions of the process. These are:

- the value and roles within the business: feed-forward systems;
- the measurement and control of the business: feedback systems.

The ability to define value by level of the business is a key enabler for the policy deployment process. In effective systems, such as Toyota's, these values are defined strictly. These values are, shown in summary, as follows:

1. The business derives value from a senior manager through their contribution to planning activities. These individuals should be future-focused and relatively unpolluted by day-to-day issues. Therefore, the worth of a senior manager is in their ability to work with the rest of the senior management team in planning the future direction of the business. This involves both the development of the business mission and the strategy but also the forward planning for policy (usually three to five years ahead). These senior managers are not best positioned to determine the means through which improvement can be achieved but can determine the predicted improvement areas for the business. This level of the business is the environmental scanning process.
2. The value of a middle manager also comes from planning activities and the integration of all departments to form cross-functional teams. The planning-forward time horizon for this level of management is less than for the senior managers – usually a year or so. These middle managers are more concerned with the means through which improvements can be achieved but also need to understand the gaps that are predicted. This level of the business is the real change agent and it is these managers who tend to know what is best practice and how it can be usefully implemented. These managers determine the velocity of change (Dimancescu *et al.*, 1997).
3. The value of the team leader and the team derives from control. These employees are bound within a local environment and therefore concern themselves with the enactment of improvement projects and the control of the process. They need to understand the needs of the business, the improvement activities and their role in the execution of these improvements.

In addition to these planning activities there is a parallel stream of feedback loops required at each level in the business in order to control the rate of progress towards the goals set. The value of this information is also a function of the role and level of the individual (and team) within the business.

1. Team leaders and teams have the most urgent need for information. The thinking cycle for the team is daily and the main review point is weekly. Therefore progress tends to have meaning when a weekly review is conducted.
2. For middle managers a daily review is too frequent to review performance and therefore a weekly approach to feedback from the teams is more valuable.

However, middle managers are interested in trends in performance in order to direct the resources to the most important improvements required. Therefore, while weekly information is transferred to the manager, the monthly trend of information (actual versus planned) is more important. As such, middle managers tend to review their performance monthly against the annual challenge set.

3. For senior managers, the review period of a month is shared with the managers but because this level of the business is concerned with benchmarking improvements against longer-term plans, the key formal review tends to be quarterly and coincides with financial reporting systems (and investment budgeting cycles). As such, the trends over the last three months are important as comparisons against the annual and three-year plans held by the business.

The roles, planning and review of individuals in the factory, set in the context of policy deployment, clearly marks a distinction between planning and execution. The policy deployment system is therefore both forward thinking and periodic in review. A second vital ingredient is the management of the lateral processes within the factory. Lateral improvement teams affect every level of the business. At the most senior level this involves the board of directors and at the shop floor level this involves quality circles and in the middle there is the need for cross-functional management.

In the West, a rich history of departmental management and 'functional ghettos' has become entrenched in the way organizations are designed (Peters, 1992), based on the logic that the greater the specialization of the factory administration then the greater the efficiency of the factory. This structuring has caused problems for businesses that operate in fast-moving or demanding markets and served to slow the response of the business through an inability to control the key processes that pass laterally through the business and the supply chain (Achrol, 1991).

These barriers do not actually exist as tangible 'walls' between departments in reality but high importance is placed on such boundaries in practice. In the absence of a mechanism that crosses these 'invisible barriers', company-wide coordination cannot be achieved meaningfully. In recent times, re-engineering efforts (Hammer and Champy, 1993) and constant 'de-layering' of businesses (Keuning and Opheij, 1994) have been techniques employed to reduce the negative impact of the functional bureaucracy but ironically tended to remove levels of middle management (the most creative level in the policy deployment process). The typical dysfunctions that these techniques have sought to correct are:

- poor communication between internal departments;
- decentralizing decision making in order to improve the responsiveness of the organisation to changes in the environment;
- the removal of administrative costs and improvements in efficiency;
- the breaking of demarcation lines between specialists and generalist managers in the business in order to improve cooperation.

Indeed, many of these 'streamlining' initiatives have been reactions to crises currently affecting the business and therefore, by default, reflections of the lack of strategic planning and historic lack of improvement activities at the factory. However, there is a further problem that these initiatives have not really addressed – most Japanese 'world class' companies remain departmentalized yet achieve the results sought of this Western organizational 'surgery'. As such, the Japanese have retained the control power of the department but also have the ability to respond quickly as a business to change (Dimancescu *et al.*, 1997). The combinations of these features have, in the West, been regarded as 'trade-offs'. The positive combination of the department and key lateral process has been achieved through the promotion of cooperation between business departments. This process has been termed cross-functional management, and was learned some 30 years ago in Japan. Once again, such professional bodies as the Japanese Union of Scientists and Engineers (JUSE) promoted the process heavily, as a means of integrating different functions within a single business.

Cross-functional management is defined by the JUSE as 'a management process designed to encourage and support interdepartmental communication and co-operation throughout a company – as opposed to command and control through narrow departments or divisions. The purpose is to attain such company-wide targets as quality, cost and delivery of products by optimising the sharing of work' (JUSE, 1988). Let us examine this definition a little further before returning to the issue of policy deployment. The definition involves several key features that are enablers for policies to be deployed properly:

- It is a method of promoting the benefits of working at the interfaces between departments in order to improve the effectiveness of the business. Thus it is deliberately intended to improve the effectiveness of the business rather than the insular goals of local efficiency.
- The improvements in quality, cost or delivery are commercial targets that enable market strategies to be effective. To improve quality is a means of differentiating products as much as it is a means of lowering the costs of production. Likewise, improving the frequency and reliability of delivery within the factory will create an effective flow of products to the distribution channels offering other forms of competitive advantage and 'order winning' behaviour such as place utility for consumers and customers.
- Quality, cost and delivery (QCD) are the only measures that unite, and act as common denominators that cross every department from purchasing to maintenance engineering and marketing to operations (Maskell 1989). Therefore, they are means of uniting different units and specializations within the business. In addition, QCD also has specific meaning for each business department and team leader in terms of the process yield, on-time deliveries to customers (including internal customers) and costs (the amount of inventory, overtime hours or costs of process changeovers).
- Quality, cost and delivery are also keys to exploiting customer service – they are for all intents and purposes what customers buy as a transaction with the business concerned.
- The sharing of work is the redeployment process whereby resources are used and shared to allow the effectiveness of the business to be released and exploited logically. This process may resolve some of the problems found in the West in terms of dislocating the activities of, say, purchasing (placing replenishment orders by production operators) with the 'value adding' role of purchasers which is to assist suppliers in reducing the total costs of supply. The sharing of work also addresses the concepts of 'empowerment' and 'job enlargement'.

In the current climate of high change and competition, the use of cross-functional management is therefore a means of retaining the efficiency of departmental management but also generating lateral processes that unite managers on the issue of 'customer service'. It is these processes that are closely correlated with long-term profitability and the need to constantly reduce the costs of making products. In addition, this technique decreases the issues associated with the creation of problems in one department and the detection (and resolution) in another. The common language of QCD is shared both vertically and, with cross-functional management, laterally within the business. The QCD approach therefore provides the means of translating corporate goals into meaningful gaps that need to be closed by the collective improvement of the key processes within the business. QCD is the principle that guides the ongoing annual challenges from year to year and can be extended to encompass other key processes (and capabilities) such as product design, supply chain partnerships and corporate citizenship. One business department cannot control these key processes and activities, they are business-wide, and therefore the power of policy deployment, integrated management and QCD creates the basic improvement fabric for a factory (Dimancescu *et al.*, 1997).

With cross-functional management used to support the senior management team, the final element of the trilogy is the quality circles and implementation teams. These individuals are brought together to execute the plans, measure key control items and improve the key business processes that yield customer and shareholder value. The shop floor improvement teams are therefore controls that serve to standardize and improve the transactions and flow of product.

THE PROCESS OF POLICY DEPLOYMENT

The policy deployment process is relatively straightforward in terms of a step-by-step list of activities (Akao, 1989). A general synopsis of the policy deployment approach reveals four generic stages of implementation:

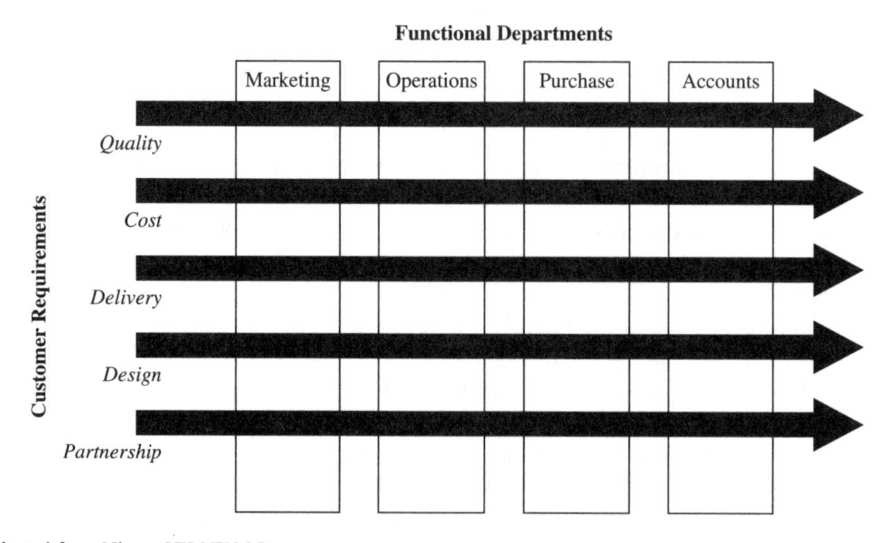

Adapted from Nissan UK NX96 Programme

Figure 1 Horizontal customer requirements and vertical organizational structures.

- a senior management scanning, positioning and evaluation process concerning the performance of the business;
- a middle management alignment activity whereby key projects are selected;
- the development of implementation teams;
- the creation of the control items and review process.

Senior management evaluation

1. The senior management team commence the process with a series of reviews concerning the business mission, other environmental scanning activities (technology, markets etc.) and stakeholder requirements (the owners, the customers, the product offering, the wider economy and competitors). This process is used to benchmark the position of the business against its future goals and by default the shortfalls in performance that need to be addressed (not the means of doing so).
2. The senior team identifies the gaps and weaknesses between the ideal future position and the current performance of the business. These are converted into key improvement challenges and decomposed into the QCD and other improvement targets required by the business over the coming year. These requirements tend to be expressed as 'challenges' such as an improvement in product quality of 20 per cent, a reduction in costs of 10 per cent, and an improvement in delivery performance of 12 per cent. To achieve this level of precision, the senior team will conduct many analyses including the future impact of changes in remuneration, tax laws, depreciation of assets, currency movements and such like. The challenges are then presented to the middle management team as a group together with any necessary clarification in a promotional activity and as a means of discussion and clarification.

The process therefore begins with a very detailed understanding of the market environment, competitor positioning and trends in terms of the future predicted rate of change in consumer demands. This is often embodied in a document entitled the 'Customer Master Plan' and involves every senior manager in the collection of information in order to position the company's current performance against the three- to five-year future position desired. In addition, this plan not only details where the company should be in five years' time but also the levels of performance needed to reach and maintain this position.

The types of issue that are reviewed by the senior management team include the marketplace, customer service and customer strategy, competitor activity, new entrants, and the emergence of new technology. In addition, changes in the corporate culture of the business (the type of person that the business needs and therefore must nurture) and what constitutes good corporate citizenship are also reviewed. These latter activities represent analyses of the 'health' and 'growth' of the human resources within the business. In addition, the three-year forward plan is deliberately used as a time block to focus thinking for all managers in the business and it also allows trends in the environment to be highlighted (such as legislation) that would have been missed by a longer planning horizon. Great long-range plans are therefore impossible in fast-moving markets and as such the annual policy deployment process means that at each point the company thinks forward three years and creates a rolling and living strategy document that is resilient to short-term and unexpected change.

The benefits of a three-year plan involve:

- the ability to predict the rate of change in customer expectations;
- the ability to predict the rate of change in competitors and the business they control;
- the use of a common time horizon such that all senior managers have a constant planning horizon upon which to focus their activities and reduce the complexity of the 'unknowns';
- the use of a common horizon that fits with the capital budgeting cycle and allows proposals to be developed for technology with longer pay-back periods.

The ability to predict changes in 'external agencies', especially those that can enact or lobby for legislation, is also analysed. The maturation of these laws is long and therefore actions can be taken by the business to comply with legislation as soon as it becomes active. A case in hand is environmentalism and green policies that will impact on every manufacturer, and those who wait for legislation to be passed will be ill-prepared for it. The Japanese car assemblers have already achieved conformance to ISO 14000, on a supply chain scale.

Cross-functional management planning

Following the senior management deployment, the middle managers engage in a period of study to understand the business requirements (such as a review of the business costs etc.) and generic adjustments to the business that could be made. This activity usually involves a process known as 'catch ball' (a reference to the children's game) but in reality is the tossing of issues between managers to highlight the ramifications of any chosen course of action (Akao, 1989). Courses of action that can be agreed immediately are confirmed by the middle managers and returned to the senior management as accepted programmes. Those issues that cannot be resolved are highlighted and members of the senior management team join the middle managers and engage in a new round of catch ball promotion. This second loop is important as unresolved issues often need authorization and would imply that a department in the business may need to be sub-optimized in order to improve the entire business. For example, JIT deliveries of small batches may increase the unit costs of purchasing and therefore to pursue JIT deliveries, the purchasing department may well have to endure a poor performance rating in the short term. The senior managers concerned can sanction this. Once the 'chosen few' initiatives have been identified, these company-wide projects become the key annual (or longer) projects to which all departments will support and engage staff. This is in addition to any focused continuous improvement effort (such as improving the quality within each manufacturing cell that is a departmental contribution and issue).

A key feature of the middle management round of analysis is the active engagement of managers in study groups in order to find out 'How' to solve the challenges set. This is, once again, conducted as a group, and implies that projects must be proposed, promoted and explained across the entire group before they will be accepted by the group and returned to senior management. As such, any improvement to reduce costs would follow a prolonged period of studying costs in the factory and their origins.

These key projects or improvement programmes would therefore be used to affect the indicators of business performance that need to be improved and it is common to

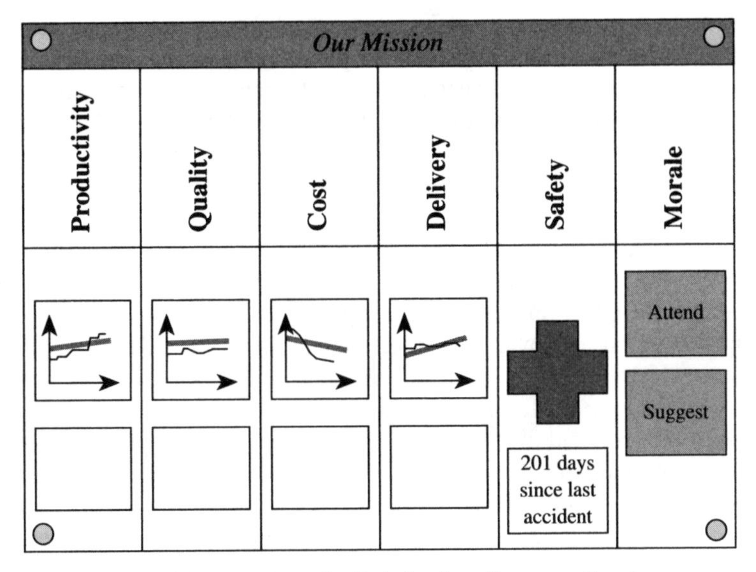

Figure 2 Local Area Performance Boards: Reinforcing Customer Service.

find these programmes listed in evaluation matrices. These matrices are themselves used to promote the 'vital few' projects that will benefit the business. The selection of 'how' is therefore a middle management team activity and spreads learning and reasoning across the middle managers who themselves will be tasked with promoting the projects (and assigning resources) within their own departments. It is important to note that for all the projects that are selected, there will be projects that are not, and this activity is important in defining the vital few projects that will release the business benefits sought. When each project is 'authorized', a senior manager is appointed as the 'champion' for the project and this person acts as the internal 'sponsor' to the middle managers throughout the duration of the project. These middle managers will themselves become sponsors of operational teams to form a linking pin structure of project management and support.

Deploying to the operational teams

Once the projects are selected, another round of promotion and catch ball is undertaken by the middle managers with the teams in the factory. This is a means of promoting awareness and clarifying issues in more detail. At this level, the 'accepted' projects and ongoing improvement targets are deployed to the team leader. In addition, project time-scales are determined and individuals assigned to projects. The measurement systems are adjusted accordingly (with the annual departmental targets declared).

The review system

Once projects have been 'signed off', a process that typically takes three months in Japan, the review system is constructed and the key control items that must be monitored are identified. This often takes the form of a bar chart (for every critical process and task) covering a year being drawn and the annual challenge trend line imposed on it. Each level of management adopts a similar process of visual reporting from the team leader to the manager and from the manager to the senior manager. However, the actual bar chart will differ slightly at each level dependent on the target improvements that are planned (as such, QCD will have a specific meaning to each team in the factory). In addition, there will be standard charts between the teams, the managers and the senior managers each showing actual versus planned performance. It is important to note that for the purchasing department, this information will be constructed for the suppliers and the improvement targets within the factory will inevitably shape the QCD targets set for suppliers.

The annual focused improvement processes can now begin. The formal reporting of progress and reviews are used to identify areas of under-performance or weaknesses. At each review point in the factory, decisions are taken and resources assigned based on need and in this respect the frequent reviews mean that the system can become autonomous in the deployment of human resources to bolster flagging projects. As such, an improvement area that lags behind the plan will attract new resources from within the department or between departments (and remedial action will be taken). The middle managers will therefore, as a team, engage in the review process and maintain a relationship with progress against the plans set (in terms of both the improvements needed and the timing of these improvements).

Annually, to close the process and begin the next round of three-year planning, the senior managers will review the actual benchmark of the business against the challenge. Any areas of success and failure will be identified and reported to a meeting of the total workforce at which point the new annual challenge is communicated. This annual review therefore 'kick starts' the new campaign and the sustainability of all the improvement processes in the factory. It is important to note that some of the projects and also areas of under-performance will be carried forward and new resources assigned to these projects in order to improve the rate of success.

The policy deployment process therefore allows the roles of each manager to be enhanced:

- For senior managers the planning activity involves the majority of their time.
- For middle managers the planning activity would account for a proportion of their time with the remainder allocated to the management of the business department.

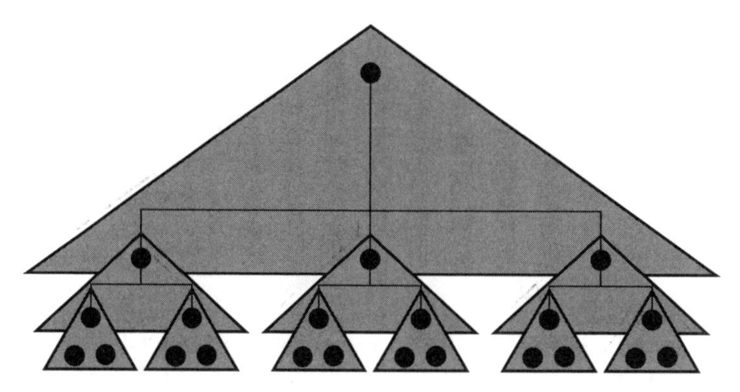

Figure 3 Likert's Linking Pin Model (1961).

- Middle managers are both team members of cross-functional groups and also team leaders of improvement activities (as part of a cross-functional initiative and/or within their department).
- For team leaders a small amount of time is spent planning change in the factory and the majority of time is used to control and improve the performance of the process.

A SUMMARY OF THE POLICY DEPLOYMENT LOGIC

The logic of the policy deployment approach is therefore relatively easy to discern and the use of this standard and formalized approach allows every individual in the business to understand their roles and contribution to the future success of the organization upon which they all depend. Therefore, policy deployment embodies:

- A 'TQM' foundation and a process that reinforces the Plan–Do–Check–Act system of management. It also involves both top-down and bottom-up planning.
- It is a holistic process that focuses on the business as a single entity and promotes the importance of the business above the importance of the department.
- The management within the business is also treated as a process and the role of the manager is much broader than the way in which managers are treated in the traditional systems.
- Focused on stakeholders (shareholders, customers and employees).
- The system is used to highlight future predicted weaknesses that affect the competitive position of the business and does not ignore these in order to compete on current strengths.
- The 'catch ball' process is a major departure from traditional Western systems and is used to align thinking, promote understanding and gain consensus about the future 'vital few' project investments to be made by the business.
- The system incorporates high levels of participation, explanation and promotion, and communication is conducted by the middle managers to their staff. As such, all managers share one 'hymn sheet' and one approach.
- Giving managers the challenge to use their collective planning skills based on facts collected about the performance of the business prevents the 'blame' culture. Instead, it is used to highlight areas of improvement that by definition will need a cross-functional management team approach. The use of the key processes (QCD) reinforces this view that the problem results from the system and market, not an individual or department.

In terms of the characteristics of an effective business strategy to enactment process, policy deployment also has the additional features that:

- It is a *formal* system of control that has distinct activities that are conducted in a logical and timely manner. This formality creates the system and allows 'normality' to be established for decision making and information processing routines within the business.
- It is *focused and prioritized* therefore allowing the actions of each element of the system to be directed towards a common purpose that is itself discriminate in determining what is important from what is not important for the business.

- It avoids the problems and risks associated with an elitist approach to the deployment of a strategy by engaging all management teams in active solicitation and participation in goal setting and understanding. This means that rather than setting objectives for individual managers, whereby the senior managers have determined the solutions that need to be implemented by middle managers, a series of discussions and *promotional* activities must take place in order to explore and clarify the change requirements of the business. *Participation* implies that this must be conducted with the middle managers as a group and thereby the solutions must be developed as a group. This latter process is one of *alignment* and creating a common focus on the 'vital few' changes that will yield superior business performance.
- It is a system that involves widespread *communication* of the need to change (couching this need in terms of customers, shareholders, and competitors and also of employees).
- It embodies a suite of key *measures*, or benchmark health checks that are used, at each level of the business, to monitor performance and take corrective actions quickly. These measures are also shared across departments and are common denominators with a known direction of improvement.
- In the current state of market, competitive and environmental change, whereby a strategy can be redundant before it has been written, the strategy must be flexible. This implies that the key features and requirements of the strategy must be enduring and relatively constant. These features include customer service, shareholder value etc., which are relatively unchanging and will change only in the metrics applied to them. The constancy of purpose is therefore emphasized.

Policy deployment and the supply chain

To finalize the analysis of the policy deployment process, it is worth considering the role of the purchasing department within this system. For most companies, the setting of direction and improvement in the supply chain is as important as any other business system. In fact, if the supply chain does not improve at the same, or a better, rate than the business itself then the strategy of the focal customer can become bottlenecked whereby internal improvements are ransomed by poor supplier performance. The policy deployment process therefore transforms the role, value and strategic integration of purchasing professionals and supply chain experts. This impact is to elevate the purchasing management to the middle and senior levels of business decision making and policy deployment. This position has, for a long while, eluded the purchasing professional and meant that the department was subordinate to either the manufacturing process or the accounts department.

Traditional arguments for the elevation of the purchaser have tended to focus on the cost implications of the supply base and the relationship between bought-out costs and the price competitiveness of the business. In the traditional system, the chosen means of supplier management have often been described as adversarial in that power wars were waged between the purchaser and the supplier. In the context of policy deployment and consensus management, the purchasing department will become an integral part of any cross-functional management team and by default the traditional adversarial purchasing regimes will conflict with the needs of such a group and the business. These suppliers, through the make or buy decision, are extensions of the focal customer factory – making what was once made in the factory – and therefore must be treated as similar to any other internal department in order to integrate and exploit the capabilities of the supply chain for improvement. The constant battles to lower piece part prices and arbitrage volume purchases, without improving the key QCD business processes, cannot be sustained if the business is seeking to improve its own creation of value or implement lean ways of working. As such, the relationship with suppliers must change from one of mistrust to one of deliberate integration and the treatment of the supplier as an extension of the focal factory.

According to Christopher (1992) 'competitive advantage is increasingly a function of supply chain efficiency and effectiveness . . . [thus] . . . greater the collaboration, at all levels, between supplier and customer, the greater the likelihood that an advantage can be gained'. Therefore, as forward-focused policies are integrated, inevitably the suppliers will be assessed in order to understand how best to exploit the efficiency and effectiveness of the supply chain. The key to integrating suppliers with the direction and improvement demands of the focal business is the purchasing department and according to Imai (one of the major contributors to the concept of continuous improvement) the policy deployment approach has changed fundamentally the role of the purchaser. He argues that the new role of the purchaser is to promote the factory policy with the supply base and includes the development of 'criteria for checking the relative strengths of the suppliers in terms of price, co-operation, quality, delivery, technology and overall management competence' (Imai, 1986). In parallel, Farmer suggests that this movement in Japan created a completely new form of purchasing specialist and 'in order to ensure the quality of response which they desired, the Japanese required the intervention of, what we might call, proactive change agents. People who influenced the manner in which supplying companies operated, the way they thought, their attitudes to quality, to cost and to relationships with their customers. In an era when life cycles were becoming shorter, they also perceived the necessity for closer, more timely collaboration in product development and management. Those roles were far removed from those performed by the traditional "hard nosed" buyer. Educational background, levels of skill . . . and involvement in the total business as a key player, were among the factors which differentiated the old role from the new' (Farmer, 1995).

The extension of the 'consensus management' approach and a new definition of what generates value from the role of the purchasing specialist therefore opens many new and exciting possibilities for purchasing to make a direct contribution to the competitive success of the business. This contribution derives, not from seeking to be different from every other manager but by seeking integration with others. In turn, for the cross-functional management processes to be effective, suppliers must be regarded as extensions of the focal factory (geographically separated operating departments). According to Newman and Rhee, 'a supplier becomes part of the team and in turn is responsible for structuring its operations to meet team requirements. Crossing traditional boundaries is acceptable because these boundaries do not exist. Traditional functional responsibility has given way to responsibility for improvement and information flow . . .' (1990).

The new organizational arrangements may therefore change fundamentally the role and perception of purchasing and allow greater levels of meaningful involvement in the management of the business. Leenders (1995) proposes that this would include:

- 'Active solicitation of supply input on major decisions being deliberated;
- Assignment of supply personnel to standing committees concerned with planning new directions or strategies for the organisations;
- Assignment of personnel to temporary task forces . . . or other organisational efforts to address specific tasks, problems or opportunities;
- Formal requests for purchasing initiatives to exploit opportunities in the marketplace;
- Non-solicited initiatives from supply which are acted upon and supported by others in the organisation;
- Ready access to information for supply personnel and vice-versa;
- Availability of extra resources . . . to develop and exploit supply opportunities;
- Requests to have supply personnel transferred or promoted within the organisation;
- Managerial support for regular reporting by supply managers;
- Management acknowledgement of and appreciation for significant supply contributions' (1995).

The process of policy deployment therefore allows purchasing activities to be promoted amongst middle and senior managers to reveal the benefits of supply chain integration. The supplier integration activity becomes a means of focusing and extending the policy deployment priorities beyond the factory to its largest single cost in a forward-focused manner. By setting stretch goals and creating a supplier evaluation system that reinforces the priorities of the focal customer, the efforts of a new 'external team' can be harnessed. The integration of suppliers also brings with it the ability to use the capabilities and improvements of suppliers to increase the growth of the focal company – a feature that was beyond the capability of pure price negotiations (Hanan, 1992). However, in order to replicate the linking pin and policy deployment structures needed to channel both information and improvement ideas a new approach to supply base management is required.

Hines (1994) proposes that the value from the new purchasing role will come from two key activities with suppliers. The first activity is supplier development and the second is supplier coordination. He contends, 'supplier development refers to the activities made by a customer to help improve the strategies, tools and techniques employed by suppliers to improve their competitive advantage, particularly by removing intra-company waste. This type of development would include the dissemination of customer strategies, so that suppliers could plan their processes more effectively, as well as the customer offering specific assistance to the suppliers in areas such as factory layout, set-up time reduction and the operation of internal kanban systems' (1994). In terms of supplier coordination, a process of information exchange, he proposes that these are 'activities made by a customer to mould their suppliers into a common way of working, so that competitive advantage can be gained, particularly by removing inter-company waste. This type of co-ordination would involve areas such as: working to common quality standards, using the same paperwork system, shared transport and employing inter-company communication methods such as EDI . . .' (1994).

However, the new approach has certain constraints and practical limitations for the 'new model of purchasing', which include:

1. the relatively few numbers of purchasers at the focal business that must engage in supplier integration activity;
2. the high numbers of suppliers that service any customer;
3. the mismatch between the traditional skill sets of purchasers and the new requirements.

In order to satisfy the resource limitations, the Japanese 'world class' manufacturers have adopted two key techniques to replicate the internal structures of policy deployment. The first is the re-tiering of the supply chain to form linking pins of suppliers as opposed to a massive number of direct supply relationships with a limited number of purchasers. In this respect, automotive companies began to buy 'systems' from suppliers as opposed to parts. These systems were effectively part-built elements of a car (a cooling system is a combination of the radiator and all the attachments to it including the fan, fan shroud, thermostat and expansion tank). As such, the new arrangement whereby previous suppliers became subservient to a key supplier meant that the number of direct interfaces with the purchaser could be logically reduced. This is opposed to supplier rationalization whereby suppliers are removed totally from the supply chain. In addition, the new tier of direct suppliers was, as Hines (1994) argues, combined into supplier associations. These associations of suppliers, work in a similar manner to internal cross-functional management processes and comprise suppliers rather than management representatives drawn from internal business departments. The associations allow suppliers, as external 'stakeholders', to become integrated in business improvement activities.

Suppliers and the QCD formula

The QCD measures are the common traded processes within a business and between a business and its supply base. These measures have a mutually reinforcing property and a logical relationship between themselves when they are correctly prioritized for the policy deployment process within a factory and the extended policy deployment approach that reaches down into the supply base.

The first priority for any supply chain initiative is to improve the quality of each link in the supply chain. Quality improvements will reduce the total cost of doing business with a supplier (in terms of fewer defects and also less need to hold safety stocks) and also improve the efficiency of the system. The traditional system that sought to reduce costs (prices essentially) and improve quality is not a sound logic. The echoes of the quality movement are therefore still valid for the logical prioritization of activities with suppliers. If quality is good then the timeliness of outputs will be important as these, particularly in batch-and-queue manufacturing, release costs through inventory reductions as safety stock is withdrawn that previously compensated for poor quality and delivery performance. However, once again the traditional logic did not stack up for cost reductions have no direct relationship with the timeliness of material flow (indeed the traditional system favoured bulk buying and constant switching of products between suppliers). Finally, there is the issue of cost reduction, which in the context of high quality and delivery reliability is a valuable activity but the value of the activity can only be extracted at this point. Cost reductions can be targeted at the processes of inventory and information flow (including the development of processes that allow the devolution of design responsibility, the management of engineering change requests and the levelling of schedules for the supply chain).

The primary goals of any business are to generate 'value' (or profitable customer service) and to continuously improve this value by attacking waste in all its forms (to lower costs). Only by generating a value can the supply chain survive and any attempt to remove weaknesses in the chain will tend to lower the total costs of the chain (which themselves must be passed to the customer but for which no value is gained). As such, the QCD formula is a force that unite the purchasing requirements of customers, internal departments within the business, purchasing agents and suppliers, combined with the 'lean logic' that customer service, at a transactional level, is measured by a cost of that service against the multiplication of quality and delivery reliability. In this respect, the management of the Q times D at a continuously reducing C is a good formula of analysis and it is identical to the way in which internal processes are focused.

Therefore, in the same manner that the weakest element of the QCD formula is targeted for improvement within the business, true supplier performance can be measured and the weakest areas highlighted for one-to-one improvement or set as the stretch goals for the supplier concerned. Therefore if a supplier achieves a 90 per cent quality performance with a 75 per cent 'on time' delivery performance then the rating for the supplier would be a true service level of 67.5 per cent. The perfect score and benchmark of every system is therefore 100 per cent – the real test of service. In this particular case, the weakness lies in both the quality and delivery process and therefore both would attract improvement goals. Only when a high level of stability is achieved can the ideals of the lean system of supply be developed. In the lean approach, inventories will be reduced, and forecasts levelled, in order to keep materials flowing without interruption.

The QCD measures therefore add the dynamic requirements of a moving strategy and policy, of a focal customer that is needed to direct improvement activities within the supply chain and the largest source of cost for most manufacturers. Ironically, the same goal of cost reduction is shared between the traditional system and the policy deployment approach but the latter does not suffer with the disadvantage of a dislocated supply chain that is beaten for price reductions. When the supplier association or group approach is used, there is the additional incentive to improve the aggregate performance of the group and allow the networking of companies to allow improvement ideas to be shared.

CONCLUSIONS: POLICY DEPLOYMENT

There is nothing mystical or magical about policy deployment yet it is, even when studied closely, almost impossible to emulate, and undesirable to copy, exactly. Copying the practices of others is believed to be a way of narrowing a gap in performance yet this perhaps devalues the strategy of the business itself. Emulation is subject to many problems, and in the UK context these problems tend to involve changes in working practices and human resource management. A history of mistrust between management and employee and between business-to-business relationships cannot be overturned in the short term. The lack of trust is itself the product of other historical failings such as poor strategy planning, or business decline that has led to redundancies in the factory, or a long history of adversarial purchasing.

Every company in the UK has the ability and crude resources with which to develop a system of policy deployment. The effectiveness of the system developed will depend on how well it is localized and promoted within the factory as much as success is determined by the calibre of people in the factory. In this respect the policy deployment process is 'universal' but the application of the technique must be contingent on the unique market circumstances of the business itself and the capabilities that have been invested in the human resources of the factory by training. Even within the LEAP programme there were wide differences between the participating companies despite operating with similar technologies and operating in the same market sector. The successful companies, those that survive and prosper, are those that have paid attention to the principles of TQM and adopted deployment methods similar to those described earlier. The unsuccessful, the businesses that continue to struggle with change management and are in market decline, have either selected poorly their customers upon whom their future depends, interpreted poorly their needs or failed to cultivate a working culture that accepts the need to improve continuously.

What prevents companies from undertaking the policy deployment process, in the UK, is an inability to put together the correct organizational ingredients in such a way that the 'system' is created. These ingredients are not simple resources to muster and there are few uniform qualities in these resources even though they may carry the same name – such as management. Most resources have to be nurtured to the point that they can be combined without 'rejection' and inevitable failure. As such, a history of under-investment in training is a major factor that inhibits the development of the policy deployment system. In addition, the 'management fad' culture of the 1980s and 1990s has created a confused environment within which all manner of cure-alls exist yet few have had a direct impact on the competitive strength of the business. Instead most initiatives have fallen far short of what was desired and in the process become discredited.

This chapter has developed the concept of policy deployment as a systematic means of developing the entire enterprise and providing a basic framework that can be used to implement logically changes in working practices. It has been portrayed as a means of focusing both internal and external improvement initiatives through the medium of consensus management and a process that is based on team work – the basic building block of the enterprise. It is policy deployment that focuses the priorities of these teams wherever they are to be found (within or beyond the factory). Without consensus there can be no alignment in this complex weave of parallel improvement activities, no shared destiny relationships and therefore no sustainable improvements. The blend of TQM values, with the planning system of the business, and a respect for the human resources in the factory (and their value roles) creates a potent combination of forces and a superstructure through which to manage the business as a single system. The forward focus of the policy deployment process also provides a means of marrying the future needs of the customer with the necessary preparations that must be undertaken by the business, and by default, its supply base.

The turbulent changes in markets such as automotive component supply makes planning difficult and has created a general trend towards the decentralization of decision making to an empowered team-based organization. The paradox is that the drive to decentralize decision making masks the need to retain a central control of the business. The policy deployment process is probably the only real process that maintains this fit between strategy and action.

REFERENCES

Achrol, R. (1991) Evolution of the marketing organisation: new forms for turbulent environments. *Journal of Marketing*, October, 77–93.
Akao, Y. (1989) *Hoshin Kanri*. Portland Or: Productivity Press.
Bicheno, J. (1994) 34 for Quality. Buckingham,: Piscie.
Dimancescu, D. Hines, P. and Rich, N. (1997) *Creating the Lean Enterprise*. New York: Amacom.
Farmer, D. (1995) Purchasing myopia – a touch of Mr. Magoo. Speech 1st World-Wide Research Symposium On Purchasing & Supply Chain Management NAPM.
Hammer, M. and Champy, J. (1993) *Re-Engineering the Corporation*. London: Brearley.
Hanan, M. (1992) *Growth Partnering*. New York: AMACOM.
Hayes, R. and Pisano, G. (1994) Beyond world class: the new manufacturing strategy. *Harvard Business Review*, Jan–Feb.
Hines, P. (1994) *Creating World Class Suppliers*. London: Pitman Publishing.
Imai, K. (1986) *Kaizen*. New York: McGraw Hill.
Japan Union of Scientists and Engineers (JUSE) in Diamanescu, D., Hines, P. and Rich, N. (1996) *Creating the Lean Enterprise.* New York: Amacon (1988) Tokyo.
Joag, S.G. and Scheuing, E. (1995) Purchasing's relationships with its internal customers. 1st World-Wide Research Symposium On Purchasing & Supply Chain Management NAPM.
Johnson, G. and Scholes, K. (1988) *Exploring Corporate Strategy*, 2nd edn. London: Prentice Hall.
Keuning, D. and Opheij, W. (1994) *Delayering Organisations*. London: Pitman Publishing.
Lamming, R. (1993) *Beyond Partnership*. London: Prentice Hall.
Leenders, Michael R. (1995) Meaningful involvement: purchasing & outsourcing. Key Note Speech, 1st World-Wide Research Symposium On Purchasing & Supply Chain Management NAPM.
Maskell, B. (1989) Performance measures for world class manufacturing. *Management Accounting*, May, 32–33.
Newman, R. and Rhee, K. (1990) A case study Of NUMMI & its suppliers. *Journal of Purchasing & Materials Management*, 26(4), 15–20.
Peters, T. (1992) *Liberation Management*. New York: Macmillan.
Rich, N., Hines, P. Jones, O. and Francis, M. (1996) Purchasing at a watershed. International Purchasing & Supply Education & Research Association, Univ. of Twente.
Womack, J.P. and Jones, D.T. (1996) *Lean Thinking*. New York: Simon & Schuster.
Womack, J.P., Jones, D.T. and Roos, D. (1990) *The Machine that Changed the World.* New York: Macmillan.

NOTES

1 The logic will be described as Quality, Cost and Delivery (QCD) as the key improvement processes that need to be managed in order to exploit commercially improvements in the factory and between companies.

Value stream management: the next frontier in supply chain management?[1]

24

Professor Peter Hines

INTRODUCTION

Imagine a company where what you did made a difference. Sadly, this may not be where you work now. Ask yourself how much of your day really makes a difference to your customers. Then think how much of your work colleagues' or staffs' time makes a difference. Is it more than 50 per cent?

Ask yourself another question: how much of your day's work would you be willing to pay for if you were the customer? How much of what you did today should you have done yesterday or postponed until tomorrow? Finally, what did you do today that will make tomorrow's job easier than today's? Maybe the answer is not much. So how are you ever going to become lean or reach world-class status? The answer, is probably that you won't. Value Stream Management and the lean approach to supply-chain improvement have been designed to get out of this inertia.

GOING LEAN

In order to go lean it is necessary to understand customers and what they value. To get your company focused on these needs you must define your internal value stream (the activities you undertake to satisfy particular customers) and later the external one (the activities your supply-chain partners undertake for these customers) as well. In order to meet customer aspirations you need then to eliminate all the waste you have just envisaged in your company in the paragraphs above. When you have done this, the next step is to find a way of setting the direction, targets and finding a way to see if your change is working or not. You then need an internal (and later external) framework to deliver value for your customers as well as a toolkit to use to make the change (see Figure 1).

If you can do this effectively you won't need to benchmark competitors to set some arbitrary and often incomparable target; perfection or the complete elimination of waste should be your goal. Sounds good, but back to the real world for a minute; if it was so easy why don't others think this way?

Sometimes we ask ourselves this question, and when we have gathered a few facts about a company, we ask them. The answer they give is usually something like 'but it can't be that simple' or 'yes, that makes a lot of sense but we never saw it that way'. Why is this?

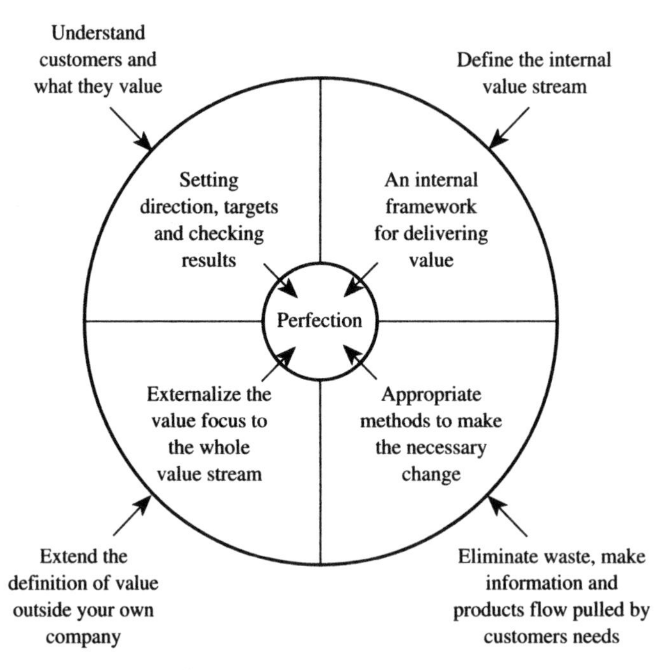

Figure 1 The central lean principles.

Here are a number of reasons:

1. Managers are too busy fire-fighting or fending off the latest crisis to spend time looking at what is really happening.
2. When advice is forthcoming it is often from consultants who make everything seem complex and often have something more than consulting to sell (have you ever wondered why many consultants have such big IT wings?).
3. Textbooks or MBA courses offering the latest 'flavour of the month' leading to 'fad surfing' whether the latest approach is of value or not.

All of these reasons mean that the existing picture is either poorly understood or is made to look very complex. Thus firms can be easily led to adopt inappropriate and costly new computer systems or physical infrastructure to cope with this complexity or the latest new concept that may be being sold at workshops or conferences. To illustrate this take a look at the following real case study with the identity of the firm withheld to protect the guilty!

THE CASE OF THE AGILE COMPUTER FIRM

This case concerns a certain UK-based American-owned computer manufacturer who had adopted a 'vanilla stocking' policy where pre-assembled laptop computers were held as finished stock and configured to order with a wide variety of software, operating systems and instruction booklets. On the face of it this agile postponement strategy seemed a sound response to a highly dynamic environment. Indeed, according to the agile textbooks this was ideal agile territory: computer industry, high variety and high unpredictability. But was all as it seemed?

On closer inspection using the value stream mapping toolkit it appeared that the demand, although variable, was reasonably predictable. Indeed, it appeared that half of the year's sales were in the last quarter of the financial year, half of the sales within each quarter were in the last month of the quarter, and half of the month's sales were in the last week.

On further investigation the cause of this demand was unrelated to the variety mix or model life cycle. In fact, the reason was solely down to how the firm rewarded its sales force and dealer network with bonuses for meeting sales targets. Worse than this, the firm, although developing a very rapid 48-hour 'quick response' dealer configuration service, had completely neglected the rest of its supply chain, but was very proud of its high-cost, completely computer controlled, football pitch-sized raw material warehouse. When a typical product was mapped using the Value Stream Mapping approach, it was found to have approximately seven months of total stock in the supply chain. On enquiry, the product life cycle of the particular computer was found to be six months!

The moral of the case is therefore that a postponement strategy may be preferable to doing nothing, but is likely to be only a stepping-stone to a more holistic lean solution. We believe it is necessary to optimize the whole supply chain, not just part of it. In this case a more appropriate solution might have involved a redesign of sales force and dealer bonuses and a reduction in information and physical production lead times upstream of the 'vanilla stock' configuration point. If this had been implemented, dealer orders could have been much flatter and could have been pulled from as far back as a supplier manufacturing de-coupling point, removing the need for several months' inventory and associated stock holding and obsolescence costs.

If you in your company or supply chain feel that any of the problems or issues discussed above are familiar to you, your next question will probably be: 'How do I get out of the vicious circle I'm in and start a virtuous circle of change? Where do I start?'

THE STARTING POINT

The starting point is to design an internal and external system that can create sustainable competitive advantage. In our travels across five continents we have found that there are four key differentiators of the leanest companies (Figure 2):

- policy deployment,
- process-based management,
- lean tools and techniques, and
- supply chain integration.

It is a rare company indeed that is excellent in all of these areas, in fact we can only think of one. However, the reason for this is that companies have often not raised the bar to aim at perfection, the reason for this is that they often don't know where to start in creating customer value and eliminating waste. The Value Stream Management

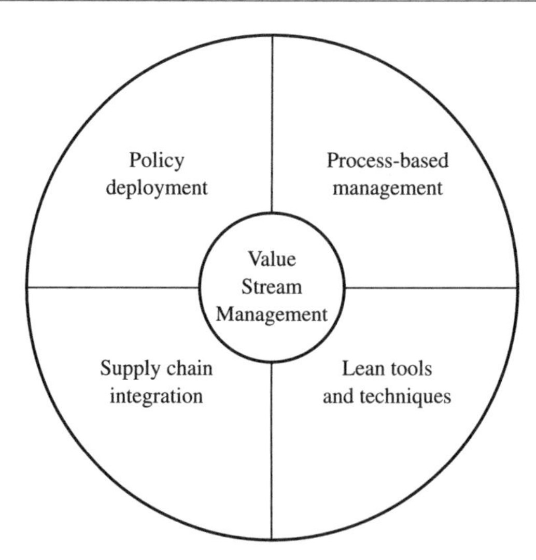

Figure 2 Going lean: the central role of Value Stream Management.

method, now trialled and perfected by application to literally hundreds of companies, is, we feel, the secret of success in unlocking your firm's innate ability to be world class.

One aspect of the Value Stream Management approach that is often neglected by companies in their quest to become lean is the strategic issue of policy deployment. The benefits and rationale of this concept are described in some detail in Chapter 23. The following section gives a brief overview of a practical approach that can be adopted by companies wishing to make a start on policy deployment.

AN OUTLINE OF A PRACTICAL APPROACH TO POLICY DEPLOYMENT

One of the main difficulties we see when companies try to apply lean thinking is a lack of direction, a lack of planning and a lack of adequate project sequencing. Knowledge of particular tools and techniques is often not the problem. In many cases lean initiatives are killed because of a lack of senior management forethought.

What is required is a simple sequence of activities by senior managers in order to gain focus involving:

1. developing critical success factors,
2. reviewing or defining appropriate business measures,
3. targeting improvement requirements over time for each business measures,
4. defining key business processes,
5. deciding which process needs to deliver against each target area, and
6. understanding which process needs detailed mapping.

We will take you through these steps before setting the scene for the top-level and subsequent detailed mapping. This process is sometimes referred to as policy deployment.

Developing critical success factors

First it is necessary to establish what are the key forces impacting your business or wider value stream. These forces may be divided into a number of categories, such as:

- **General business environment**
- **Industry specific**
- **Customer specific**
- **Company specific**

The best way to do this is as a brainstorming exercise using a flip chart or Post-It notes facilitated by a team leader.

The next step is to develop critical success factors against these key forces. Critical success factors are a limited number (10 or fewer) of key areas where 'things must go right' in order for the business to succeed and flourish. They should be directly linked to the specific factors impacting your company or value stream and particularly be influenced by the most important specific factors. An example is shown in Table 1.

Reviewing or defining appropriate business measures

For most companies a set of top-level (often financial based) business measures will already be in place. However, these may not be aligned to the critical success factors. This is very important as existing measures will drive aspirations and ultimately performance. So here we are checking if your strategic goals and operational measures are aligned.

Let's try to develop a set of balanced measures against the critical success factors, bearing in mind a balance between financial measures and customer or supply chain oriented measures. We can then check if they adequately measure what is important (Table 2)

This example may be regarded as a reasonably aligned set of measures. Although the measures may not be the absolute optimum set, they are good enough to pilot, although a later review may be of value, perhaps at the end of the first year.

Targeting improvement for each business measure

Targeting the actual required improvement rate is the next stage, one that many

Table 1 Example

Key force	*Examples of key specific factors*	*Possible critical success factors*
General business Environment	1. Recession	1. Turnover growth
Industry specific	2. New competitors	2. Maintain or grow market share
Customer specific	3. Main customer in decline	3. Find new customers
	4. High cost-down pressures	4. Dramatically reduce costs
	5. Severe quality improvement targets	5. Dramatically improve quality
	6. New product requirement targets	6. Develop new products
Company specific	7. Exacting holding company	7. Keep holding company happy

Table 2 Measures against critical success factors

Key Operational performance measures:	*Strategic level critical success factors*						
	Turnover growth	*Improve market share*	*Find new customers*	*Reduce costs*	*Improve quality*	*Develop new products*	*Holdings company happy?*
Return on capital				Maybe		Negative (short term)	Yes
Net cash				Yes (short term)			Maybe
Stock turn				Yes			Maybe
Overall equipment effectiveness				Yes	Yes		Maybe
Total cost reduction	Yes	Yes	Yes	Yes	Maybe		Yes
Total turnover	Yes	Maybe	Maybe			Maybe	Maybe
Market share	Yes	Yes	Maybe			Maybe	Maybe
Sales to new customers	Yes	Yes	Yes			Maybe	Maybe
Product quality	Yes	Maybe	Maybe	Maybe	Yes	Maybe	Maybe
New product sales	Yes	Yes	Yes	Maybe	Maybe	Yes	Maybe

Table 3 Targeting the required improvement rate

Measure	*Now*	*Target end year 1*	*Target end year 2*	*Target end year 3*	*Target end year 5 vision*
Return on capital	2.4%	4.4%	6.4%	8.4%	12.4%
Net cash	(£2.4m)	(£2.2m)	(£1.8m)	(£1.2m)	£1.0m
Stock turn	8.3	12	16	26	46
Overall equipment effectiveness	43.4%	50.0%	60.0%	70.0%	85.0%
Total cost reduction	(4.5%) last year	5.0%	5% additional	5% additional	5% additional
Total turnover	£10.4m	£12m	£16m	£20m	£35m
Market share	4.5%	5%	6%	8%	15%
Sales to new customers	6.0%	10%	15%	20%	25%
Product quality	8,300 ppm	4,000 ppm	1,000 ppm	400 ppm	50 ppm
New product sales	5.0%	5.0%	8.0%	13.0%	25.0%

companies fail to undertake. Where companies do this they usually set only one target, perhaps for six months' time. However, for an effective lean conversion programme a more realistic timescale is three to five years within a long term vision, with staged targets for every 6 or 12 months (Table 3).

Again the first time this targeting exercise is tried the result will probably not be the optimum, but it is likely to provide a good direction and can be improved perhaps on an annual basis.

The above process has set a broad direction for the company over the next three years, inside a long-term vision. What we now need to work out is how are we going to achieve this. The starting point is gaining an understanding of key business processes.

Defining key business processes

A key business process can be defined as:

Patterns of interconnected value-adding relationships designed to meet business goals and objectives.

Processes have a series of inputs and a number of steps, tasks or activities that convert these inputs into a number of outputs. They typical run across several departments in a business or indeed businesses and in general are designed to encourage and support inter-departmental communication and cooperation throughout the company or value stream.

In our use of the term 'process' we are referring to a very limited number (up to 10) key activity groups that are required to deliver value to the business or value stream. The fewer you define, the easier they will be to manage. Remember that these processes are not everything a company does, but they are the core or key activities it undertakes and must get right.

Don't fall into the BPR trap of defining 100+ business processes. Brainstorm many, but settle on a few. Once you have listed and agreed on between four and ten key processes make sure each has a definition. This will prevent later confusion.

In our example this brainstorming and analysis may have yielded the processes shown in Table 4.

Table 4 Processes resulting from analysis

Key business process	*Definition*
1. Order fulfilment	The taking of orders, processing the orders, production planning, production, delivery to customer and payment management.
2. Sales acquisition	The winning of new business with new or existing clients.
3. Product life-cycle management	The management of customer needs for new products, developing new products, introducing them into the market and retiring old products.
4. Technology, plant and equipment management	The development, management and maintenance of operating equipment (including IT).
5. Human resource development	The development, management and maintenance of employees.
6. Strategy and policy deployment	The strategic management of the company, focusing of change and management of critical success factors.
7. Supplier integration	The integration of suppliers into the other key business processes.
8. Continuous improvement	The continuous radical or incremental improvement of all other processes.

Deciding which process needs to deliver against each target area

Here we are going to decide which key business process area is likely to give us the targeted improvements as the result of some development activity. In order to do this a simple correlation of the target areas and key business processes is required. To do this, ask if the business process is likely to yield benefit to each target area if improved. Record Yes, Maybe or No. Do not answer Yes unless there is a direct link.

We will now do this for our example (Table 5).

Once you have completed this exercise it will start to become clearer where attention needs to be focused on improvement activity.

Understanding which process needs detailed mapping

In order to decide which process or processes need detailed mapping it is necessary, as we have already done, to decide which are likely to yield the greatest gains against the target areas. In our case example, we may divide the different processes into three categories:

1. Processes focusing overall direction but not directly impacting on targets or **strategic processes**
2. Processes directly impacting on targets or **core processes**
3. Processes indirectly impacting on targets or **support processes**

In our example the processes can be classified as shown in Table 6.

The strategy and policy deployment sets the direction, the five core processes are required to deliver the targeted results aided by the two support processes. At this point it may be useful to try to gauge or estimate where the targeted improvements are likely to come from within the core processes. It is not possible, for now, to be categorical in this until greater data can be obtained. This will be possible after mapping each of the core processes.

To keep things simple at this point, just pick one time scale over which to target the required performance improvements. In this case we will take the five-year horizon. We will then estimate how much of the targeted gains should come from each core process area (Table 7).

The next choice is which of these processes to map first and in what order to map the later processes. In many instances it is best to start with the order fulfilment process as it is easy for all to understand and it is central to the operations of most companies and value streams. In other cases, and depending on the relationship with key customers, the sales acquisition process might be mapped first. However, inexperienced mappers should not be used with a customer before having piloted the approach internally.

In our example an effective route may be to map:

1. Order fulfilment, then
2. Sales acquisition, then
3. Supplier integration, then
4. Product life-cycle management and finally
5. Technology, plant and equipment management.

How to go about mapping at the overview and detailed levels has been the subject of various earlier chapters in this book.

Table 5 Correlation of target areas and key business processes

Measure	*Key Business Processes*							
	Order fulfilment	*Sales acquisition*	*Product life-cycle management*	*Technology, plant and equipment management*	*Human resource development*	*Strategy and policy deployment*	*Supplier integration*	*Continuous improvement*
Return on capital	Maybe	Yes	Maybe	Yes	Maybe	Maybe	Maybe	Yes
Net cash	Yes	No	Maybe	Maybe	Maybe	Maybe	Yes	Yes
Stock turn	Yes	Maybe	Maybe	Maybe	Maybe	Maybe	Yes	Yes
Overall equipment effectiveness	Maybe	No	No	Yes	Maybe	Maybe	Maybe	Yes
Total cost reduction	Yes	Maybe	Yes	Yes	Maybe	Maybe	Yes	Yes
Total turnover	Maybe	Yes	Yes	Maybe	Maybe	Maybe	Maybe	Yes
Market share	Maybe	Yes	Yes	Maybe	Maybe	Maybe	Yes	Yes
Sales to new customers	Maybe	Yes	Maybe	No	Maybe	Maybe	Maybe	Yes
Product quality	Yes	No	Yes	Yes	Maybe	Maybe	Yes	Yes
New product sales	Maybe	Yes	Yes	Maybe	Maybe	Maybe	Maybe	Yes
Total	4 Yes 6 Maybe 0 No	5 Yes 2 Maybe 3 No	5 Yes 4 Maybe 1 No	4 Yes 5 Maybe 1 No	0 Yes 10 Maybe 0 No	0 Yes 10 Maybe 0 No	5 Yes 5 Maybe 0 No	10 Yes 0 Maybe 0 No

Table 6 Classifying processes

Strategic processes	*Core processes*	*Support processes*
1. Strategy and policy deployment	1. Order fulfilment	1. Human resource development
	2. Sales acquisition	2. Continuous improvement
	3. Product life-cycle management	
	4. Technology, plant and equipment management	
	5. Supplier integration	

Table 7 Core process contributions

Measure	*Core Processes*					
	Total 5-year targeted improvement	*Order fulfilment*	*Sales acquisition*	*Product life-cycle management*	*Technology, plant and equipment management*	*Supplier integration*
Return on capital	10%		7%		3%	
Net cash	£3.4m	£2.4m				£1m
Stock turn	37.7	30				7.7
Overall equipment effectiveness	41.6%				35%	6.6%
Total cost reduction	25%	5%		10%	5%	5%
Total turnover	£24.6m		£10m	£14.6m		
Market share	10.5%		3%	7%		0.5%
Sales to new customers	19%		19%			
Product quality	8,250 ppm	1,250 ppm		2,000 ppm	2,000 ppm	3,000 ppm
New product sales	20%		10%	10%		

Policy deployment: summary of steps

- Developing critical success factors,
- reviewing or defining appropriate business measure,
- targeting improvement requirements over time for each business measure,
- defining key business processes,
- deciding which process needs to deliver against each target area, and
- understanding which processes need detailed mapping.

This is essentially a senior management process, perhaps involving line managers responsible for the key business processes.

VALUE STREAM MANAGEMENT: A SUMMARY OVERVIEW

Value Stream Management is therefore the vital missing piece or core of a lean transformation, as we outlined in Figure 2. It is a strategic and operational approach to the data capture, analysis, planning and implementation of effective change within the core cross-functional or cross-company processes required to achieve a truly lean enterprise. In summary its main stages are:

1. Map out the existing position both within and between your company and its supply-chain partners to identify where and quantify how great are the opportunities for improvement of productivity, inventory reduction, time compression and quality improvement. And remember that your long-term target should be perfection.
2. Identify what is critical to the success of your company. Check whether your existing measures are likely to support such success. Are there too many, are they functional in nature or are they just inappropriate? Think of a more appropriate set of measures that will help support success. Then set targets for a five-year period against these measures and cascade them throughout the company so that you achieve 100 per cent employee alignment.
3. Develop an understanding of what the key processes are in your business; it may come as a surprise that success requires more than just fulfilling orders for customers. Indeed, for most companies there will be between four and ten key processes. So what are they? Well, they are the key cross-functional or even cross-

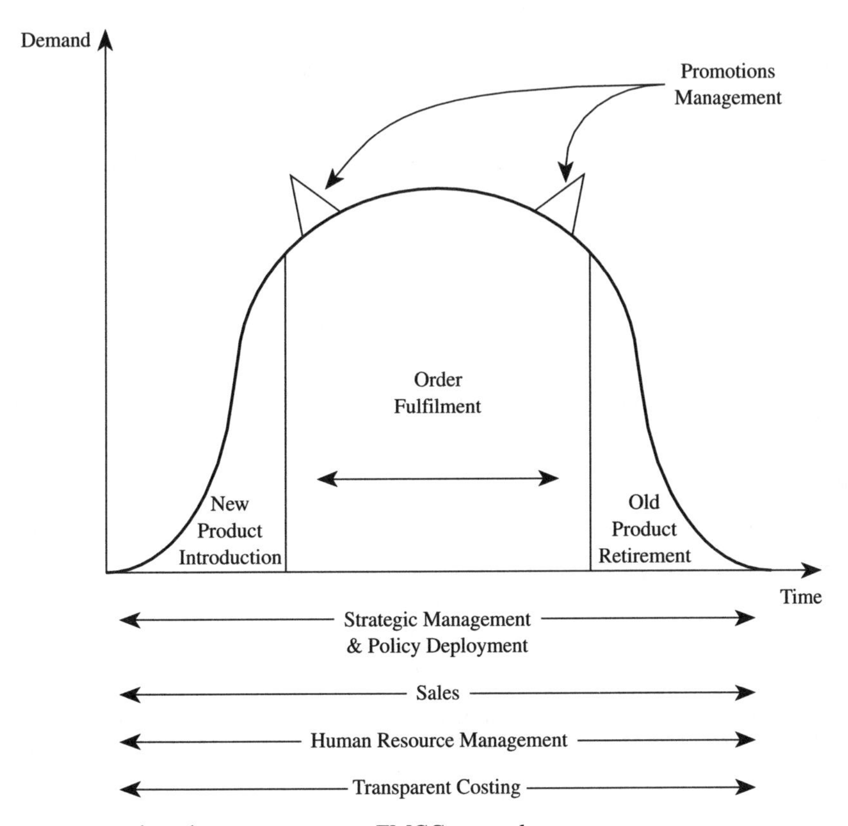

Figure 3 Process-based management: an FMCG example.

company activities that a company has to get right in order to succeed against its own critical success factors. As a result, these processes are unique to a particular company. A fast-moving consumer goods (FMCG) example is shown in Figure 3 within the context of a product life cycle.
4. Apply a series of appropriate tools and techniques once the opportunities, strategy and processes have been established, but not in a piecemeal way as in the computer firm above. These improvements may be based on the shop floor (the traditional lean production environment), the office, the warehouse or lorry but can equally well be within the R&D laboratory within the New Product Development process or the accounting department applying Activity Based Costing.
5. Repeat the first four steps in the wider supply chain. Undertake joint mapping with customers and suppliers, just as we have done within the Lean Processing Programme (LEAP). Then share or develop strategies and gain operational alignment. Which measures would be best used to optimize the whole supply chain? What are the key supply-chain processes, how will the gains be shared? Then decide upon an operating approach and structure to address these key processes such as a Supplier Association (see Hines, 1994). Then apply an appropriate range of tools and techniques to aim at perfection. This is likely to include demand alignment so that products and information can flow at the pull of the customer.

If you can achieve all of this you will find that you have greatly reduced complexity, developed supply-chain transparency and will be on the way towards a truly lean enterprise. At this point ask yourself a few more questions:

Does what you do make a difference?

How much of your day's work would you be willing to pay for if you were the customer?

How much of what you did today should you have done yesterday or postponed until tomorrow?

What did you do today that will make tomorrow's job easier than today's?

Judging by the firms we have worked with on the Lean Processing Programme (LEAP), the answers to your questions will be very different than at the starting point.

REFERENCES

Peter Hines, *Creating World Class Suppliers: Unlocking Mutual Competitive Advantage.* Pitman Publishing, London, 1994.

Peter Hines & David Taylor, *Going Lean: A Guide to Implementation.* Lean Enterprise Research Centre, Cardiff, 2000

NOTES

1 This chapter is based an extended version of an article which was first published in *Logistics & Transport Focus*, 1(3), Sept, 1999 and is published here with the kind permission of the Institute of Logistics.

From Current State to Future State: the steel to component supply chain

25

David Brunt

INTRODUCTION

The chapter attempts to consolidate the key learning points from LEAP in the context of 'Learning to See' style 'Material and Information flow (Value Stream) maps' as described by Rother and Shook (1). The application of the 'Learning to See' mapping tool is discussed firstly at the level of the single plant, before being expanded across the whole supply chain – from steel, through steel service centres to first-tier component manufacturers. Firstly, the value stream mapping technique is briefly revisited before describing the process of developing a 'Current State' map for a single firm.[1] In the third part of the paper, the considerations required to develop a 'Future State' map for a single firm are discussed. Fourthly the technique is rolled out to cover the whole value stream for a component and a technique for implementing improvements suggested. Finally the approach is discussed in terms of the critical success factors required to implement identified improvements and conclusions are drawn from the work.

VALUE STREAM MAPPING

In the book *Lean Thinking*, Womack and Jones (2) describe the value stream as all the actions (both value added and non-value added) currently required to bring a product through the main flows essential to its introduction. These can be described as:

1. the design of the product – from its conception through to its launch in the market place;
2. the flow of the product – from raw materials to delivery to the customer;
3. the flow of information necessary to trigger and support these flows.

It is argued that the benefits of removing waste will be most fully realized by taking a view of the big picture – the value stream – so that the whole rather than individual processes can be optimized.

During three years of research the LEAP team attempted to build on previous work carried out at the Lean Enterprise Research Centre to develop a methodology for mapping and analysing value streams in order to identify waste and attempt to remove it. A critical assessment of the mapping tools described in Chapter 3 by Hines and Rich (3) suggested that a key weakness with the initial mapping approach was its lack of visuality. That is, the ability of a researcher or manager with knowledge of lean tools and techniques to explain the current dynamics of the firm being analysed and to communicate an action plan that would be understood by all the key stakeholders. As a result, the LEAP team developed the mapping approach to include visual process flow charting (using brown paper) as described earlier in this book in Chapter 7 (4).

However, while this approach addressed the issue of visuality the team found that the method could be improved further as key data with which to create a lean system was still missing from the tool. In addition, the use of a 'Brown Paper' in conjunction with the Seven Value Stream mapping tools meant that the researcher or practitioner needed to constantly switch from the macro-level 'Brown Paper' to the micro-level tools such as the process activity map or demand amplification map.

CURRENT AND FUTURE STATE MAPS

The Current and Future State mapping technique originated in Toyota as 'Material and Information Flow Mapping'. Rother and Shook explain that the technique 'is used by Toyota Production System practitioners to depict current and future, or 'ideal' states in the process of developing implementation plans to install lean systems'. The method examines the flows of material and information in order to focus attention on the whole value stream for a product rather than discrete production processes.

There are four key steps to material and information flow value stream mapping:

1. Select a product family and for that product family.
2. Draw a 'Current State' map.
3. Develop a 'Future State' map.
4. Construct a work plan of activities in order to get from the Current to the Future State.

As part of the LEAP methodology specific product families had been mapped back from first-tier component suppliers through steel service centres to the steel producer. For example, sub-frames or body-in-white components supplied to different automakers had been mapped through these firms. This case uses a sub-frame supplied to a UK-based vehicle manufacturer as an example.

CURRENT STATE MAPPING

Having selected a product family the next step was to draw a 'Current State' map for the sub-frame (at a first-tier component supplier). The map was constructed from 'door-to-door' production flow inside the firm, see Figure 1. (A list of icons are shown in Figure 5.) This was constructed with data collected using the Hines and Rich value stream mapping tools. The process activity and quality filter maps were particularly useful for creating a 'Current State' map.

Within the drawing of the 'Current State' map there are four distinct stages:

1. Gather details about the customer's requirements.
2. Detail the physical flow with all processes, data boxes and inventory triangles.
3. Map the supply of materials.
4. Map the information flows and determine push and pull system.

The customer, a vehicle maker, requires 12,000 components per month which are delivered in returnable containers each holding 14 components. The customer

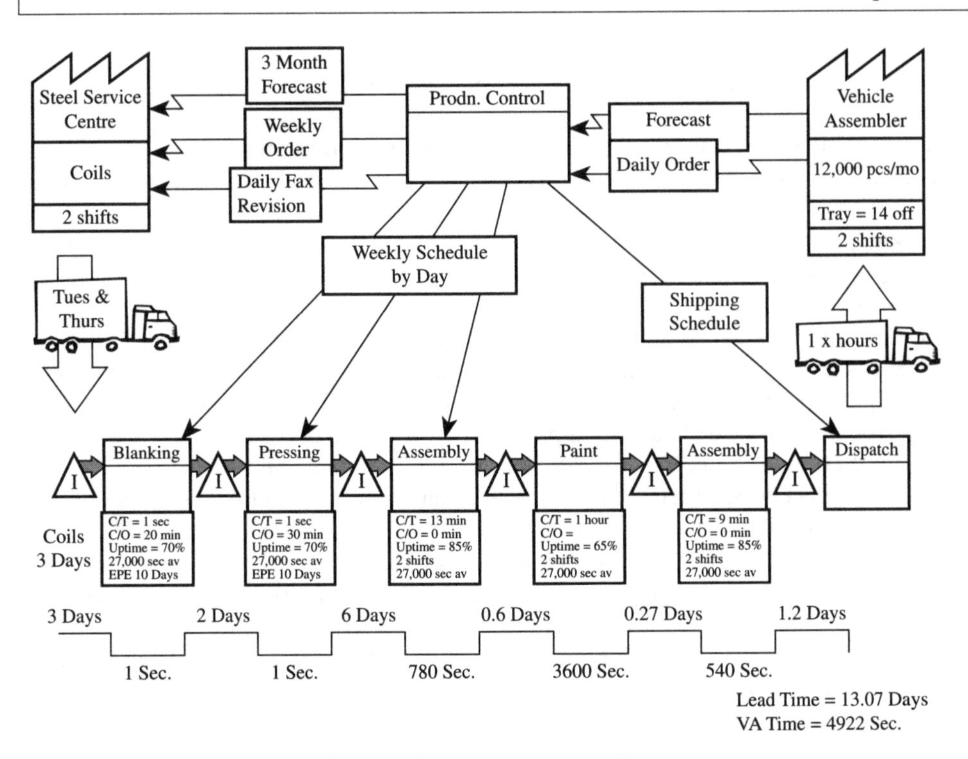

Figure 1 Current State map of a first-tier component supplier.

operates a two-shift operation. To make the components, steel is supplied in coil form. The first process at the component manufacturer is to make a blank (steel cut to a shape). The blank is then sent to a store, where it waits for approximately two days before being taken out of the store and moved to the press shop to be formed. The 'Current State' map provides a 'snap-shot' in time. The amount of time the component is stored takes into account the inventory quantity stored and the customer demand (how quickly this inventory will be consumed by the customer). Therefore an inventory quantity of 1000 pieces and a customer demand of 500 pieces/day equates to two days of inventory. In some cases more than one blanking or pressing operation may be required; however, the component chosen requires only one blank and one press operation. Following the press operation the finished pressings are stored once more (six days of stock) before being built into an assembly, painted and finished. There is approximately 1.2 days of finished goods.

In order to know how many sub-frames to make and when, the information flows are mapped. These show that the customer sends a five-month forecast of its requirements into the future. On a day-to-day basis, daily orders are sent to the first tier firm indicating the time of day parts are required. The component supplier uses an MRP system to plan its production. A three-month forecast is sent to the steel service centre (its supplier of steel to make the sub-frames) and weekly orders are sent to firm the requirements. In addition, any revisions are sent by fax daily.

Internally, a weekly schedule by day is developed for the blanking shop, press shop and assembly line. A shipping schedule is also developed. Rother and Shook suggest that 'as you figure out how each process knows what to make for its customer (the following process) and when to make it, you can identify a critical piece of mapping information: material movements that are pushed by the producer, not pulled by the customer. 'Push' means that the process produces something regardless of the actual needs of the downstream customer process and 'pushes' it ahead.' In the case of the first tier component supplier in the 'Current State' map a push system operates from blanking, through pressing to assembly. Estimating (scheduling what the next process needs) is usually very difficult as schedules can change (as noted by the revisions the firm sends daily to the steel service centre) and production difficulties (quality losses and downtime losses) mean that processes rarely produce to plan. In addition, many processes are not balanced on a one-for-one basis. For example, a press can make a single part much more quickly than an assembly takes to weld (although the press may probably not be dedicated to a product family). As a result there is a tendency to build buffers into the process in the form of extra components.

The final action required to create the 'Current State' map is to draw a time line on the bottom of the map. The time line mirrors the amount of value added and non-value added time required making the product. For the first-tier component supplier, the value-added proportion of time to manufacture one sub-frame was 4922 seconds (82 minutes). However, the time that the steel used to make the sub-frame was on the site, from it coming on site to leaving as a completed component, assuming steel goes first-in first-out through each inventory point, is 13.07 days (98 hours on a 7.5 hour day). This means that value is being added for only 1.38 per cent of the time that the steel is on site.

FUTURE STATE MAPPING

While drawing the 'Current State' map enables a pictorial understanding of the operations in the business to be developed, the real utility of the tool is the ability to develop a vision of how good the value stream for the product could be. While a number of the companies in the LEAP project had a plan for future business and its development, few companies had a picture of the whole plant and the value stream.

In order to develop the 'Future State' map the 'Current State' map was analysed using the following guidelines developed by Rother and Shook:

1. Produce to the TAKT time based on the demand and the available working time of processes closest to the customer.
2. Develop continuous flow wherever possible.
3. Use 'supermarkets' to control production where continuous flow is not possible upstream.
4. Try to send the customer schedule to only one production process.
5. Distribute the production of different products evenly over time at the 'pacemaker' process.
6. Create an 'initial pull' by releasing and withdrawing small, consistent increments of work at the pacemaker process.
7. Develop the ability to make 'every part every day' in the processes upstream of the pacemaker process.

The guidelines were then turned into questions that can be divided into four areas. These are questions concerned with:

1 Demand
   a. What is the TAKT time?
   b. Should the company build to a 'finished goods supermarket' or directly to shipping?
2 Material flow
   c. Where can the company use continuous flow processing?
   d. Where do we need to use supermarket-based pull systems?
3 Information flow
   e. At what single point in the production chain will the company schedule production?
   f. How will the company level the production mix at the pacemaker process?
   g. What increment of work will the company consistently release and take away at the pacemaker process?
4 Supporting improvements
   h. What supporting process improvements are necessary (key improvement initiatives and critical success factors for implementation of the 'Future State' map)?

A 'Future State' map for the first-tier component assembler is shown in Figure 2. The starting point for the creation of the 'Future State' map is the understanding of TAKT time. The TAKT time is expressed in the equation:

$$\text{TAKT} = \frac{\text{Available Working Time}}{\text{Customer Demand}}$$

It can be seen from the diagram that the vehicle assembler requires 12,000 pieces per month and that the available work time is 27,000 seconds (7.5 hours after breaks have been removed). This means that in order to meet the customer demand within normal working hours, the first tier firm needs to produce a sub-frame every 45 seconds in its assembly process. The TAKT time cannot be changed by the first-tier supplier. Therefore the company must work out how to meet the demands of its customer.

It can be seen from the 'Future State' map that sub-frames are built into a finished goods supermarket. The supermarket links an upstream process to its downstream customer via a pull system. Thus the customer goes to the supermarket and withdraws what it needs when it needs it. The upstream process then produces to replenish what has been withdrawn. This is a more pragmatic solution than producing directly to shipping right away (in essence a more 'lean' solution.) Producing into a finished goods supermarket allows the firm to manage small fluctuations in demand from the customer and allows a safety margin while the firm moves from a push to a pull system. As the customer buys in multiples of 14, this is the choice of order size. Therefore when a stillage with 14 sub-frames is removed, a signal to make another 14 is triggered.

The 'Future State' map shows where the firm can introduce continuous flow. It is suggested that this can take place through assembly, painting and final assembly. At a more detailed level the Process Activity Map can be used to highlight the cycle times of each of operations which can then be compared with the TAKT time. If the cycle time is above the TAKT time then it is not possible to have continuous flow without making improvements first. In addition, maintenance issues and quality losses should be captured, as these may need to be solved before continuous flow is possible.

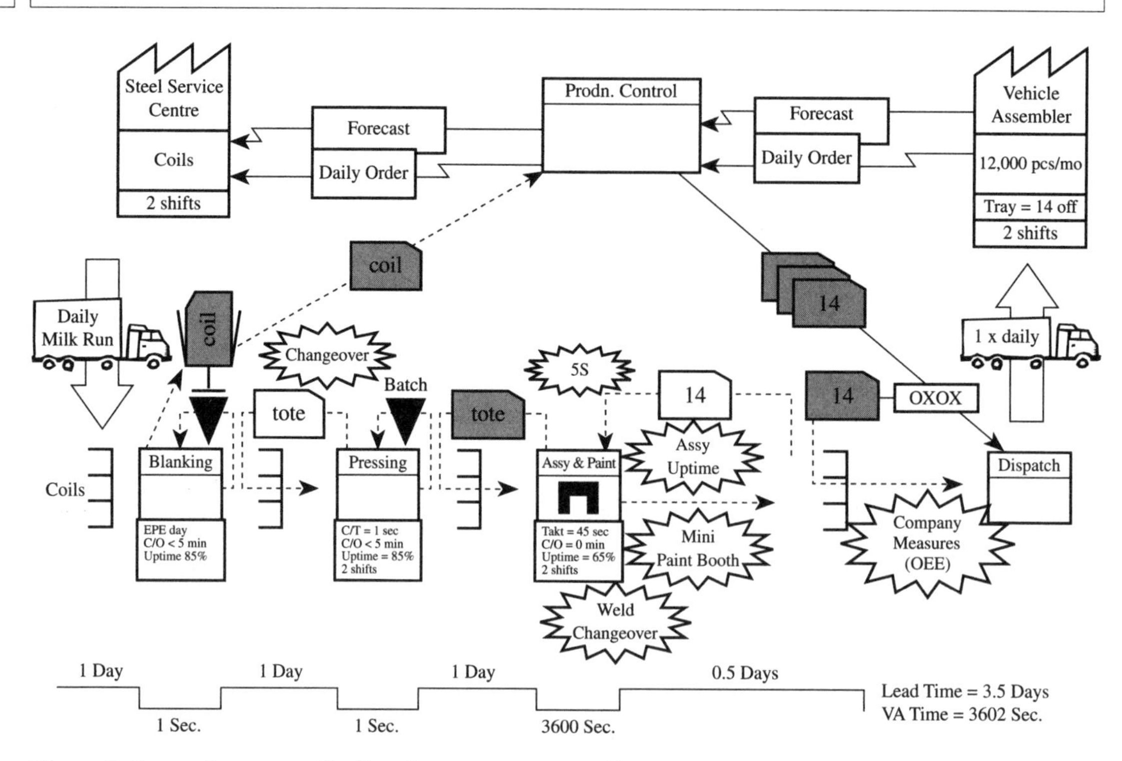

Figure 2 Future State map of a first-tier component supplier.

Where the first tier supplier cannot flow the sub-frame, supermarket pull systems will need to be installed. In the 'Future State' map these are located at the pressing and blanking processes. The use of these pull-based systems on products with regular demand allows the upstream processes to be linked together. When the assembly cell has used its pressings, it withdraws pressings from supermarket. The removal of the pressings (either visually or through a kanban card system) triggers the production of another batch of parts. To produce the parts the press operation withdraws blanks which, in turn, triggers the blank operation to withdraw coils to replenish what has been taken away. As both pressing and blanking are batch operations exact quantities linked to the end-customer demand (14 pressings and 14 blank parts) would not be feasible. Therefore a series of signals would be sent back until the press and blank operations had enough demand from their customer (assembly) to produce a batch of each of these components.

The 'Future State' differs from the 'Current State' in terms of the information flow. In the 'Future State' map a single point along the physical flow is used to schedule manufacture of sub-frame components, rather than separate schedules being made for individual processes. This single point is referred to as the 'pacemaker' process. Downstream of this single point, processes need to occur in a flow. Therefore the logical place to pace the production of sub-frames is the assembly cell.

When looking in detail at sub-frame manufacture, it becomes apparent that there are a number of variants. These depend on engine type and specification of suspension (for example, sport models have a different suspension set-up to other models). As a result the firm will need to understand the demand of the different variants and establish supermarkets for each of the 'runner' sub-frames (5). By levelling the production mix, developing the ability to distribute the variants evenly over the available time, the firm will be able to smooth the work flow more evenly. In order to do this, changeover times may need to be worked on.

The 'Future State' map is also designed to communicate the TAKT time through all the activities in the company. This means that work needs to be paced by developing the ability to release and take away a set amount of components at a designated process along the physical flow – the pacemaker process.

Finally, to enable the 'Future State' to be implemented, support improvements need to be highlighted and documented on the map. These are shown in 'lightning bursts' which can then be used as the basis for an action plan. For example, in Figure 2, 'weld changeover' refers to the need to reduce the changeover time at the weld operation by using a single-minute exchange of dies implementation.

MAPPING THE WHOLE VALUE STREAM

Rother and Shook suggest that 'as your lean experience and confidence grow you can expand outward, from the plant level toward the complete molecules-to-end-user map'. Having mapped a single firm in the LEAP value stream, it is possible to extend the 'Current State' map to cover the supply and manufacture of steel that makes up part of the sub-frame. The principles used to map the first-tier component supplier are replicated. Figure 3 shows the 'Current State' map for the value stream from steel production to component supplier.

It can be seen from Figure 3 that looking at the first-tier component supplier only reveals a partial view of the supply chain. The map shows the activities from hot-rolling steel through to delivery to the vehicle assembler. In summary, the value-added elements of making the sub-frame (from the completion of hot-rolled steel to dispatch of the completed assembly) take 121 minutes. However, if a piece of steel were to be processed on a first-in, first-out basis through the pipeline then there would be an additional 47 to 69 days of delays in the system. It should be noted that this does not include time for the hot-rolling and steel-making processes which could add a further 12 to 28 days.

In addition to clearly showing the ratio of value added to non-value added time, the map also shows the interrelationship of information and material flows at a macro and company level. At the macro level, in the case of the product being mapped, the vehicle producer has a relatively steady demand pattern. However, this demand pattern is distorted through the first tier and steel service centre before getting to the steel company. At the company level, each of the firms sends schedules to individual processes along the physical flow, pushing rather than pulling material through the system. The demand distortions are mapped separately through Demand Amplification Mapping as described in Chapters 9 and 10.

In order to create a vision of how this system can be improved a 'Future State' map that crosses these three firms is displayed in Figure 4. It can be seen that the pull system discussed within the 'Future State' map for the first-tier supplier has been extended backwards up the value stream. While more detail is required to fine-tune the system, it can be seen that implementing this version of the 'Future State' map would lead to an improvement in the lead time through the system from 47 to 69 days down as low as

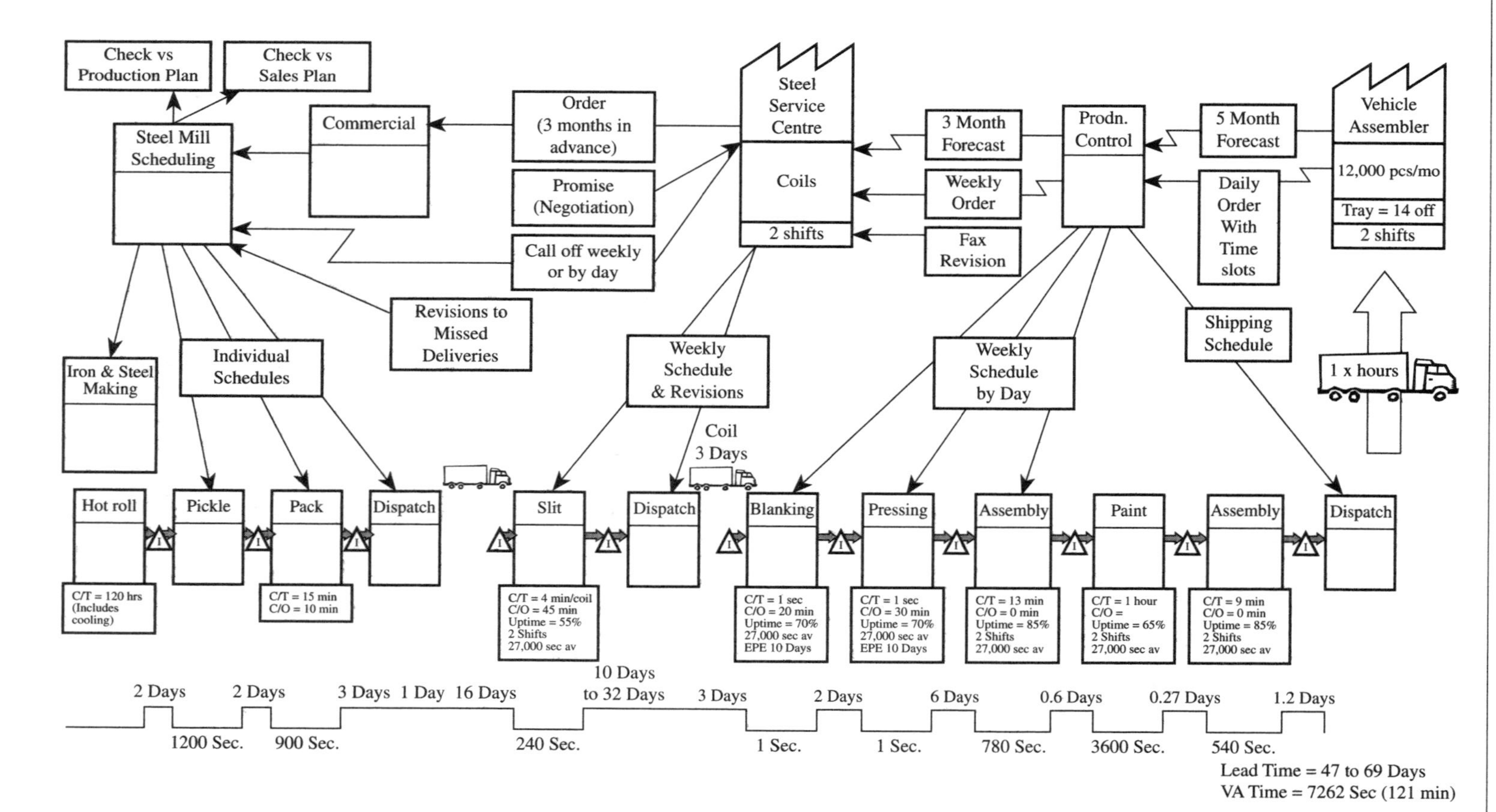

Figure 3 The 'Current State' value stream from steel to component supplier.

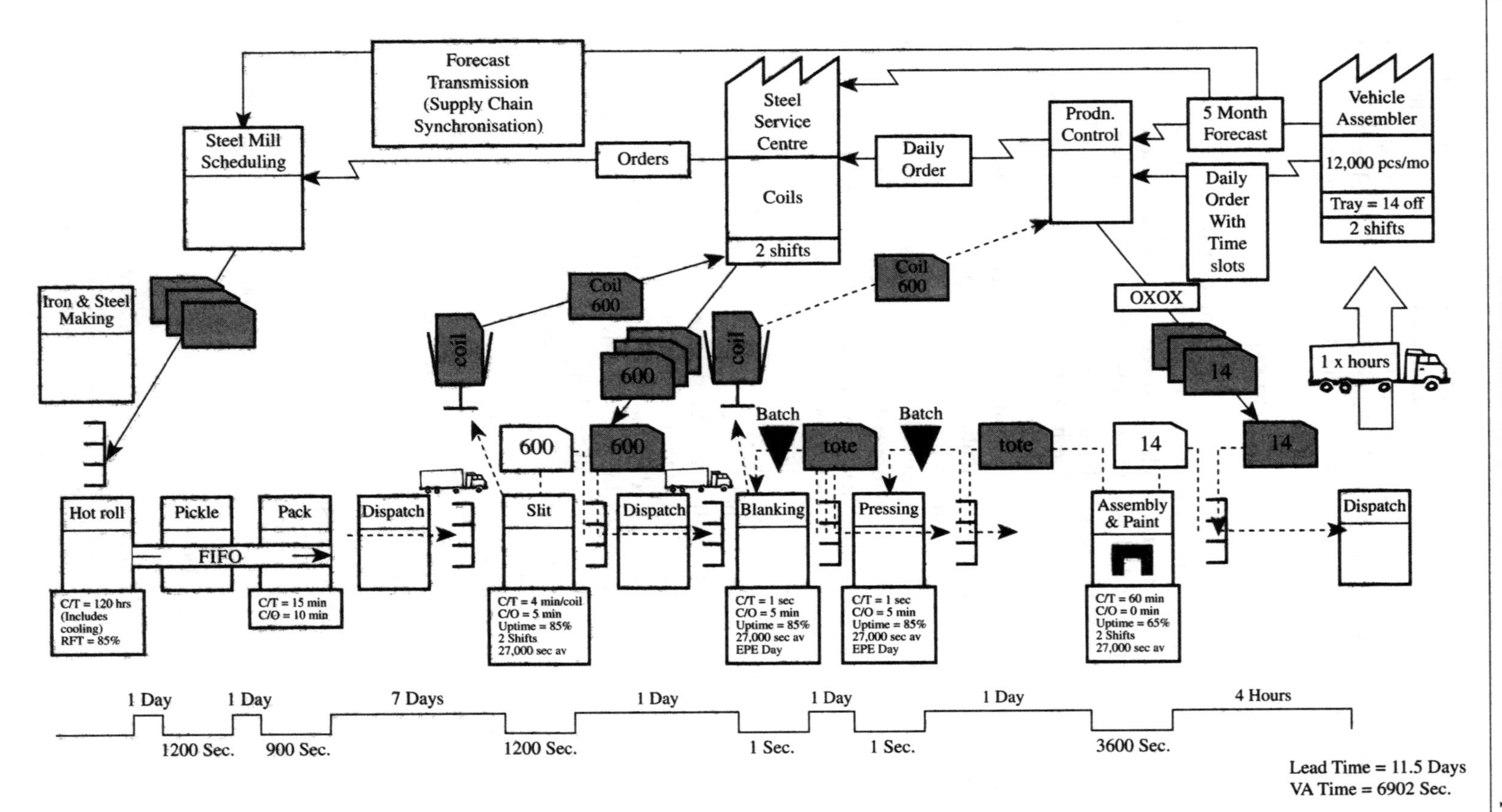

Figure 4 The 'Future State' value stream from steel to component supplier.

11.5 days. In addition, the information flows across the three firms are much simplified. The vehicle assembler's demand data is communicated to all levels in the supply chain, so that each firm receives true information about what is being produced. This activity is supported at an operational level by the kanban system which will allow management to focus more on integration issues rather than the day-to-day planning changes.

MOVING FROM CURRENT TO FUTURE STATE

Making the transition from 'Current State' to 'Future State' will give huge benefits to each of the firms displayed in the maps. However, a number of critical success factors are required in order to manage such improvement projects. These issues are particularly important when they take place across a number of companies. The first issue is the calculation of business performance benefits and how those benefits will be divided between the firms making the improvements. These calculations need to be agreed by all participating firms.

Secondly, the finance and company measurement systems used by the individual firms need to be clearly understood so that any conflicting measures can be removed before embarking on the improvements necessary. For example, if one company in the chain is measuring on volume per employee and this measure is not balanced by a set of operational measures, then the firm will want to overproduce regardless of the demand for the product. These conflicts between measurement and intended actions will probably result in items in the 'Future State' map not being implemented.

Thirdly, an assessment of other value streams is necessary so that an understanding of where these value streams overlap can be created. This is particularly important when considering capacity constraints or bottlenecks that occur within the supply chain.

Finally, while looking at the upstream value stream is of benefit, it is important to consider the total value stream – from the customer through to steel supply. It is only when the total value stream is analysed that it is possible to truly rethink the delivery of value to take out more waste/cost. In the automotive industry much of the work in becoming lean has been concerned with applying lean tools and techniques at the factory level. In many instances most of these improvements have focused on the physical flow on the shop floor. However, in the constant pursuit of competitive advantage, the auto companies need to extend this lean logic to re-assess the value they offer to the customer (end-user). As the environment changes, supply chains will need to change as well. Thus understanding value and the value stream working backwards from the customer not forwards from the steel company will be required.

SUMMARY AND CONCLUSIONS

This chapter has attempted to consolidate the key learning points from the Lean Processing Programme in order to discuss the application of 'Learning to See' style 'Material and Information flow (Value Stream) maps'. The application of the mapping tool has been expanded across the whole supply chain – from steel, through steel service centres to first tier component manufacturers. Current and Future State maps have been highlighted to illustrate the benefits of a lean system pictorially and a method of constructing an action plan has been discussed.

In summary, analysis of part of the value stream shows that, while many companies profess to be 'getting lean', most have only scratched the surface. This shows the true power of the Toyota system and highlights that 'the more waste that is removed, the more new waste is seen'.

NOTES

1 Initial Value Stream maps were created with the help of Chris Butterworth at British Steel. I would like to thank him for his assistance in this research.

REFERENCES

(1) Rother, M. and Shook, J. (1998), *Learning to See, Value Stream Mapping to add value and eliminate muda.* Brookline: The Lean Enterprise Institute.
(2) Womack, J. P. and Jones, D. T. (1996), *Lean Thinking, Banish waste and create wealth in your corporation.* New York: Simon & Schuster.
(3) Hines, P. and Rich, N. (1997) The seven value stream mapping tools. *International Journal of Operations and Production Management*, 17(1), 46–64.
(4) Brunt, D. C. (2000) Company activities viewed as part of the value stream – creating big picture maps of key processes. *In Manufacturing Operations and Supply Chain Management – the Lean Approach.* London: International Thomson Business Press.
(5) Bicheno, J. (1998) *The Lean Toobox.* Buckingham: Picsie Books.

APPENDIX

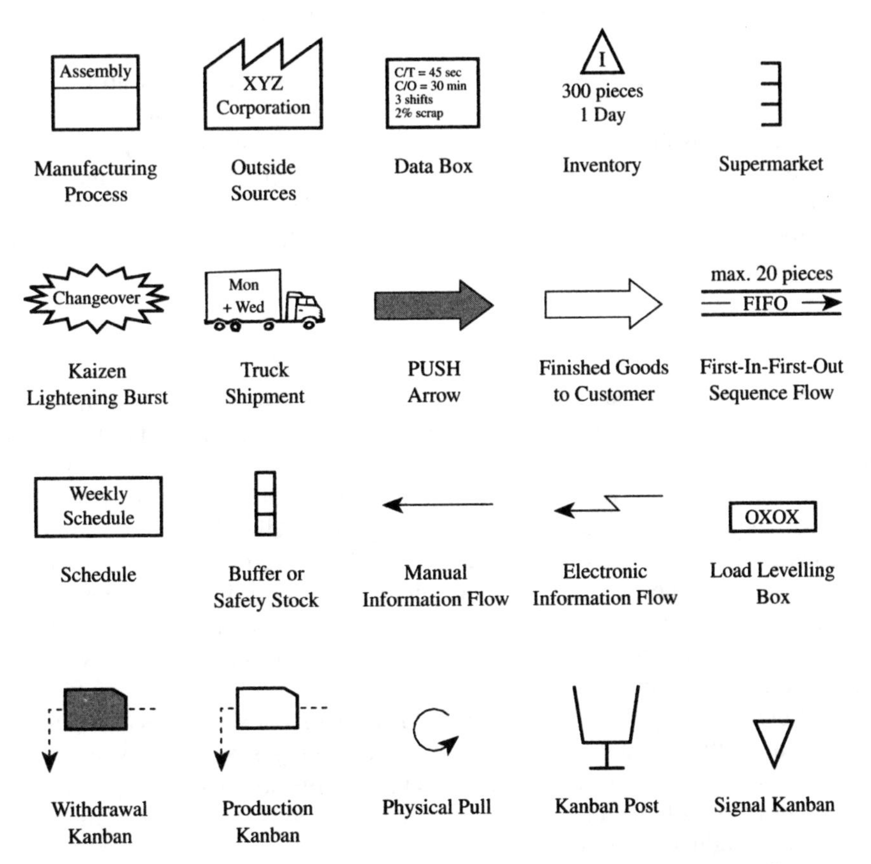

Figure 5 Symbols used in the value stream map from steel to component supplier.

26 The challenge: building cars to order – can current automotive supply systems cope?

Matthias Holweg and Daniel T. Jones

In this article, Matthias Holweg and Professor Daniel Jones of the Lean Enterprise Research Centre at Cardiff Business School investigate whether current vehicle manufacturer order fulfilment systems are able to exclusively deliver 'built-to-order' cars, as part of a demand-driven production system.

Building cars to order could revolutionize the way the car industry works and provide the potential solution to overproduction, the resulting stockpiles of finished vehicles sitting in airfields, and the current low industry profitability – itself induced by high discounts needed to support current 'push'-based selling.

This article is based on initial findings of the research of the *3DayCar Programme*, a three-year joint initiative which commenced in 1999, involving Cardiff Business School, The University of Bath and the International Car Distribution Programme (ICDP), and sponsored by a consortium of car manufacturers, suppliers, logistics companies, IT service providers and the UK government. The aim of the *3DayCar Programme* is to develop a framework in which a vehicle could be built and delivered to customer specifications in minimal lead times – with three days' order-to-delivery time as the ultimate goal.

INTRODUCTION

Recent news about the car industry has tended to focus on three major topics. First, the waves of mergers and acquisitions, which appear to aim to build critical mass to support further R&D costs and economies of scale. Second, the over-capacity problem, estimated at 5.8 million units in Europe alone. And third, concern over supposedly high consumer car prices (especially in the UK). However, a fourth topic has emerged in the midst of these issues: 'build-to-order' or demand-driven production. The question asked – inspired by successful implementations of customer-driven assembly operations in other industry sectors (such as in the often praised Dell Computers supply chain) – is whether the customer still plays a central role in the current system.

Initial evidence suggests that the car industry does not seem to have adopted the principle of putting customer demand into the driving seat of the vehicle supply system. At present, the manufacturers attempt to forecast likely customer demand (both in terms of volume and specification) many months in advance, and figure out an attractive product mix in terms of both balancing labour in the assembly plant and financial terms.

Actual customer orders that are received are either fitted into the plan laid out by the production programme months ahead, or the forecast orders in the system are amended to customer requirements – to the extent the production flexibility allows. However, neither approach is able to provide the customer with an order-to-delivery lead time of less than 40–60 days for a custom-built vehicle.

Currently, the auto industry is excessively focused on the easily measured performance in the assembly plant and the volume of sales in the market, and is failing to deliver specific customer-ordered vehicles within a satisfactory timeframe. Current production programming – and hence scheduling – seems rather guided by line balancing and financial objectives to achieve a profitable mix, than by customer service measures. This has been prioritized at the expense of market responsiveness – with the result that although plants are often efficient, a vast number of cars are produced that do not match customer requirements, ending up as finished stock at dealers or central stocking locations throughout the country. However, due to increasing product variety and market volatility, forecasting demand has become exceedingly difficult and wrong guesses immediately result in redundant stocks and costly stock-clearing initiatives.

CURRENT SYSTEMS FAILURE

The new car buying process tends to be a very frustrating experience for customers. In the UK, for example, a typical volume car customer has to wait, on average, 48 days for his or her custom-built vehicle from the factory to arrive – in some cases it is even longer than 60 days.

For an increasing number of customers, this delivery time is beyond their tolerance limit, so inevitably, many compromise on specification to obtain a car within an acceptable time frame. In the UK, for example, ICDP research in 1997 showed that 26 per cent of customers were not willing to wait longer than 7 days, and only 19 per cent were willing to wait 30 days. Across Europe, this tolerance level varied, with Germany showing the most patient customers. Yet even in Germany, for only 71 per cent of the customers was 30 days waiting time an acceptable proposition.

However, even if customers decide to order a custom-built vehicle from the factory, they are likely to experience even further delays, as the delivery date given at the time of the purchase of the car cannot be kept in 24 per cent of the UK new car purchases. In general, dealers across Europe (being aware of this unreliability) add on average nine additional days to the promised date given by the manufacturer, to give a more realistic delivery date to the customer.

The current system of vehicle supply, it can be argued, not only fails to deliver the right product within an acceptable time frame to the customer, but also puts strain on the manufacturers and dealer networks. The industry has burdened itself with a distribution system characterized by massive levels of finished stocks in order to ensure that it can supply automobiles that are reasonably similar in specification to those demanded by the customer. The magnitude of the stocking problem is illustrated by the situation in the US, which has an annual sales volume of around 17 million light vehicles, and a supply pipeline of 45 days, amounting to just over two million vehicles. This results in multi billion-dollar stock financing costs – not to mention cost of insurance, damage and quality defects.

Yet holding stock not only ties up capital, but also incurs all sorts of other costs. For an average UK dealer with an annual sales of 400 vehicles, insurance costs for new car stock will be £5,000 – in other words, £125 per new car sale are direct costs the customer has to pay for the dealer's stock.

Furthermore, both the sales volume targets given by manufacturers to dealers, and the incentives granted to persuade customers to settle for alternative specification, result ultimately in low dealer and manufacturer margins. Dealer margins on new car sales, which theoretically range from 10 to 20 per cent, in practice are quite often close to zero.

From the manufacturer's viewpoint, it might even be argued that only the profit made by financial and aftersales/repair operations actually justify building the vehicles in the first place. In the case of Ford of Europe, for example, 25 per cent of the profits over the lifetime of a vehicle come from financing and insurance, 29 per cent from after-sales and bodyshops, yet only 1 per cent originates from car assembly, and 3 per cent from car retailing. Similar proportions can be observed for most manufacturers.

Consequently, the current system not only frustrates customers (and dealers) with long lead times and unreliable delivery dates, it also incurs high cost for the manufacturers in terms of cost of capital and vehicle distribution. For the dealers, the current system results in low margins on new car sales and a further strain on residual values, especially if airfields full of stock finally need to be cleared by heavy discounting to push them into the marketplace.

The effects of this massive discounting (up to 30–40 per cent below the recommended retail price) and sales incentives on the whole system are not yet fully understood. Clouded by accounting practices and dispersion into manufacturing, sales and distribution figures, the manufacturer does not see what costs actually occur by producing and distributing a particular vehicle. In practice, the profit for the manufacturing operation is already accredited as the vehicle leaves the factory. Yet the actual cash intake from the customer can take up to a year, while the car remains subject to stock financing agreements and potential incentivized sales between the National Sales Company and the dealers.

Therefore, the feedback on what costs are actually caused by producing vehicles to stock and then discounting and pushing them into the market remains invisible. So the critical question still remains to be answered by the 3DayCar Programme: which is more profitable – to push cars into a market at a discount, or to produce cars to order, pulled by the customer?

It might even be argued that the current system operates in a vicious circle, whereby the quest for volume and market share tends to foster forecast-driven

Table 1 Stock levels in the marketplace

Region	*Manufacturer category*	*Average stock levels in days of sales*
Europe	Volume	55
	Japanese	80
	Specialist	40
USA	Volume	45
	Japanese	45–110

production (Figure 1). As a result, the long order lead times and the pressure to sell the existing stock encourages dealers to sell the existing stock, generally involving discounts to cater for alternative specification. In return, this results in lower margins and profitability, demanding even higher economies of scale. Additionally, push-based selling prevents manufacturers from seeing actual demand in the market, further fostering forecast-driven production.

FAILURE RECOGNITION

Public announcements by the industry suggest that there is some recognition of this failure. Concerned about the inefficiency of an intensive dealer network, aware of the growing use of incentive payments to manage demand and, like so many mature industries, equally terrified and enthralled by the possibilities of e-commerce, it is slowly realizing that there may be a better way of doing things.

What started with groundbreaking news in the *Wall Street Journal* back in August 1999 was an article offering the shattering news that Toyota were about to break the mould in vehicle supply by producing their Camry Solara Coupe in five days on a 'made-to-order' basis. Compared to the industry average, this was an incredible claim – but so it turned out to be. What Toyota aims to achieve is that (by means of a novel ordering system recently installed) it would be possible to *change* the specification of a car up to five days before production.

However, even car makers such as Ford, GM and Renault have recently announced their intentions and projects to reduce order-to-delivery times, yet initiatives so far have been focused mainly on the organization downstream of the factory gate. While there are undoubtedly gains to be made in distribution, it is inevitable that the manufacturers will soon realize that they cannot improve their ability to match cars to customer requirements without realigning the order fulfilment process to support demand-driven production.

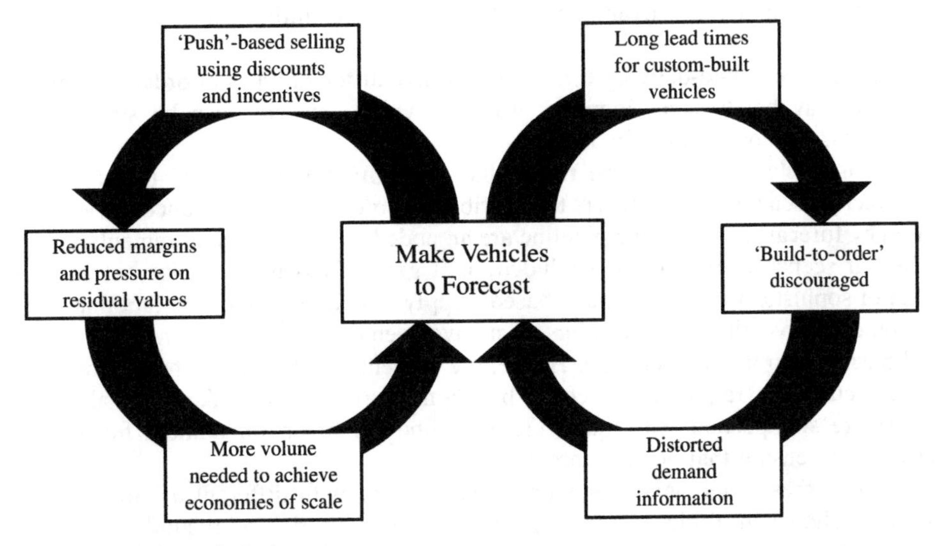

Figure 1 'Vicious circles'.

Renault's 'Projet Nouvelle Distribution' is probably amongst the first to have recognized this challenge, and the objective of a stock reduction of 300,000 vehicles across Europe by the end of 2000 sounds promising.

IS 'BUILD-TO-ORDER' THE SOLUTION?

Attempting to supply the customer with a product of the right specification within an acceptable time frame sounds likes a sensible business premise. In this context, 'build-to-order' seems an obvious approach – to have a demand-driven production system which aims to provide custom-built vehicles in a minimal lead time. One would imagine that the car industry, no longer enjoying the luxury of having demand exceeding its ability to supply (as it did for the last decades), would be governed by these objectives anyway. While the grand scale of investment needed to deliver a product to the market at an economically competitive price creates certain constraints, providing the customer with a car that meets his or her exact requirements as quickly as possible has unmistakable logic.

If companies could provide custom-built vehicles to order, as opposed to make them to forecast, it could solve the major deficiencies of the current system:

- **Redundant stocks** would not occur, as cars would only be manufactured to customer order, relieving manufacturers and dealers of the stock financing burden – and the airfields full of cars would disappear.
- Cars would be sold without **discounts**, as there is no need to grant discounts for alternative specification or to clear stock – hence allowing for reasonable margins for both manufacturers and dealers.
- **Customer service levels** would rise, as right specification and acceptable lead time are the major objectives of the order-to-delivery system.

However, despite this obvious logic, concerns are uttered, particularly from the manufacturers, whether a 'build-to-order' system can replace the current 'make-to-forecast' or order amendment system. To discuss the potential pitfalls of the 'build-to-order' system, a closer examination of what the order fulfilment process means is needed.

In fact, the order-to-delivery time relates to five different types of order fulfilment, as there may be different loops or ways in which new cars can be supplied to customers, as shown below in Table 2.

In this context, it should be noted that the term 'build-to-order' is sometimes incorrectly used by manufacturers to describe the order amendment function (loop 4), whereby forecast orders in the pipeline are amended to customer requirements. In the 3DayCar scenario, this loop is excluded, as it generally is nothing else but another level of sophistication of the 'push-based' supply system; consider the fact that if no customers arrive, the forecast orders (that have been decided months ahead) are built and pushed into the marketplace. Hence, the percentage of orders actually amended to real customer requirements could be claimed as 'built-to-order', but all other orders are still pushed. The main reason behind the order amendment function is basically to ensure that volume targets are met.

Each of these approaches, or loops, comes along with different advantages and risks, as shown in Table 3. For loops 1–3, an obvious risk of redundant stock is present, as the vehicles are already built. Also, a 'specification risk' occurs, as those

Table 2 Order fulfilment loops

Loop	*Order-to-delivery approach*		*Order-to-delivery time (UK data)*
Loop 1	Dealer stock	The car is bought from the stock at the visited dealer.	Instantly available
Loop 2	Dealer transfer	The car is located at another dealer in the country, and transported to the dealer. The additional cost occurring is > £100 for a dealer 'swap' within the UK.	3 days
Loop 3	Distribution centre	The vehicle is sourced from a central stock location, controlled by the manufacturer. Generally the dealer does not hold any new cars in his or her own stock, so most sales would be made from the DC itself.	4 days
Loop 4	Order amendment	Orders are laid out as forecast in the first place, and once the customer specifies his or her order, these unsold 'pipeline' orders are amended to customer requirements.	Variable, 11 days on average
Loop 5	Build-to-order	This implies that the order is entered as a new order into the system. This happens only in 32% of the new vehicle purchases in the UK at the moment, which have an average order-to-delivery lead time of 48 days	40–60 days

Table 3 Risk profiles of order fulfillment loops

Sales sourcing	*Stock redundancy risk*	*Alternative specification risk*	*Discounting risk*	*Lead-time risk*	*Capacity utilization risk*
Dealer	++++	++++	++++	0	++
Dealer transfer	+++	+++	+++	+	++
Distribution centre	++	++	++	+	++
Order amendment	+	+	+	++	++
Build-to-order	**0**	0	0	++++	++++

[0: No risk, +: Low risk, ++: Moderate risk, +++: High risk, ++++: Very high risk]

cars in stock might not be the right spec for the customer. Potential stock redundancy and wrong specification then relate to the overall risk that discounting might have to be used to sell those cars.

As customers are not prepared to wait, a potential risk of lost sales occurs, if the order-to-delivery time exceeds the customer's waiting tolerance. The customer not willing to wait might instead buy from a different brand offering shorter order-to-delivery times. This risk is called 'lead time risk' or 'lost sales risk'.

Also, as the provision of vehicle production capacity is one of the major costs incurred, manufacturers tend to strive for the most efficient utilization of their production and assembly facilities. And this is where the 'build-to-order' is most often criticized. Manufacturers fear for the efficiency of their plants, as 'real' customer orders might not arrive in a sequence that most suits the production schedules of the plants.

However, there seems to be some misunderstanding: in the long run, 'build-to-order' has the same capacity utilization risk as a forecast-driven production system – if there is no demand, there is no justification for build in either system. 'Build-to-order' is as sensitive to pricing and incentivizing as 'make-to-forecast', with the simple difference that in the 'build-to-order' scenario the production volume would need to be supported – as opposed to clearing existing stock from the airfields. The actual risk of 'build-to-order' is short-term volatility – what happens if no orders come in the first week of the month, but all arrive in the second week?

This fear is justified, as under the current reactive management there is no way of catering for short-term volatility. However, the flaw is not to be seen in the 'build-to-order' approach, but in the manufacturers' abilities to manage demand. Car makers these days do not understand and manage their demand, but simply react to incoming orders, and increase marketing efforts if the market share target seems under threat.

In contrast, a 'build-to-order' system would require a proactive management of demand and a segmentation of demand. Why not use non-urgent orders, such as demonstrator and showroom cars for dealerships, or cars for use by the company's own employees or even large fleet orders – which generally provide visibility for several weeks ahead – to buffer the service for those customers who require short delivery times of their custom-built vehicle?

A buffer of those orders would then enable the manufacturer to overcome short-term fluctuations. Again, this seems obvious, but as some assembly plants do not see whether a vehicle is a customer order or a stock car, this would be a leap forward.

In conclusion, the 'build-to-order' approach offers the best risk profile, and in combination with demand management to cater for short-term volatility, seems the obvious strategy to adopt. The only risk occurring is the actual lead times involved, as potentially sales might be lost if the order lead time exceeds the customer's waiting tolerance. Hence the central question: are current vehicle supply systems capable of providing short enough lead times to support 'build-to-order'?

ARE CURRENT VEHICLE SUPPLY SYSTEMS ABLE TO SUPPORT 'BUILD-TO-ORDER'?

To answer this question, 3DayCar researchers analysed the order fulfilment processes of major European manufacturers – basically 'stapling ourselves to a customer order' as it is being processed, and the vehicle is manufactured and delivered.

At each step of the process the system-related delays were researched, and the minimum time delays recorded. As the order processing time depends on several factors which could falsify the results (for example, a waiting list for a high demand product), the vehicle supply *system capability* was analysed, which reflected the *optimal* throughput time for an order – any waiting queue, rework or part unavailability will further extend the evaluated order-to-delivery time.

The minimum throughput times refer to the minimum time required for an order to stay at each stage, determined, for example, by only once a week data transmission, overnight runs of computer systems, or the physical layout and line speed of the assembly track. The average order fulfilment times therefore tend to be even longer, yet depend on the demand and supply situation for each product, but do not provide a basis for comparison.

To understand why custom-built vehicles require 40 days or more to be delivered, Figure 2 shows a simplified version of the 'spaghetti-world' of different IT systems, departments and processes the orders have to fight their way through. Each of these steps shown are major processes related to the order fulfilment, with the red path highlighting the way customer orders flow through the process.

The major steps are:

- **Order entry**, which is a check whether the orders are feasible to be built, and if they are, transfers them into the order bank.
- **Order bank**, which holds all unsold orders until they are scheduled for production. The order bank does not fulfil any real purpose apart from providing a 'comfort' buffer for the manufacturer to build a schedule which allows for efficient production.
- **Order scheduling**, which picks the orders from the order bank and assigns them to build periods (generally weeks) at the different plants. The scheduling tool takes parts availability, market and dealer fair share allocations and mix constraints into account, most of which are decided in the production programming meeting.

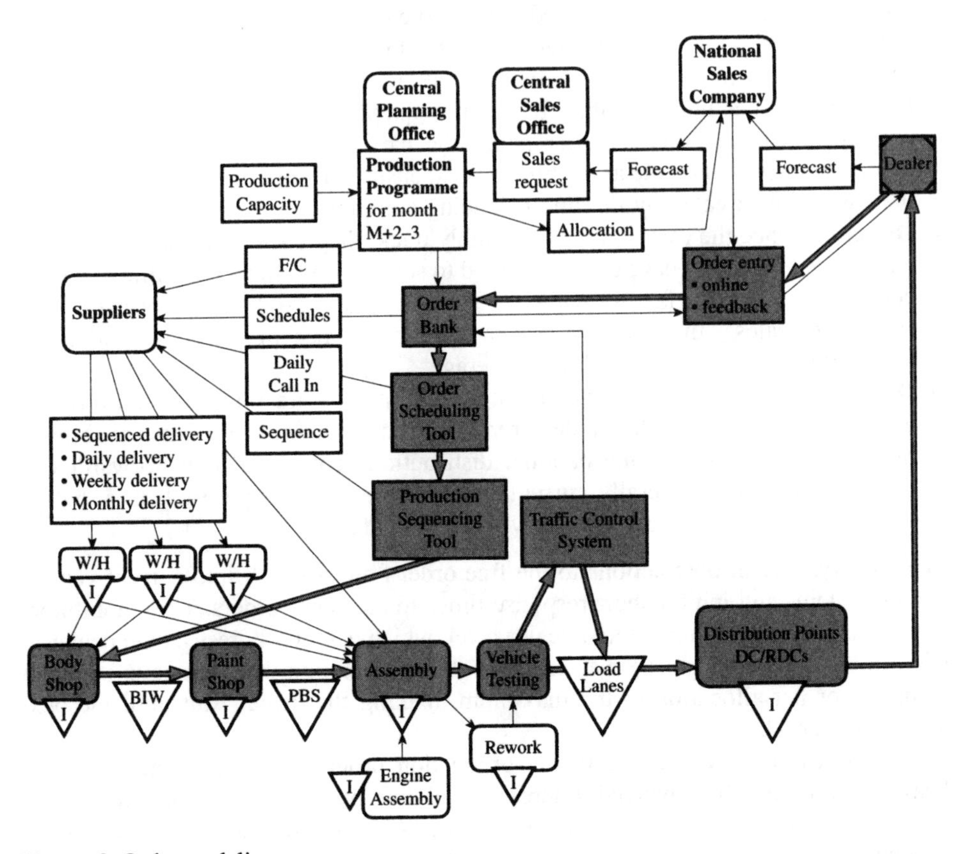

Figure 2 Order-to-delivery process.

Once scheduled, the orders are generally held for two weeks, although nothing happens to them. The general excuse used here is to blame the suppliers, claiming that this 'drop-dead' period is required by suppliers to schedule their production. However, our research indicates that today's component suppliers are generally capable of supporting much shorter scheduling cycles – and that these scheduling procedures are as much a legacy as the IT systems that execute them.

- **Order sequencing,** where the scheduled orders for a build week are re-shuffled into a sequence of build orders for the assembly plants. The sequencing tool needs to take build constraints into account, as for example, the assembly line balance might only cope with 50 per cent estate cars – hence every other car will be a less labour-intensive three-door or four-door model. In any case, only after the orders are sequenced do suppliers actually receive their final call-off of what is required, as only then is it actually defined what parts will be needed. This is another reason why holding scheduled orders is of very limited use, and explains why the schedules issued by the manufacturers show such a high degree of fluctuation!
- **Manufacturing**. Once sequenced, the orders are sent to the body shop, where the order is generally identified with the physical floorpan, which then becomes a complete body. After the body shop, the body enters the body-in-white storage, the only purpose of which is to achieve efficiency in the paint shop by accumulating bodies that are meant to be sprayed in the same colour. Paint batching surprisingly is still an issue, although the actual savings by doing so are hardly more than £1 per car. The downturn of batching is that the initial production sequence is distorted, hence it becomes unpredictable for all subsequent operations as to what cars are coming down the process. After paint, the cars are generally reshuffled again before they are sent on the assembly track, to ensure the mix of cars is aligned with the constraints created by the line balancing activities.
- **Despatch**. Once the cars leave the assembly track, they undergo different tests and checks, and generally have to be reworked to some extent. In fact, as little as 33 per cent of cars leave the factory without having been reworked at all. However, once 'passed to sales', the cars are driven into load lanes in the plant and await transportation – for a whole day on average.
- **Distribution.** Several different distribution strategies are operated in Europe – either direct distribution to the dealer via distribution hubs, or the cars are shipped into several regional or one national distribution centre, and then forwarded to dealers as required. Generally, current distribution systems require at least 4–5 days to transport a car from a UK factory to any UK dealer.

In summary, several obstructions to the free order flow are built into current vehicle supply systems and inhibit short response times to customer orders. These inhibitors are allocations both at market and dealer level, which essentially restrict the dealer to stick to the allocation given, even if some other dealer in a different country is not using his or her allocation to the maximum, making the dealer wait until the next allocation period.

The allocations are decided in the production programming meeting, which is heavily influenced by financial interests – giving priority to markets with high profitability, i.e. the UK, Germany and France. Also, the decisions concerning the product mix obstruct the order flow. It is much more profitable to produce a rich mix of specifications, than just to build 'standard' cars. If, however, more 'standard CL' vehicles are ordered, they would simply be delayed to allow for production of more GTIs and GLXs to be built – whether there are customer orders or not!

Deeply hidden in the departmental functionalism, current IT systems also are a serious inhibitor to responsive order fulfilment. The reason is the batch layout of the system, which was necessary due to technological constraints at the times those systems were introduced; current IT systems require 4–5 days to process an order – 'today's problems were yesterday's solutions'.

It does not come as a surprise that the order-to-delivery process for a custom-built order, on average, needs more than 40 days to complete. Analysing the distribution of where these days are lost is shown in Figure 3.

Manufacturing, surprisingly, is clearly not the issue in terms of delays. The actual assembly operation takes only 6–8 hours, the complete production hardly more than 22 hours – plus additional time for testing and rectification. Most time, 85 per cent in fact, is lost in the information flow, whereas the actual manufacturing operation hardly exceeds 1.5 days.

In the past, time compression initiatives in the automotive industry have concentrated on the assembly area, which we clearly see as 'shop floor myopia'. In the physical production process probably only minutes are to be gained in assembly – maybe hours in case of de-coupling body and paint shop from assembly. The problem clearly lies in information management. The major gains for time compression are to be gained in the information flow, where the savings can be measured in weeks.

However, there is one issue with current manufacturing practices. As mentioned above, due to frequent rescheduling and resequencing it is not foreseeable in which sequence the vehicles are coming off the line. As about 65 per cent of all vehicles spend some time being reworked, it becomes impossible to predict the output of the plant. The problem with this arises in distribution, where efficient utilization of the resources requires advanced planning and scheduling. Yet, as the sequence is unreliable, the truckloads are only consolidated after the vehicles have come off the line – simply waiting an additional day in the plant.

In summary, it can be argued that the current system is not designed to provide built-to-order vehicles to customers. Not surprisingly, the ICDP research findings in

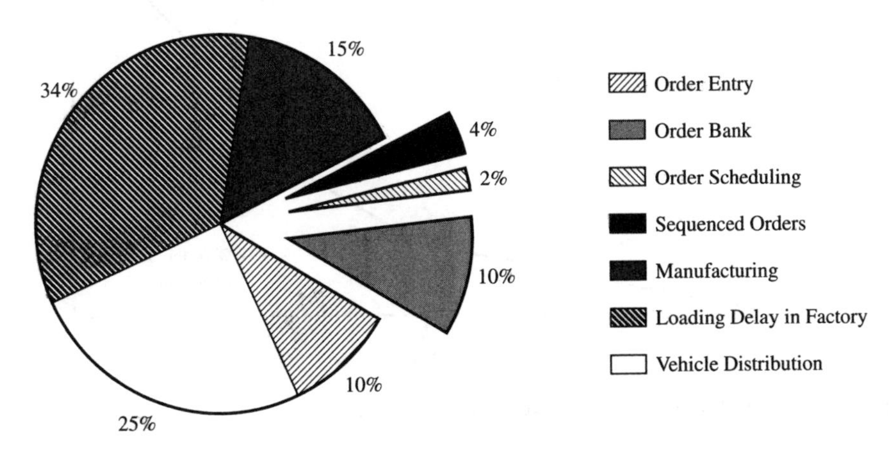

Figure 3 Time delays in the order fulfilment process for custom-built orders.

the UK show that 68 per cent of customers are served from existing stock or amended orders in the pipeline, yet only 32 per cent are new orders put into the system. And it is only this ratio that can provide cars within the tolerance of the customer's expectations.

Figure 4 shows the current order fulfilment – dealer stock, central stock, order amendment and custom-built vehicles – weighted by their average sales and compared to the average waiting tolerance of customers. What can be seen is that, at the moment, lead time is not an issue, as most sales are from existing stock. The inventory and discounts granted cover the manufacturers and dealers against their inability to provide custom-built vehicles in a short period of time.

But what if manufacturers were to adopt a 'build-to-order' strategy – using current systems? If the order fulfilment philosophy were to change under current systems, the result would be devastating for the customer service level, leaving a big performance gap, as illustrated in Figure 5.

The supply system is unable to provide vehicles within the expected lead time of customers, hence the manufacturers face the risk of lost sales, as customers might buy a different brand with better availability.

The conclusion that has to be drawn is that current vehicle ordering and supply systems cannot support a higher degree of 'built-to-order' vehicles, as they are not capable of delivering responsive order fulfilment. If the number of cars built to order were raised, customer service levels would further drop. So current systems have to rely on high levels of finished vehicle stock to provide a reasonable service to customers.

Redesigned systems are necessary if vehicle manufacturers are to embrace this new philosophy to provide custom-built vehicles within an acceptable time frame for the customer. Piecemeal improvement, sometimes promoted as the way ahead, is simply futile, as the whole concept behind it is based on a 'push' or wholesale supply system, which also has left its legacy in the IT systems that have grown 'organically' alongside over the years.

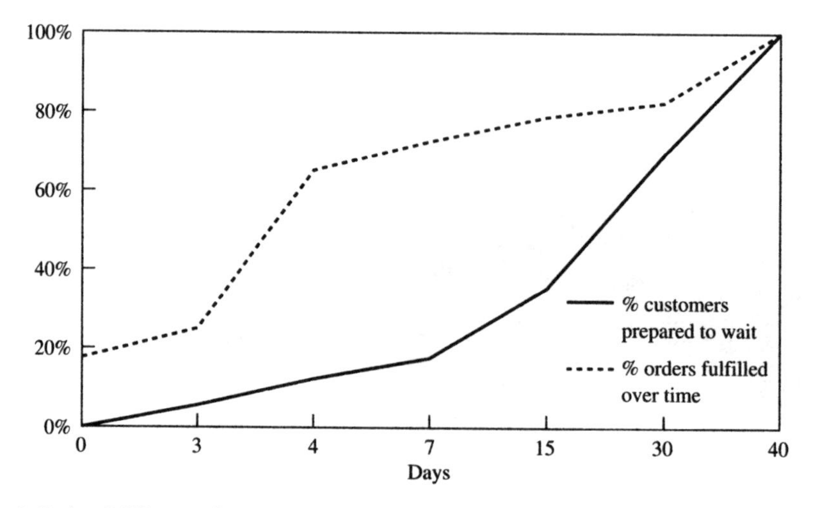

Figure 4 Order fulfilment times vs. preparedness to wait.

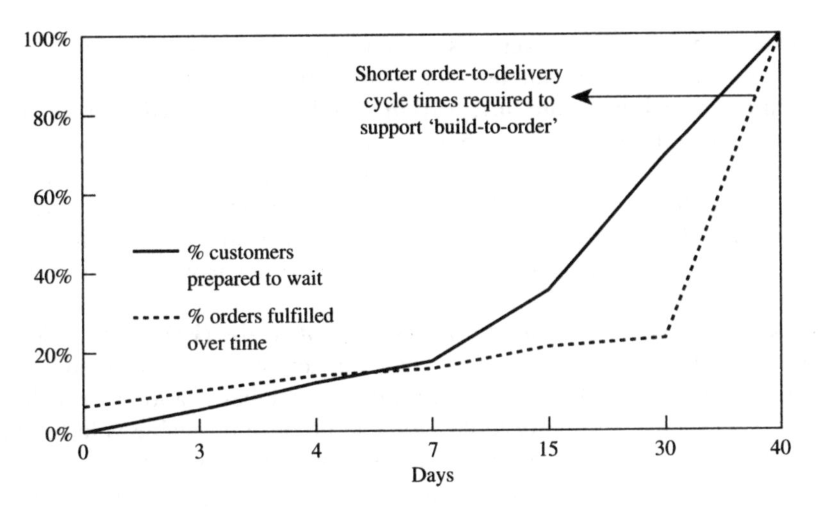

Figure 5 Implications of increased 'build-to-order' content.

THE FIVE CHALLENGES

The 3DayCar research clearly shows that current vehicle supply systems are a frustrating experience for customers, and put heavy financial strain on supply, manufacturing and distribution of new vehicles. We strongly believe that a 'build-to-order' approach has the potential of alleviating current constraints in the system at the risk of potential decreased capacity utilization in the short term and less efficient assembly operations, by catering for more flexibility in the process.

The authors believe that there are five major challenges that need to be overcome to turn the legacy of 'building to forecast' into a responsive 'build-to-order':

1. **Abandon 'push-based' system mindset**. A new mindset with new key performance measures is needed, promoting customer service and total costing, as opposed to volume and market share. Total costing of the complete order-to-delivery process is needed to visualize sunk cost in the current system, which are not yet visible. The challenge is to resist overproduction in order to maintain margins and residual value, which are both essential to maintain a strong brand. A build-to-order culture needs to be planted, avoiding overproduction and discounting / incentive schemes.
2. **Enable demand-driven production**: Separate tactical allocation decisions from the operational order scheduling, and enable daily scheduling processes, or even direct order booking into the production sequence to ensure minimal order-to-delivery times. To achieve this, the organizational layout needs to be changed from a 'departmental chimney' structure to a cross-functional approach.
3. **Understand real demand** – and provide the appropriate service. The challenge is both to understand current demand structures and customer expectations, and to manage these expectations. This knowledge is essential to support the demand-driven production system. Differentiation in order treatment is imminent, although heavily resisted by the manufacturers. However, with changes in the vehicle

ownership model – manufacturers converting into a service mobility provider, rather than just being a manufacturer – this point will gain momentum.
4. **Information visibility and integration:** 'build-to-order' will not be achieved without integration of both suppliers and logistics providers. For both, the provision of appropriate demand visibility is crucial, hence an online access to the order scheduling system would be the logical thing to do. Also, adversarial behaviour and short-term bidding needs to be replaced by long-term partnering. With the growth of 'mega-suppliers', changes in the power base in the supply chain are foreseeable in the near future.
5. **Break dependency on current economies of scale (EOS).** A major future challenge will be to escape the constraints of steel stamping and painting. Exploring other body structure and assembly techniques is a long-term challenge, yet will determine the ability to develop and produce profitable volume cars in a market with steadily decreasing life cycles and increasing variety. The standard steel monocoque will need to be replaced by modular spaceframe or composite bodies, embracing modular assembly and supply strategies. Modules should be standardized across models and maybe even brands. Also, complexity and variety reduction will further alleviate R&D cost coverage requirements.

The systems research within the 3DayCar programme is currently developing a direct order booking system, whereby an order will be directly entered into the assembly sequence. Furthermore, the customer order will not be identified with the physical vehicle before the start of the assembly line (de-coupling), treating the body and paint shops as internal suppliers to the actual assembly operation.

The direct order booking system will be validated using an holistic simulation model. However, initial findings indicate that each manufacturer will require an individual solution towards an optimized order-to-delivery approach, with particular strategies or hybrid approaches being more suitable for one than for the others.

To achieve a build-to-order system requires not only a redesigned ordering and supply system, but first of all a significant change in company philosophy. And changing the mindset might even prove to be even more of a 'legacy' than redesigning the outdated IT systems, as 'build-to-order' challenges the most established measures in the car industry – capacity utilization and market share.

So far, the car industry has been getting away with ignoring customer demand by producing against forecasts and supplying from stock. We believe that in the light of overcapacity and competitive pressure in the world automotive industry this approach has reached its limit – and a 'build-to-order' strategy might prove to be just the cutting edge required to survive in today's markets.

Appendix 1
LEAP publications list

The following publications resulted directly from the activities of the LEAP project:

1. Hines, P. (1997) Mapping the value stream, *Lean Summit,* UK.
2. Taylor, D. (1997) A methodology for approaching supply chain improvement, *Second World Logistics Conference, London.*
3. Bicheno, J. (1998) Mapping muda in the automotive supply chain, *Proceedings of the SAPICS 20th International Conference, Cape Town,* pp. 201–203.
4. Bicheno, J. (1998) 'Kaizen and Kaikaku: Can they be made continuous?' Harry Boer and Jose Gleskes (eds), University of Twente, pp. 55–62.
5. Bicheno, J. (1998) Automotive value stream mapping. *Proceedings of POMS 98, Cape Town,* pp. 52–61.
6. Bicheno, J. (1998) 5S. *Control,* August, 10, ISSN 02661713.
7. Bicheno, J. (1998) The Kano Model. *Control*, September, pp. 27, ISSN 02661713.
8. Bicheno, J. (1998) Automotive value stream mapping. *POMS South Africa Meeting: An International Meeting of the Production and Operations Management Society,* Graduate School of Business, University of Cape Town, 29 June–2 July, pp. 52–61.
9. Bicheno, J., Holweg, M., Taylor, D., Sullivan, J. and Butterworth, C. (1998) *The Lean Leap Supply Chain Game*. Cardiff Business School/Picsie Books.
10. Bicheno, J., Taylor, D., Holweg, M., Hines, P., Brunt, D., Rich, N., Butterworth, C. and Sullivan, J. (1998) Learning supply chain dynamics through participative gaming. *4th International Workshop on Games in Production Management, IFIP/University of Ghent, Belgium.*
11. Brunt, D. and Butterworth, C. (1998), Waste elimination in Lean Production – a supply chain perspective. *Proceedings of the ISATA Conference, Germany,* pp. 275–282.
12. Brunt, D., Rich, N. and Hines, P. (1998) Aligning continuous improvement along the value chain. *Proceedings of the 7th Annual International IPSERA Conference, London,* pp. 80–88.
13. Brunt, D., Sullivan, J. and Hines, P. (1998) Costing the value stream – an improvement opportunity. *Proceedings of the LRN 2nd Annual Conference, Cranfield,* pp. 223–231.
14. Butterworth, C. (1998) *Supply Chain Management*. London: Logistica.

15. Butterworth, C., Norgrove, S. and Wheeler, C. (1998) Mapping the whole value stream. *Lean Summit,* UK.
16. Hines, P. (1998) New directions: value stream mapping. *SCDP Meeting,* Unipart.
17. Hines, P. (1998) Value stream mapping. *Lean Summit,* UK.
18. Hines, P. and Rich, N. (1998) Outsourcing competitive advantage. *International Journal of Physical Distribution and Logistics Management,* 28(7), 524–546, ISSN 0960-0035.
19. Hines. P., Rich, N., Bicheno, J., Brunt, D., Taylor, D., Butterworth, C. and Sullivan, J. (1998) Value stream management. *International Journal of Logistics Management,* 9(1), 25–42, ISSN 0957–4903.
20. Hines, P., Rich, N., Brunt, D., Taylor, D., Bicheno, J., Butterworth, C. and Sullivan, J. (1998), The value stream mapping tools – a critique of the methodology. *Proceedings of the 5th International Conference of EurOMA, Dublin,* pp. 231–236.
21. Munro-Faure, M., Atkinson, D. and Bicheno, J. (1998) Denso Manufacturing UK: Improvement through teamworking. In *Quality Improvement: Teamwork Solutions from the UK and North America,* Munro-Faure, L., Teare, R. and Scheuing, E. (eds), London: Cassell, pp. 17–40, ISBN 0304703117.
22. Rich, N. and Hines, P. (1998) Value stream mapping: a macro view. *Lean Summit,* USA.
23. Taylor, D. (1998) Redesigning an elephant: parallel incremental transformation strategy: an approach to the development of lean supply chains. *Proceedings of the LRN 2nd Annual Conference, Cranfield,* pp. 189–209.
24. Taylor, D. (1998) Parallel incremental transformation strategy: an approach to the development of lean supply chains. *Proceedings of the AMA Annual Conference, Vienna,* pp. 116–125.
25. Taylor, D. (1998) Benchmarking makes you average: the lean alternative. *Logistics 98, The Oriana.*
26. Bicheno, J. (1999) Implementing 'lean' principles: kaizen and kaikaku. *Logistics Focus,* 7(3), 12–17, ISSN 13506293.
27. Brunt, D. (1999) Value stream mapping tools: application and results. *Logistics Focus,* 7(2), 24–31, ISSN 13506293.
28. Brunt, D. and Butterworth, C. (1999) The seven new wastes. *Logistics in the Information Age: Proceedings of the 4th International Symposium on Logistic, Florence,* pp. 399–402, ISBN 8886281374.
29. Butterworth, C. (1999) Partnership: the future of steel in the motor car. MIRA Yearbook.
30. Butterworth, C. and Bicheno, J. (1999) Kaikaku in action: a case study of the Achievements in just 2 days. *Financial Times Automotive Components Analysis,* February.
31. Esain, A. (1999) New product introduction – the triangulation tools for time compression and competitive advantage: a process industry case study. *Proceedings of the 5th International Conference on Concurrent Enterprising, The Hague,* pp. 473–480, ISBN 0951975986.
32. Esain, A. (1999) Benchmarking and improvement of the strategic supply base. *Proceedings of the 2nd International Conference on Managing Enterprises, Newcastle, Australia*, pp. 95–100, ISBN 072591078X.

33. Esain, A. (1999) New product introduction – the triangulation tools for time compression and competitive advantage – a process industry case study. *Logistics in the Information Age: Proceedings of the 4th International Symposium on Logistics, Florence,* pp. 331–336, ISBN 8886281374.
34. Esain, A. (1999) New product introduction and the lean processing programme (LEAP). *Logistics Focus,* 1(2), ISSN 1466836X.
35. Esain, A. and Sullivan, J. (1999) How strategy affects achievement, innovation and continuous improvement in new product introduction. *6th International Product Development Management Conference, Cambridge.*
36. Hines, P. (1999) Supply chain management: from lorries to macro-economic determiner. *Proceedings of the LRN 2nd Annual Conference, Cranfield,* pp. 1–23.
37. Hines, P. (1999) Future trends in supply chain management. In *Global Logistics and Distribution Planning,* Water, D. (ed.), (3rd edition), London: Kogan Page, pp. 39–62, ISBN 0749427795.
38. Hines, P. (1999) Value stream management: next frontier in the supply-chain? *Logistics Focus,* 1(3), 36–39, ISSN 1466836X.
39. Hines, P., Sullivan, J. and Holweg, M. (1999) Waves, beaches and breakwaters: new insights into supply chain dynamics. *Proceedings of the 15th International Conference on Production Research: Manufacturing for a Global Market,* Hillery, M.T. and Lewis, H.J. (eds), University of Limerick, Ireland, pp. 571–575, ISBN 1874653569.
40. Hines, P., Sullivan, J., Holweg, M. and Rich, N. (1999) Waves, beaches and breakwaters: new insights into supply chain dynamics. *Logistics in the Information Age: Proceedings of the 4th International Symposium on Logistics, Florence*, pp. 67–72, ISBN 8886281374.
41. Holweg, M. and Bicheno, J. (1999) The automotive steel supply chain: what happens, what could happen? *Managing Operations Networks: Proceedings of the EurOMA 6th International Annual Conference, Padua*, pp. 221–228, ISBN 8886281390.
42. Rich, N. (1999) Supply-chain management: the measurement 'Wall', *Logistics Focus,* 7(4), 26–31, ISSN 13506293.
43. Sullivan, J. and Bicheno, J. (1999) The application of value stream management to muda reduction in a first tier automotive component manufacturer. *Managing Operations Networks: Proceedings of the EurOMA 6th International Annual Conference, Padua*, pp. 311–321, ISBN 8886281390.
44. Taylor, D. (1999) Parallel incremental transformation strategy: an approach to the development of lean supply chains. *International Journal of Logistics Research and Applications,* 2(3), 305–323, ISSN 1367 5567.
45. Taylor, D. (1999) Eradicating demand amplification: a step towards the lean supply-chain. *Logistics Focus,* 1(1), 52–61, ISSN 1466836X.
46. Taylor, D. (1999) A methodology for the elimination of demand amplification effects across multi-levels of a supply chain. A case study from automotive assembler to steel manufacturer. *Logistics in the Information Age: Proceedings of the 4th International Symposium on Logistics, Florence*, pp. 365–372, ISBN 8886281374.
47. Taylor, D. (1999) Supply-chain improvement: the lean approach. *Logistics Focus,* 7(1), 14–20, ISSN 13506293.

48. Taylor, D. and Butterworth, C. (1999) From demand amplification to synchronised supply. *Proceedings of the 8th International Annual IPSERA Conference, Belfast and Dublin*, pp. 741–756.
49. Taylor, D. and Brunt, D. (1999) Balancing flow and demand. *Proceedings of The Lean Summit Europe*, *Cannes*, Sept., pp. 523–527.
50. Taylor, D. (2000) An approach to the measurement and analysis of demand amplification across multi levels of the supply chain. *The International Journal of Logistics Management*, 11(1).
51. Taylor, D. (2000) Demand amplification: has it got us beat? *The International Journal of Physical Distribution And Logistics Management*, Vol 30(6), pp. 515–533
52. Hines, P. and Taylor, D. (2000) Going lean: a guide to implementation. *Lean Enterprise Research Centre*, *Cardiff Business School.*
53. Hines, P., Bicheno, J. and Rich, N. (2000), *End to end lean*. Portland, Oregon: Productivity Press.
54–61. The seven articles describing the LEAP project which were first published in *Logistics Focus* (Nos 26, 27, 34, 38, 42, 45, 47 above) are to be translated into Portuguese and published in Brazil in the journal *Movimentacao & Armazenagem/Log* during 2000.
62. Hines, P. and Taylor, D. (2000) Enxugando A Empresa: Un guia para implementacaio. Instituto IMAM, Sao Paula, Brazil.

Appendix 2 Books giving further guidance on the concepts and implementation of lean approaches

Lean Thinking: Banish Waste and Create Wealth in your Corporation
James P. Womack and Daniel T. Jones
Simon & Schuster, New York, 1996

The Machine That Changed the World
James P. Womack, Daniel T. Jones and Daniel Roos
Rawson Associates, New York, 1990

Going Lean: A Guide to Implementation
Peter Hines and David Taylor
Available from:
Lean Enterprise Research Centre
Aberconway Building
Colum Drive
Cardiff CF10 3EU, UK
e-mail *lovells@cardiff.ac.uk*

The Lean Toolbox, 2nd edition
John Bicheno
Picsie Books, Box 622, Buckingham MK18 7YE
Email *picsie@axiom.co.uk/picsie/*

End to End Lean
Peter Hines, John Bicheno and Nick Rich
Productivity Press, Portland, Oregon, 2000

Learning to See: Value Stream Mapping to Add Value and Eliminate Muda
Mike Rother and John Shook
The Lean Enterprise Institute, Brookline, Mass, 1998 (*www.lean.org* or email at info@lean.org)

Value Stream Management: The Development of Lean Supply Chains
Peter Hines, Paul Cousins, Nick Rich, Daniel Jones and Richard Lamming (eds)
FT Management, London, 2000

The Lean Enterprise: Designing & Managing Strategic Processes for Customer Winning Performance
Dan Dimancescu, Peter Hines and Nick Rich
AMACOM, New York, 1997

Appendix 3
Sources of assistance for lean development

RESEARCH ASSISTANCE

At LERC there are a number of ongoing research programmes within both manufacturing and service environments either on a group or individual basis. If you would like to discuss your specific requirement please contact:

Lean Enterprise Research Centre
Aberconway Building
Colum Drive
Cardiff CF10 3EU, UK

In the first instance contact either Shirley Lovell or Nick Rich:

e-mail *lovells@cardiff.ac.uk*
richn@cardiff.ac.uk
or visit our web site on: *http://www.cf.ac.uk/uwc/carbs/lerc*

EDUCATIONAL ASSISTANCE

A range of educational courses are run at LERC, including:

MBA in Supply Chain Management
MSc in Lean Operations
The Automotive Retail Management Programme
Short courses on specialist topics
Tailored short courses for individual companies

For further information please contact:

Claire Gardner
Education Manager
Lean Enterprise Research Centre
Aberconway Building
Colum Drive
Cardiff, UK
e-mail: *gardnerc@cardiff.ac.uk*

or visit our web site on: *http://www.cf.ac.uk/uwc/carbs/lerc*

Index